Nonlinear Wave Equations: Analytic and Computational Techniques

CONTEMPORARY MATHEMATICS

635

Nonlinear Wave Equations: Analytic and Computational Techniques

AMS Special Session
Nonlinear Waves and Integrable Systems
April 13–14, 2013
University of Colorado, Boulder, CO

Christopher W. Curtis
Anton Dzhamay
Willy A. Hereman
Barbara Prinari
Editors

American Mathematical Society
Providence, Rhode Island

2010 *Mathematics Subject Classification.* Primary 35P25, 47A40, 35Q51, 35Q53, 35Q55, 35Q35, 37K40, 37K15, 65M70, 65M85.

Library of Congress Cataloging-in-Publication Data

Nonlinear wave equations : analytic and computational techniques : AMS Special Session on Nonlinear Waves and Integrable Systems : April 13–14, 2013, Boulder, CO / Christopher W. Curtis, Anton Dzhamay, Willy A. Hereman, Barbara Prinari, editors.

pages cm. – (Contemporary mathematics ; volume 635)

Includes bibliographical references.

ISBN 978-1-4704-1050-6 (alk. paper)

1. Nonlinear wave equations. I. Curtis, Christopher W., editor. II. Dzhamay, Anton, 1970– editor. III. Hereman, Willy A., 1954– editor. IV. Prinari, B., 1972– editor.

QA927.N65945 2015

512′.353–dc23 2014033711

Contemporary Mathematics ISSN: 0271-4132 (print); ISSN: 1098-3627 (online)

DOI: http://dx.doi.org/10.1090/conm/635

∞ The paper used in this book is acid-free and falls within the guidelines
established to ensure permanence and durability.
Visit the AMS home page at http://www.ams.org/

10 9 8 7 6 5 4 3 2 1 20 19 18 17 16 15

Contents

Preface

The field of nonlinear waves is an active mathematical research area with a long tradition and storied history. Starting with the observation of John Scott Russell in 1834 of the great "wave of translation," now called a *soliton*, nonlinear dispersive wave equations have proven over the years to be fundamental for modeling nonlinear wave phenomena in such diverse fields as fluid and gas dynamics, nonlinear optics, low-temperature physics, biology, and more. The discipline can be succinctly described as a study of waves usually resulting from a precise balance between nonlinearity and dispersion. While dispersion attenuates in physical problems, it does so weakly and typically over long spatial and temporal scales. Coupled with an intricate process of wave mixing due to nonlinearity, the analysis and computation of nonlinear dispersive waves becomes challenging and complicated. Furthermore, computing solitons and describing their elastic collisions requires mathematical methods that are quite distinct from those used to analyze and solve linear partial differential equations.

Major progress in the study of soliton equations came with the discovery of the "Inverse Scattering Transform" (IST) by Gardner, Greene, Kruskal, and Miura in 1967. As the name suggests, the IST combines techniques from inverse problems and scattering theory to solve a special class of nonlinear dispersive equations, called completely integrable systems, of which the Korteweg-de Vries (KdV) and nonlinear Schrödinger (NLS) equations are prototypical examples. The IST method can be viewed as a nonlinear analog of the Fourier transform. Not only does the IST allow one to solve the initial-value problem for nonlinear integrable equations for fairly general initial conditions, it has also provided unprecedented insight into how nonlinear and dispersive effects influence one another. Research in integrable systems has developed into a rich discipline at the crossroads between analysis, geometry, algebra, and mathematical physics. Recently discovered connections with the study of Riemann–Hilbert problems and Riemann surfaces are examples of such intersections.

Unfortunately, not every nonlinear dispersive system can be solved with the IST. Notably, the problem of modeling free surface irrotational waves in incompressible and inviscid fluids is beyond the scope of the IST. As a "parent" model for several of the most important problems in integrable systems though, one might question how phenomena seen in integrable systems manifest themselves in this more complicated, yet physically more realistic setting. Likewise, within the last several years, techniques from integrable systems have found their way into research of the free surface problem. Despite such efforts to solve a problem with a nearly 200 year history, the benefits and limitations of these novel lines of attack are still unclear.

With the purpose of bringing together researchers working on these different problems, a special session called "Nonlinear Waves and Integrable Systems" was organized at the AMS Western Sectional Meeting in Boulder, Colorado, in the spring of 2013. The organizers were fortunate in having such a diverse group of speakers agree to present their research and commit to write a research paper for this volume which attempts to capture the scope of the special session. While by no means exhaustive, this volume addresses and explains many of the major analytic and computational techniques used across several sub-disciplines of the study of nonlinear wave equations.

The volume begins with an article by Ben Herbst, Garrett Nieddu, and A. David Trubatch who return to questions from the genesis of the field of integrable systems. The authors study whether discretizations of continuous integrable systems inherit their dynamic properties, in this case "recurrence," from the integrability of the continuous model.

Next, a series of papers on techniques related to the IST is presented. Francesco Demontis, Cornelis van der Mee, and Federica Vitale study the scattering data of a defocusing Zakharov–Shabat system with non-zero boundary conditions. This problem is of particular relevance to the study of dispersive shock-waves, a subject of much recent interest.

The next two papers address questions related to the Novikov–Veselov (NV) equation, a two-dimensional generalization of the KdV equation. The first paper, co-authored by Ryan Croke, Jennifer L. Mueller, Michael Music, Peter Perry, Samuli Siltanen, and Andreas Stahel, is devoted to the use of the IST to study the NV equation. The authors illustrate the use of powerful techniques from inverse problems and scattering theory to determine for what types of initial conditions the NV equation is well posed. In the second paper, the stability of particular solutions to the NV equation is studied by Ryan Croke, Jennifer L. Mueller, and Andreas Stahel.

One last paper concerned with integrable systems is written by Gregory Lyng who studies the semi-classical limit of the focusing NLS equation showing the role of Riemann–Hilbert problems in investigating integrable systems.

The subsequent series of papers addresses questions on the free surface problem in fluids. The first paper by Jon Wilkening looks at a classic question on the correspondence of phenomena found in the KdV equation and the free surface problem from which the KdV equation is derived.

In the next paper, Katie Oliveras and Bernard Deconinck investigate modulation instabilities in the free surface problem. Modulational instabilities were first found in the NLS equation, which is derived from the free surface problem. Using a formulation of the free surface problem first proposed by Ablowitz, Fokas and Musslimani (AFM) – an approach strongly motivated by methods from integrable systems – the authors extend the results from the NLS equation to a more physically realistic model.

In the same spirit, Katie Oliveras and Vishal Vasan show how to use the AFM formulation to determine the height of free surface waves from pressure measurements along the "sea-floor". This work has practical importance for oceanographers, and represents a significant improvement over existing methodologies.

Finally, Jon Wilkening and Vishal Vasan investigate several competing numerical schemes for simulating the free surface problem. The merits and drawbacks of the AFM approach for numerical solutions are examined in detail, and superior alternatives to this approach are provided and explained.

The editors wish to thank the authors who contributed to this volume. Without their diligent efforts this volume would not exist. Likewise, we wish to express our gratitude to the anonymous reviewers who helped to make this volume a strong contribution to the nonlinear waves and broader mathematical community. Finally, we would like to thank all of the participants in the special session who helped to motivate the creation of this volume.

Christopher W. Curtis
Anton Dzhamay
Willy A. Hereman
Barbara Prinari

List of Special Session Participants

Douglas E. Baldwin
University of Colorado Boulder

Michael Benfield
North Carolina State University

Gino Biondini
State University of New York at Buffalo

Sarbarish Chakravarty
University of Colorado Colorado Springs

Christoper W. Curtis
University of Colorado Boulder

Francesco Demontis
Università di Cagliari

Anton Dzhamay
University of Northern Colorado

Willy A. Hereman
Colorado School of Mines

Mark A. Hoefer
North Carolina State University

Matthew Johnson
University of Kansas

Stephane Lafortune
College of Charleston

Hector E. Lomeli
University of Texas Austin

Gregory Lyng
University of Wyoming

Kenichi Maruno
University of Texas Pan-American

Jennifer L. Mueller
Colorado State University

Katie Oliveras
Seattle University

Peter A. Perry
University of Kentucky

Barbara Prinari
University of Colorado Colorado Springs

Harvey Segur
University of Colorado Boulder

Thomas Trogdon
University of Washington

Vishal Visan
Pennsylvania State University

Jon Wilkening
University of California, Berkeley

Contemporary Mathematics
Volume **635**, 2015
http://dx.doi.org/10.1090/conm/635/12677

Recurrence in the Korteweg-de Vries equation?

Ben Herbst, Garrett Nieddu, and A. David Trubatch

ABSTRACT. The Zabusky-Kruskal lattice (ZK) was derived as a finite-difference approximation of the Korteweg-de Vries equation for the purpose of numerical simulation of that PDE. Like the Fermi-Pasta-Ulam lattice from which it was ultimately derived, ZK was also observed to exhibit near-recurrence of its initial state at regular time intervals. The recurrence has not been completely explained, though it has been attributed to the solitons or, less specifically, the integrability of the KdV continuum limit The attribution of recurrence to the integrablity of the continuum limit (i.e., KdV) leads naturally to the hypotheses that simulations (i) on smaller grid separations, (ii) with discretizations that are integrable as spatially discrete systems or (iii) with higher-order discretizations should all exhibit stronger recurrences than observed in the original simulations of ZK. However, systematic simulations over a range of grid scales with ZK, as well as an integrable discretization introduced here and a spectral discretization of KdV are *not* consistent with these hypotheses. On the contrary, for the ZK and an integrable finite-difference discretization of KdV, recurrence of a low-mode initial state is observed to be strongest and most persistent at an intermediate scale. We conclude that the observed recurrences is a lattice property and not a reflection of the integrable dynamics of KdV.

1. Introduction

When Zabusky and Kruskal [**36**] performed numerical simulations of the Korteweg-de Vries equation (KdV),

$$u_t + uu_x + \delta^2 u_{xxx} = 0, \tag{1.1}$$

with periodic boundary conditions ($u(0,t) = u(2,t)$), they were looking for the recurrence of the single-mode initial condition

$$u(x,0) = \sin(\pi x)\ ,\ 0 \le x \le 2\ , \tag{1.2}$$

at regular time intervals. The recurrence of the initial condition was expected because, in this instance, KdV (1.1) was considered an asymptotic limit of the nonlinear oscillator lattice first investigated by Fermi, Pasta and Ulam (FPU) [**15**]. In addition to observing repeated (partial) recurrences, Zabusky and Kruskal also famously observed the elastic nonlinear interaction of localized waves and coined the term *soliton* to denote their particle-like coherence [**36**].

2010 *Mathematics Subject Classification.* Primary 34C60, 37J35.

The second and third authors gratefully acknowledge support for this work by NSF DMS 1009517.

The story of the first investigations by FPU –as well as the follow-on discoveries by Kruskal, et. al – has been told many times, so it will not be repeated here. Moreover, the subsequent literature on the both the FPU problem and KdV is too vast and varied to summarize adequately in this space. For our purposes, the essential elements are as follows: FPU considered their nonlinear lattice systems to be "toy" models of a general, macroscopic, physical system in which nonlinear terms cause interaction between the normal modes of the linear system (i.e., Fourier modes). In accordance with the postulates that underly statistical derivations of macroscopic thermodynamics, FPU expected to see a single-mode initial condition evolve to a state in which the energy was evenly distributed across all the Fourier modes ("thermalization"). Instead, numerical simulations yielded repeated approximate recurrences of the single-mode initial state at evenly-spaced time intervals.

Given the discovery of recurrence by FPU and the derivation of KdV as an asymptotic model of the FPU lattice (cf. e.g, [**31**]), Zabusky and Kruskal did not identify recurrences observed in simulations of KdV [**36**] as a novelty. Instead, for Zabusky and Kruskal, the novel surprising result observed in simulations of KdV was the elastic interaction of solitons (in the approximate form of the classical sech^2 traveling-waves of KdV). The discovery of the solitons (more precisely, their elastic interactions) inspired the development of the Inverse-Scattering Transform (IST) method for the solution of KdV [**16**], which in turn led to the characterization of KdV as a completely integrable Hamiltonian system (cf. e.g., [**13**]). Despite the original suspicions of Kruskal [**27**], the theory of solitons and the IST is, of course, not uniquely applicable to KdV, and has been extended to a broad array of physically important and mathematically rich nonlinear evolution equations on the line (cf. e.g., [**3**]), as well as systems with periodic and quasiperiodic boundary conditions (cf. e.g., [**11, 23, 29**]).

The logical circle is completed by invoking the solitons of KdV as an explanation for the recurrence observed by Zabusky and Kruskal (e.g., [**30, 36**]). By extension, the recurrence observed in simulations of the FPU lattice is then explained on the grounds that FPU dynamics are governed by KdV (e.g., [**8, 32**]). Even at first glance, the connections are non-trivial: the individual sech^2 solitons, and multisoliton solutions, are solutions of KdV with vanishing boundary conditions on the whole line ($u \to 0$ as $x \to \pm\infty$), while solutions of KdV with periodic boundary conditions are more complex (cf. e.g., [**11, 23, 29**]). Nevertheless, if an individual soliton is narrow compared to the spatial period, the soliton does not "see" the periodic boundary condition and exists as an approximate solution (the limit of a periodic "cnoidal" wave). Indeed, the correspondence between periodic and whole-line solutions can be developed even for a spatially periodic solution whose support is not small compared the spatial period [**30**]. Even so, the periodic boundary conditions are an important ingredient for the soliton view of recurrence: to achieve (approximate) recurrence, the unidirectional solitons of whole-line KdV "wrap around" in order to nearly recover their original arrangement as the system evolves.

The system integrated by Zabusky and Kruskal [**36**] was not, strictly speaking, KdV (1.1) itself, but rather the lattice

(1.3)
$$\frac{\mathrm{d}}{\mathrm{d}t}u_n = \frac{(u_{n-1}+u_n+u_{n+1})(u_{n-1}-u_{n+1})}{6h} + \delta^2\frac{u_{n-2}-2u_{n-1}+2u_{n+1}-u_{n+2}}{2h^3}$$

which, by inspection, is an $O(h^2)$ discretization of KdV. The system of equations for the lattice can be written in vector-matrix form as

$$\frac{\mathrm{d}}{\mathrm{d}t}\boldsymbol{u} = -\left(\boldsymbol{A}_3\boldsymbol{u}\right) \times \left(\boldsymbol{D}_c\boldsymbol{u}\right) - \delta^2 \boldsymbol{D}_{3c}\boldsymbol{u}$$

where: $\boldsymbol{u}$ is an N-component vector of the values u_n, the matrices $\boldsymbol{A}_3$, $\boldsymbol{D}_c$ and $\boldsymbol{D}_{3c}$ represent the differences in (1.3) and $\times$ represents a component-by-component vector product. (Of course, a scheme is also required to discretize the time steps. However, consideration of the time-stepping method used in [**36**] is not essential, as the results described there, and here, can be obtained with any of a number of sufficiently accurate ODE integration schemes.) Further simulations [**17**] with the same lattice (1.3), but with a two-mode intial condition and a different value of δ^2, also showed repeated near-recurrences of the initial state.

The averaging coefficient, $\boldsymbol{A}$, in the nonlinear term of the lattice (1.3) leads to the exact preservation of the quantities

$$m_1 = \sum_n u_n \qquad m_2 = \sum_n u_n^2,$$

with the imposition of periodic or vanishing (on a bi-infinite lattice) boundary conditions. These conserved quantities correspond to the preserved quantities

$$m_1 = \int u \, dx \qquad m_2 = \int u^2 \, dx$$

for KdV under the same boundary conditions. In particular, the conservation of m_2 on the lattice precludes runaway growth of the solution as it evolves.

An alternative explanation for the observed near-recurrence of the low-mode initial condition is to assert that the evolution of the ZK lattice (1.3) is governed by the properties of KdV (1.1) as a completely integrable (infinite-dimensional) Hamiltonian system, but without reference to the solitons of KdV (1.1) on the line. In rough summary, for KdV with periodic boundary conditions, the "finite gap" quasiperiodic (in time) solutions are dense in the space of solutions [**25, 26**], and these quasiperiodic solutions are almost periodic (in time).

While a generic perturbation of an integrable system destroys the integrability, the well-known KAM theorem describes the persistence of quasiperiodic orbits of finite-dimensional integrable systems under Hamiltonian perturbations (cf. [**9**] and references therein). In this regard, we note that, even though the ZK lattice (1.3) is not integrable as a system of ODEs [**24**], it can be understood as a perturbation of KdV, which, as noted, is a completely integrable Hamiltonian PDE. In this regard, there has been significant work to extend KAM to infinite-dimensional systems, including especially KdV with peridoc boundary conditions [**20, 22**].

The Nekhoroshev theorem [**28**] complements the KAM theorem by providing a bound, valid for exponentially-long times, on the divergence of orbits of the perturbed system from the quasiperiodic orbits of the unperturbed (integrable) system. (See [**10**] for a unified treatment of KAM and Nekhoroshev theorems.) In the standard Nekhoroshev theorem, both (i) the bound on the divergence of the orbit of the perturbed system from the orbit of the integrable system and (ii) the expression for the maximum time of the validity for that bound include the a term in which the perturbation parameter is raised to a power of $\frac{1}{N}$, where N is the dimension of the system. If we consider the ZK lattice (1.3) as a perturbation of KdV with spatial period L, we have $L = Nh$, where N is the number of grid points and h is the both the grid spacing and the perturbation parameter. Then, because

$$h^{(\frac{1}{N})} = h^{(\frac{h}{L})} \to 1 \,, \quad \text{as} \quad h \to 0 \,, \tag{1.4}$$

the bound provided by the standard Nekhoroshev theorem becomes ineffective in the continuum (PDE) limit. Notwithstanding this complication, there has been some progress in the development of Nekhoroshev-like theorems for infinite-dimensional systems (e.g., [**6**, **7**, **14**, **33**]. Hence, it would be reasonable to hypothesize Nekhoroshev-like bounds on the orbits of discretizations of KdV.

The supposition that recurrence in the ZK lattice (1.3) is a manifestation of the solitons and/or integrability of KdV leads one naturally to hypothesize that *stronger* and *more-persistent* recurrences than described in [**36**] will occur:

H1 as the lattice spacing h approaches zero (with $Nh = L$ fixed), and the lattice converges to KdV;
H2 with a higher-accuracy discretization of KdV, such as a spectral or pseudo-spectral discretization;
H3 with a discretization that is itself integrable as a system of ODEs.

In this article, we present evidence obtained by numerical simulation that the above hypotheses are *false*. Consequently, we infer that the underlying structure which leads to recurrence in the lattice is not a result of either the (approximate) existence of solitons nor of integrability more generally.

The original expectation of FPU (that they would observe the equipartition of energy among the modes in nonlinear lattices) is, in some sense, diametrically opposite from the results of their simulations (repeated recurrence of a single-mode state). However, while recurrence precludes equipartition, the absence of recurrence is, by itself, insufficient to imply equipartition. Instead, a system can evolve to a state in which energy is shared unequally between the modes (or a subset of the modes), but there is no recurrence of the single-mode state. Therefore, the question of the mechanism of recurrence is distinct from, though likely related to, the question of equipartition in KdV and FPU lattices.

Following the original work of Zabusky and Kruskal [**36**], we consider simulations of the ZK lattice (1.3) with the single-mode initial condition (1.2) restricted to an evenly spaced lattice,

$$u_n(0) = \sin \pi h n \,, \quad n = 0, \ldots, N-1 \,, \tag{1.5}$$

where $Nh = L = 2$. Moreover, following the original simulations, we take $\delta = .022$ in (1.3). In an examnation of 1, Section 2 describes the results of simulations of ZK (1.3) with $70 \leq N \leq 300$. To examine Hypothesis 1 we consider the spectral discretization of (1.1), as described in Section 3. To shed light on the role of integrability in the recurrence (1), we construct a semi-discretization (i.e., a spatial discretization) of KdV which is integrable as a lattice/ODE system (indeed, a family of integrable systems that converges to KdV as the grid-spacing parameter $h \to 0$) in Section 4. For all three spatial discretizations, the time evolution was computed with the variable-step-size "ode5r" routine in the "odepkg" [**35**] extension of Octave [**12**]. The ode5r routine of odepkg uses the radau5.f Fortran code [**19**] to compute the time evolution.

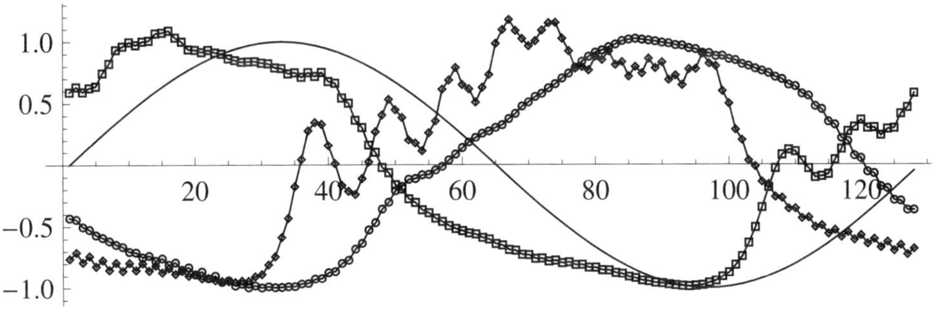

FIGURE 1. Evolution of ZK lattice (1.3) showing repeated, approximate recurrences of the lowest Fourier mode at regular time intervals. For this simulation, $\delta = .022$ and $N = 128$ with the single-mode initial condition (1.5). Legend: circles $t = 9.956$; squares, $t = 19.910$; diamonds, $t = 29.856$.

2. Zabusky-Kruskal Lattice

The approximate recurrence to the low-mode condition (with a phase shift) is plainly visible in plots of the simulation (Figure 1). However, the recurrence is more striking in Fourier space (Figure 2). Moreover, the Fourier-space view of the evolution of (1.3) clearly shows that the regular recurrence of the low-mode state is a transient phenomenon.

The strength and persistence of the recurrence (or, inversely, the decay rate of the recurrence peaks) depends strongly on the number of nodes of the lattice (inversely proportional to the discretization length, h). Specifically, the recurrences are strongest and most persistent for a middle range of N (centered on $N \approx 100$) values. Notably recurrence *weakens* as $N \to \infty$ and the lattice approaches KdV. In contrast, the *time* of recurrences, is not observed to be dependent on N. (See Figure 3.)

These simulations suggest, surprisingly, that the mechanism of recurrence is a property of the ZK lattice itself, not simply a reflection of the properties of the continuum limit, KdV. In other words, recurrence in KdV occurs because KdV is "close" to the ZK lattice, and not vice-versa. Given the narrow range (in N) of the strongest and most persistent recurrence, it is was a stroke of good luck that Zabusky and Kruskal considered simulations in the middle of this set values.

3. Spectral Discretization

The spectral discretization of KdV (1.1) is obtained by truncation of the Fourier-series expansion of the the PDE [**2**]. That is, we write KdV with periodic boundary conditions in terms of the Fourier series (in x) of u to obtain

$$\frac{\mathrm{d}}{\mathrm{d}t}\hat{u}_k = -\frac{i2\pi k}{L}\sum_{\ell}\hat{u}_{k-\ell}\hat{u}_\ell - \delta^2\left(\frac{i2\pi k}{L}\right)^3\hat{u}_k\,,$$

where

$$u = \sum_k \hat{u}_k e^{i\left(\frac{2\pi k}{L}\right)x}\,,$$

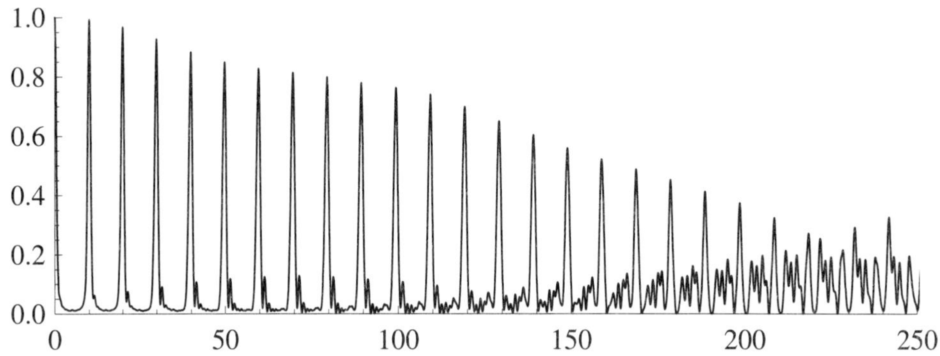

FIGURE 2. Evolution of the magnitude of the first Fourier mode of ZK (1.3) lattice showing repeated, approximate recurrences at regular time intervals. For this simulation, $\delta = .022$ and $N = 128$ with the single-mode initial condition (1.5).

and then truncate the convolution sum for some maximum $|k|$. In vector-matrix form, the spectral discretization is

$$\frac{\mathrm{d}}{\mathrm{d}t}\boldsymbol{u} = -\boldsymbol{D}_f\boldsymbol{F}^{-1}\left[(\boldsymbol{F}\boldsymbol{u}) * (\boldsymbol{F}\boldsymbol{u})\right] - \delta^2\boldsymbol{D}_f^3\boldsymbol{u}\ , \tag{3.1}$$

where: $\boldsymbol{D}_f = \boldsymbol{F}^{-1}\boldsymbol{\Omega}\boldsymbol{F}$ is the Fourier differentiation matrix; $\boldsymbol{F}$ is the matrix form of the discrete Fourier transform; $\boldsymbol{\Omega}$ is the (diagonal) matrix of wavenumbers; and $*$ is the convolution operation,

$$(\boldsymbol{u} * \boldsymbol{v})_k = \sum_j u_{k-j}v_j\ ,$$

where the subscript is the index label of the component.

In comparison with the $O(h^2)$-approximation given by the ZK lattice (1.3), the spectral discretization is accurate beyond all orders in h [**18**, **21**] (and is exponentially accurate for spatially analytic data [**34**]).

The distinction between the spectral discretization and a *pseudo*-spectral discretization (collocation), in which the convolution term in (3.1) is replaced with a pointwise product, should not be overlooked. The pseudospectral discretizations of KdV are nonlinearly unstable, even for relatively fine discretizations, due to aliasing. Hence, we instead consider the stable numerically spectral method (3.1) as a high-accuracy discretization of KdV (1.1).

The spectral scheme shows repeated, decaying recurrences reminiscent of the result for the ZK scheme (1.3). However, for $N = 128$ (equivalently, 128 Fourier modes) the spectral discretization (3.1) evolves like the ZK lattice with a much larger N value. (See Figure 4.) As expected, the spectral scheme converges to the common continuum limit, KdV (1.1), much faster (in N) than the ZK scheme. As a consequence, there is no intermediate lattice scale for which the recurrence is stronger than for N large. Instead, as N increases, the scheme has a relatively abrupt transition from N values with no discernible recurrence to the limiting recurrence behavior. (See Figure 3.) Previous investigation of the spectral method for KdV [**1**] did not reveal recurrence due to the small number of modes used (as limited by computing capacity available at that time).

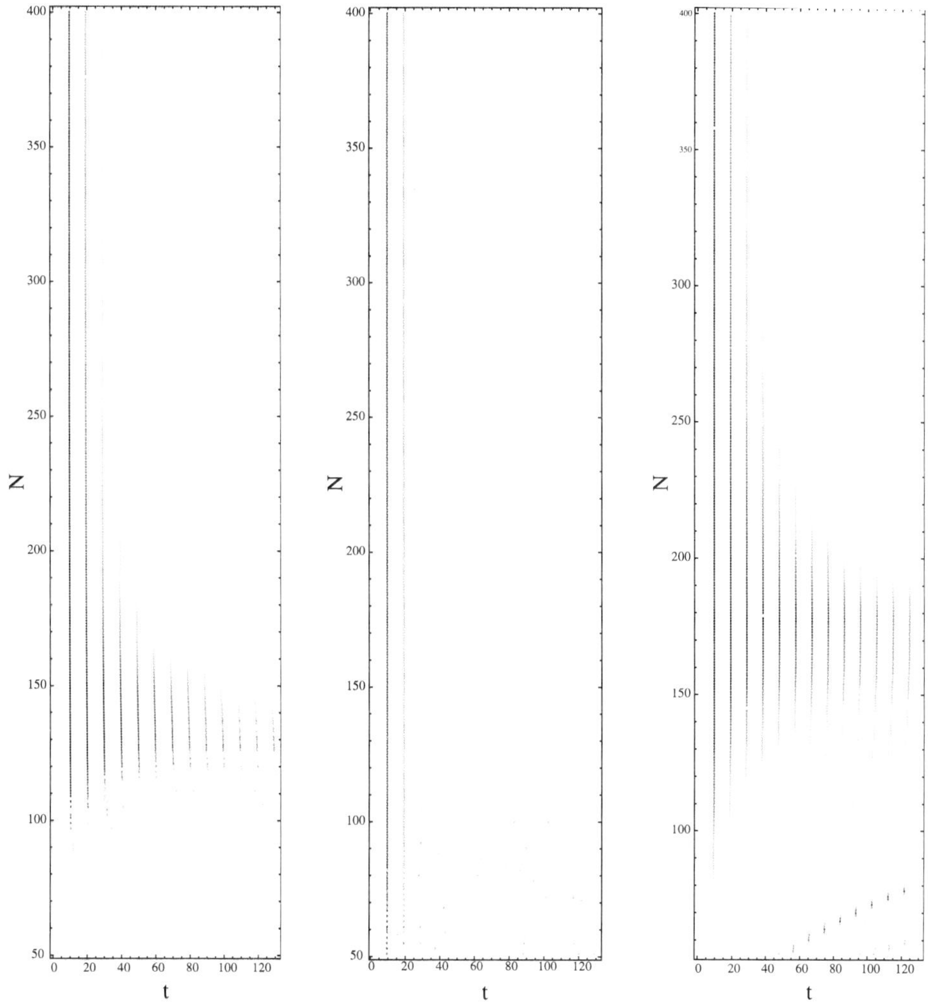

FIGURE 3. Temporal peaks of the first Fourier mode by the number of nodes (N), for the ZK lattice (1.3) (left), Spectral Discretization (3.1) (center), and Integrable Discretization (4.6) (right), all with $\delta = .022$ and single-mode initial condition (1.5). The intensity of the peak is indicated by the gray level of the point: $|c_1| = 0$, white to $|c_1| = 1$, black.

4. Integrable Discretization

We construct an integrable discretization of KdV (1.1) by (i) discretizing the linear scattering problem of the zero-curvature condition associated wiht KdV (1.1) and then (ii) choosing a time-dependence matrix so that the evolution equation that results from the discrete compatibility condition has KdV as a continuum limit. This approach is an adaptation of the method used to construct integrable discretizations of nonlinear Schrödinger equation [**4**] as well as matrix and vector generalizations [**5**].

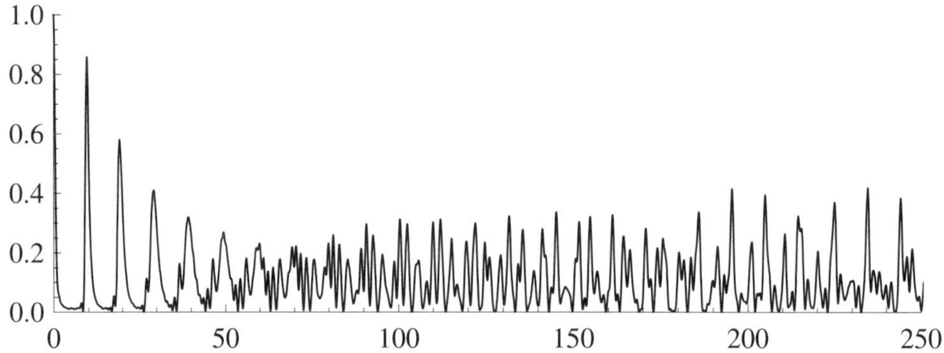

FIGURE 4. Evolution of the magnitude of the first Fourier mode of the spectral scheme (3.1), with $\delta = .022$, $N = 128$ and single-mode initial condition (1.5), showing repeated, approximate recurrences at regular time intervals with rapid decay of the recurrence strength.

We recall that KdV (1.1) is equivalent to the to the zero-curvature condtion,

$$\boldsymbol{S}_t = \boldsymbol{T}_x + \boldsymbol{T}\boldsymbol{S} - \boldsymbol{S}\boldsymbol{T}\,, \tag{4.1}$$

where

$$\boldsymbol{S} = \begin{pmatrix} ik & u \\ -\frac{1}{6\delta^2} & -ik \end{pmatrix}, \qquad \boldsymbol{T} = \begin{pmatrix} -\frac{1}{6}u_x - \frac{i}{3}uk & \frac{1}{3}u^2 - \delta^2 u_{xx} - 2i\delta^2 u_x k + 4\delta^2 uk^2 \\ \frac{1}{18\delta^2}u - \frac{2}{3}k^2 & \frac{1}{6}u_x + \frac{i}{3}uk - 8i\delta^2 k^3 \end{pmatrix},$$

and k is a space- and time-independent parameter. The zero-curvature condition (4.1) is equivalent to the equality of mixed derivatives ($\boldsymbol{\psi}_{xt} = \boldsymbol{\psi}_{tx}$) for simultaneous solutions of

$$\boldsymbol{\psi}_x = \boldsymbol{S}\boldsymbol{\psi}\,, \qquad \boldsymbol{\psi}_t = \boldsymbol{T}\boldsymbol{\psi}\,.$$

These two linear equations are used to construct the Inverse-Scattering Transform, which makes integrability operational, but we do not discuss this here.

The spatially discrete equation

$$\boldsymbol{\psi}_{n+1} = \boldsymbol{S}_n\boldsymbol{\psi}_n, \text{ where } \quad \boldsymbol{S}_n = \begin{pmatrix} z & U_n \\ \alpha & z^{-1} \end{pmatrix}\boldsymbol{\psi}_n\,, \tag{4.2}$$

with

$$U_n = hu_n\,, \; \alpha = \frac{h}{6\delta^2}\,, \; z = 1 + ikh + O(h^2) = e^{ikh}\,, \tag{4.3}$$

is a forward-difference discretization of the linear differential equation (in x) above. With the time-dependence equation

$$\frac{\mathrm{d}}{\mathrm{d}\tau}\boldsymbol{\psi}_n = \boldsymbol{T}_n\boldsymbol{\psi}_n\,, \tag{4.4}$$

where

$$
\begin{aligned}
T_{1,1} &= \tfrac{1}{2}z^{-4} - \alpha U_{n-1}z^{-2} + \alpha(U_n - U_{n+1}) - \alpha^2 U_n(U_{n-1} + U_n + U_{n+1}) \\
&\quad + (2 + \alpha U_n)z^2 - \tfrac{1}{2}z^4 \\
T_{1,2} &= -U_{n-1}z^{-3} - (U_{n-2} - 2U_{n-1} - \alpha U_{n-1}(U_{n-2} + U_{n-1}))\, z^{-1} \\
&\quad + (2u_n + U_{n+1} + \alpha U_n(U_n + U_{n+1}))\, z + U_n z^3 \\
T_{2,1} &= -\alpha z^{-3} + \alpha\,(1 + \alpha(U_{n-1} + U_n))\, z^{-1} + \alpha\,(1 + \alpha(U_{n-1} + u_n))\, z - \alpha z^3 \\
T_{2,2} &= -\tfrac{1}{2}z^{-4} + (2 + \alpha U_{n-1})z^{-2} + \alpha(U_{n-2} - U_{n-1}) + \alpha^2 U_{n-1} + \alpha U_n z^2 + \tfrac{1}{2}z^4\,,
\end{aligned}
$$

the compatibility condition (counterpart of the zero-curvature condition),

$$\frac{\mathrm{d}}{\mathrm{d}\tau}\boldsymbol{S}_n = \boldsymbol{T}_{n+1}\boldsymbol{S}_n - \boldsymbol{S}_n\boldsymbol{T}_n\,,$$

is equivalent to

(4.5)

$$
\begin{aligned}
\frac{\mathrm{d}}{\mathrm{d}\tau}U_n = (1 + \alpha U_n)\,\{\alpha\,[U_{n-1}(U_{n-2} - U_n) &+ U_{n+1}(U_n - U_{n+2}) \\
&+ (U_{n-1} + 2U_n + U_{n+1})(U_{n-1} - U_{n+1})] \\
&+ U_{n-2} - 2U_{n-1} + 2U_{n+1} - U_{n+2}\}\;.
\end{aligned}
$$

Under the rescaling (4.3) and $\tau = \frac{1}{h^4}t$, equation (4.5) becomes

(4.6)

$$
\begin{aligned}
\frac{\mathrm{d}}{\mathrm{d}t}u_n = \left(1 + \frac{h^2 u_n}{6\delta^2}\right)\Bigg[\frac{u_{n-1}(u_{n-2} - u_n)}{12h} &+ \frac{(u_{n-1} + 2u_n + u_{n+1})(u_{n-1} - u_{n+1})}{12h} \\
&+ \frac{u_{n+1}(u_n - u_{n+2})}{12h} \\
&+ \delta^2\frac{u_{n-2} - 2u_{n-1} + 2u_{n+1} - u_{n+2}}{2h^3}\Bigg]\,,
\end{aligned}
$$

which converges to KdV (1.1) as $h \to 0$. We note that, reminiscent of the ZK discretization (1.3), the nonlinear terms of (4.6) are equivalent to averages of elementary discretizations of the nonlinear term of KdV (1.1).

The integrable scheme (4.6) manifests a recurrence behavior similar to that of the ZK lattice (1.3): the strongest and most-persistent recurrence occurs in a middle range of N values, the solutions converge to the KdV limit (with weaker recurrences) as N gets large and recurrences are weak or non-existent for N below a threshold. Moreover, as with ZK (1.3), the time of recurrences, when they do occur, does not depend on N. However, the range of N where recurrence is strongest and most persistent is not the same as for the ZK scheme. (See Figure 3.) The strongest and most persistent recurrence in the integrable scheme is more pronounced than that of the ZK scheme (which can be seen by comparison of Figure 2 and Figure 5) .

5. Conclusion: Recurrence as a Property of the Lattice

Taken together, the systematic simulations described in the preceding sections suggest that recurrence of the low-mode initial condition observed by Zabusky and Kruskal [**36**] is a property of nonlinear lattices and *not* primarily a manifestation of properties of the PDE limit (i.e., KdV (1.1)). On the other hand, a reduction

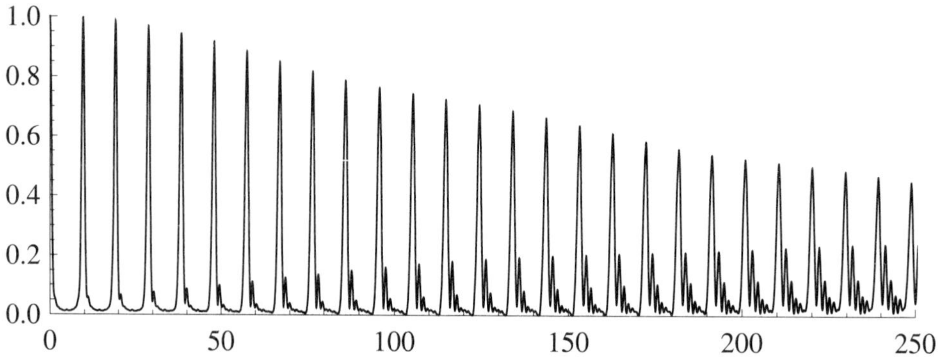

FIGURE 5. Evolution of the magnitude of the first Fourier mode of the integrable scheme (4.6), with $\delta = .022$, $N = 170$ and single-mode initial condition (1.5), showing repeated, approximate recurrences at regular time intervals.

of the number of degrees of freedom in the system leads to a complete breakdown of recurrence as the number of degrees of freedom falls below a threshold. Hence, recurrence can not be attributed to a restriction of the phase space. In the 'middle' range between the breakdown of recurrence (N small) and dynamics dominated by the PDE limit (N large), the Zabusky-Kruskal (1.3) and integrable (4.6) lattices exhibit their strongest recurrence, due to some as-yet-undetermined property of the lattices. We emphasize that the existence of this middle-N region of maximal recurrence in the integrable scheme demonstrates that integrability itself is not decisive in determining the recurrence behavior of a lattice.

The representation of the state of the lattice systems by Fourier coefficients can be interpreted as an (isometric) change of coordinates. The convention of using the Fourier coefficients and the low-mode initial condition to investigate recurrence goes back to FPU [**15**] and is grounded in the fact that the magnitudes of these coefficients are the power in the normal modes of the linear part of system. However, for the configurations considered here (and previously by Zabusky and Kruskal [**36**]), the linear term is not dominant and the Fourier modes interact strongly. Therefore, the Fourier coefficients are not intrinsic to the observed dynamics. Instead, the integrable system has a natural set of coordinates, the action-angle variables. In action-angle coordinates, the mechanism of recurrence may be clearer. However, the development of the action-angle coordinates is probably best realized through the development of the semi-discrete Inverse-Scattering Transform for the pair (4.2),(4.4), which is beyond the scope of this work.

References

[1] Kanji Abe and Takashi Abe, *Recurrence of initial state of the Korteweg-de Vries equation*, Phys. Fluids **22** (1979), no. 9, 1644–1646, DOI 10.1063/1.862824. MR542702 (80e:65110)

[2] Kanji Abe and Osamu Inoue, *Fourier expansion solution of the Korteweg-de Vries equation*, J. Comput. Phys. **34** (1980), no. 2, 202–210, DOI 10.1016/0021-9991(80)90105-9. MR559996 (81a:65113)

[3] M. J. Ablowitz and P. A. Clarkson, *Solitons, nonlinear evolution equations and inverse scattering*, London Mathematical Society Lecture Note Series, vol. 149, Cambridge University Press, Cambridge, 1991. MR1149378 (93g:35108)

[4] M. J. Ablowitz and J. F. Ladik, *Nonlinear differential-difference equations*, J. Mathematical Phys. **16** (1975), 598–603. MR0377223 (51 #13396)
[5] M. J. Ablowitz, B. Prinari, and A. D. Trubatch, *Discrete and continuous nonlinear Schrödinger systems*, London Mathematical Society Lecture Note Series, vol. 302, Cambridge University Press, Cambridge, 2004. MR2040621 (2005c:37117)
[6] Dario Bambusi, *Nekhoroshev theorem for small amplitude solutions in nonlinear Schrödinger equations*, Math. Z. **230** (1999), no. 2, 345–387, DOI 10.1007/PL00004696. MR1676714 (2000h:35146)
[7] D. Bambusi and N. N. Nekhoroshev, *Long time stability in perturbations of completely resonant PDE's*, Acta Appl. Math. **70** (2002), no. 1-3, 1–22, DOI 10.1023/A:1013943111479. Symmetry and perturbation theory. MR1892373 (2003f:37141)
[8] Dario Bambusi and Antonio Ponno, *On metastability in FPU*, Comm. Math. Phys. **264** (2006), no. 2, 539–561, DOI 10.1007/s00220-005-1488-1. MR2215616 (2007a:37087)
[9] *Smooth ergodic theory and its applications*, Proceedings of Symposia in Pure Mathematics, vol. 69, American Mathematical Society, Providence, RI, 2001. Edited by Anatole Katok, Rafael de la Llave, Yakov Pesin and Howard Weiss. MR1858533 (2002d:37005)
[10] Amadeu Delshams and Pere Gutiérrez, *Effective stability and KAM theory*, J. Differential Equations **128** (1996), no. 2, 415–490, DOI 10.1006/jdeq.1996.0102. MR1398328 (97e:58202)
[11] B. A. Dubrovin and S. P. Novikov, *Periodic and conditionally periodic analogs of the many-soliton solutions of the Korteweg-de Vries equation* (Russian, with English summary), Ž. Èksper. Teoret. Fiz. **67** (1974), no. 6, 2131–2144; English transl., Soviet Physics JETP **40** (1974), no. 6, 1058–1063. MR0382877 (52 #3759)
[12] J. W. Eaton et al., *GNU Octave*, `http://www.octave.org`.
[13] Ludwig D. Faddeev and Leon A. Takhtajan, *Hamiltonian methods in the theory of solitons*, Reprint of the 1987 English edition, Classics in Mathematics, Springer, Berlin, 2007. Translated from the 1986 Russian original by Alexey G. Reyman. MR2348643 (2008f:37145)
[14] Erwan Faou and Benoît Grébert, *Quasi-invariant modified Sobolev norms for semi linear reversible PDEs*, Nonlinearity **23** (2010), no. 2, 429–443, DOI 10.1088/0951-7715/23/2/011. MR2578486 (2011j:37156)
[15] E. Fermi, J. Pasta, and S. Ulam, *Studies of Nonlinear Problems I*, Tech. Report LA-1940, Los Alamos Scientific Laboratory of the University of California, 1955.
[16] C. S. Gardner, J. M. Greene, M. D. Kruskal, and R. M. Miura, *Method for solving the Korteweg-de Vries equation*, Physical Review Letters **19** (1967), 1095–1097.
[17] K. Goda, *Numerical studies of recurrence of the Korteweg-de Vries equation*, Journal of the Physical Society of Japan **42** (1977), 1040–1046.
[18] David Gottlieb and Steven A. Orszag, *Numerical analysis of spectral methods: theory and applications*, Society for Industrial and Applied Mathematics, Philadelphia, Pa., 1977. CBMS-NSF Regional Conference Series in Applied Mathematics, No. 26. MR0520152 (58 #24983)
[19] E. Hairer and G. Wanner, *Solving ordinary differential equations. II*, 2nd ed., Springer Series in Computational Mathematics, vol. 14, Springer-Verlag, Berlin, 1996. Stiff and differential-algebraic problems. MR1439506 (97m:65007)
[20] T. Kappeler and J. Pöschel, *KdV & KAM*, Ergebnisse der Mathematik und ihrer Grenzgebiete. 3. Folge. A Series of Modern Surveys in Mathematics [Results in Mathematics and Related Areas. 3rd Series. A Series of Modern Surveys in Mathematics], vol. 45, Springer-Verlag, 2003.
[21] H. Kreiss and J. J. Oliger, *Methods for the approximate solution of time-dependent problems*, GARP publications series, Global Atmospheric Research Programme - WMO-ICSU Joint Organizing Committee, 1973.
[22] S. B. Kuksin, *Nearly integrable infinite-dimensional Hamiltonian systems*, Lecture Notes in Mathematics, vol. 1556, Springer-Verlag, 1993.
[23] Peter D. Lax, *Periodic solutions of the KdV equation*, Comm. Pure Appl. Math. **28** (1975), 141–188. MR0369963 (51 #6192)
[24] Decio Levi and Christian Scimiterna, *Integrability of the Kruskal-Zabusky discrete equation by multiscale expansion*, Nonlinear and modern mathematical physics, AIP Conf. Proc., vol. 1212, Amer. Inst. Phys., Melville, NY, 2010, pp. 66–80. MR2648953 (2011f:37137)
[25] H. P. McKean and E. Trubowitz, *Hill's operator and hyperelliptic function theory in the presence of infinitely many branch points*, Comm. Pure Appl. Math. **29** (1976), no. 2, 143–226. MR0427731 (55 #761)

[26] H. P. McKean and E. Trubowitz, *Hill's surfaces and their theta functions*, Bull. Amer. Math. Soc. **84** (1978), no. 6, 1042–1085, DOI 10.1090/S0002-9904-1978-14542-X. MR508448 (80b:30039)
[27] R. Miura, Personal Communication.
[28] N. N. Nehorošev, *An exponential estimate of the time of stability of nearly integrable Hamiltonian systems* (Russian), Uspehi Mat. Nauk **32** (1977), no. 6(198), 5–66, 287. MR0501140 (58 #18570)
[29] S. P. Novikov, *A periodic problem for the Korteweg-de Vries equation. I* (Russian), Funkcional. Anal. i Priložen. **8** (1974), no. 3, 54–66. MR0382878 (52 #3760)
[30] A. R. Osborne and L. Bergamasco, *The solitons of Zabusky and Kruskal revisited: perspective in terms of the periodic spectral transform*, Phys. D **18** (1986), no. 1-3, 26–46, DOI 10.1016/0167-2789(86)90160-0. Solitons and coherent structures (Santa Barbara, Calif., 1985). MR838309 (87j:35342)
[31] Richard S. Palais, *The symmetries of solitons*, Bull. Amer. Math. Soc. (N.S.) **34** (1997), no. 4, 339–403, DOI 10.1090/S0273-0979-97-00732-5. MR1462745 (98f:58111)
[32] A. Ponno and D. Bambusi, *Korteweg-de Vries equation and energy sharing in Fermi-Pasta-Ulam*, Chaos **15** (2005), no. 1, 015107, 5, DOI 10.1063/1.1832772. MR2133458 (2005k:37160)
[33] Jürgen Pöschel, *On Nekhoroshev estimates for a nonlinear Schrödinger equation and a theorem by Bambusi*, Nonlinearity **12** (1999), no. 6, 1587–1600, DOI 10.1088/0951-7715/12/6/310. MR1726666 (2001e:37100)
[34] Eitan Tadmor, *The exponential accuracy of Fourier and Chebyshev differencing methods*, SIAM J. Numer. Anal. **23** (1986), no. 1, 1–10, DOI 10.1137/0723001. MR821902 (87c:65048)
[35] T. Triechl, *OdePkg*, `http://octave.sourceforge.net/odepkg`.
[36] N. J. Zabusky and M. D. Kruskal, *Interaction of solitons in a collisionless plasma and the recurrence of initial states*, Physical Review Letters **15** (1965), 240–243.

Department of Applied Mathematics, University of Stellenbosch, Stellenbosch, South Africa

Department of Mathematical Sciences, Montclair State University, Montclair, New Jersey 07043

Department of Mathematical Sciences, Montclair State University, Montclair, New Jersey 07043
E-mail address: `david.trubatch@montclair.edu`

Contemporary Mathematics
Volume **635**, 2015
http://dx.doi.org/10.1090/conm/635/12679

On the Location of the Discrete Eigenvalues for Defocusing Zakharov-Shabat Systems having Potentials with Nonvanishing Boundary Conditions

F. Demontis, C. van der Mee, and F. Vitale

Abstract. In this article we prove that the discrete eigenvalues of the Zakharov-Shabat system belong to certain neighborhoods of the endpoints of the spectral gap and the discrete eigenvalue of the free Hamiltonian.

1. Introduction

The nonlinear Schrödinger (NLS) equation is a well-known physically and mathematically significant nonlinear evolution equation extensively studied for over forty years. For example, the NLS equation has been derived in the modeling of ocean water waves [**2**, **21**], Bose-Einstein condensation [**18**], and optical fibers [**10**, **11**].

In this work we consider the *defocusing* NLS equation, i.e.,

$$iq_t + q_{xx} - 2|q|^2 q = 0\,, \tag{1.1}$$

[subscripts x and t denote partial differentiation throughout] with nonzero boundary conditions (NZBCs)

$$q(x,t) \to q_\pm(t) = q_0 e^{2iq_0^2 t + i\theta_\pm}, \quad \text{as } x \to \pm\infty, \tag{1.2}$$

where i denotes the complex unit, $q_0 > 0$ and $0 \le \theta_\pm < 2\pi$ are arbitrary constants. It is well-known that Eq. (1.1) is associated to the so-called Zakharov-Shabat (ZS) system:

$$\frac{\partial X}{\partial x}(x,k) = (-ik\sigma_3 + Q(x))X(x,k), \qquad x \in \mathbb{R}, \tag{1.3}$$

where

$$\sigma_3 = \begin{pmatrix} 1 & 0 \\ 0 & -1 \end{pmatrix}, \qquad Q(x) = \begin{pmatrix} 0 & q(x) \\ q^*(x) & 0 \end{pmatrix}, \tag{1.4}$$

$q(x)$ is the potential, k is a complex spectral parameter and the asterisk denotes the complex conjugate, by means of the inverse scattering transform.

Recently, the defocusing NLS (1.1) with NZBCs has been the subject of renewed interest because of its applications to Bose-Einstein condensates [**8**, **9**] and dispersive shock waves in optical fibers [**19**]. This justifies our effort to investigate some questions connected with this subject. In particular, we focus our attention on some aspects which arise when the Inverse Scattering Transform (IST) (see [**1**, **2**, **20**] for

2000 *Mathematics Subject Classification.* Primary 34A55; Secondary 78A46.

a general review of this method) is applied to the equation (1.1) with NZBCs (1.2). In fact, the IST for (1.1) with NZBCs (see [**3**, **4**, **7**, **12**, **13**, **17**, **22**]) is much more complicated than for (1.1) with decaying potentials, in particular with regard to the analyticity properties of the eigenfunctions of the scattering problem (1.3) and the corresponding scattering data. A step forward in that direction was recently made in [**6**] where it was proved that the direct scattering problem is well defined for potentials q such that $q - q_\pm$ belongs to the functional class $L^{1,2}(\mathbb{R}^\pm)$ ($L^{1,p}(\mathbb{R}^\pm)$ consists of all functions $f(x)$ satisfying $\int_{\mathbb{R}} dx\,(1+|x|)^p\,|f(x)| < \infty$). For this reason, we will assume that $q - q_\pm \in L^{1,2}(\mathbb{R}^\pm)$.

It is also worthwhile to study if an *area theorem* for the defocusing NLS (1.1) with NZBCs can be proved, which means establishing existence and location of discrete eigenvalues (see Section 2 for the definition of a discrete eigenvalue) of the scattering problem (1.3) as a function of the area of the initial profile of the solution of (1.1). It is well known that for the focusing NLS with vanishing boundary conditions such a result already exists [**2**, **14**–**16**]: in fact, there are no discrete eigenvalues of (1.3) if the L^1-norm of the potential is smaller than $\pi/2$. Only recently [**5**], the non existence of an analogous result for equation (1.1) with NZBCs has been proved. In [**5**], the authors showed that no area theorem is possible for the defocusing NLS with NZBCs by providing explicit examples of box-type initial conditions where at least one discrete eigenvalue exists. In the present paper, we analyze a class of potentials more general than that considered in [**5**] (we only require that $q - q_\pm \in L^{1,2}(\mathbb{R}^\pm)$) and establish the conditions (equations (4.8) and (4.13) in Section 4) which the potentials have to satisfy in order that a particular, but well specified subset of $(-q_0, q_0)$ does not contain any discrete eigenvalue.

The paper is organized as follows. In Section 2 we review the basic facts on the direct scattering problem for (1.1) with NZBCs (1.2) and discuss an explicit example (different from that considered in [**5**]) which establishes the presence of at least one discrete eigenvalue in $(-q_0, q_0)$. In Section 3 we explicitly construct the resolvent of the free hamiltonian of the spectral problem originating from (1.3) and, finally, in Section 4, adapting the technique used in the vanishing case in [**14**–**16**], we state our main results, namely Theorems 4.3 and 4.4.

2. Preliminaries

In this section we study the direct scattering problem for (1.3) by using the same notations adopted in [**6**] to which we refer the interested reader for details. Moreover, we discuss a significant example which shows that in the spectral gap $(-q_0, q_0)$ there may exist a discrete eigenvalue if $q_+ \neq q_-$.

To study the direct scattering problem of (1.1) with NZBCs (1.2), a new spectral parameter

$$\lambda = \sqrt{k^2 - q_0^2}$$

is introduced which is a conformal mapping from the Riemann k-surface $\mathbb{K}$ onto the Riemann λ-surface Λ. Here $\mathbb{K}$ consists of two sheets, $\mathbb{K}^+$ and $\mathbb{K}^-$, which both coincide with the complex k-plane cut along the semilines $\Sigma = (-\infty, -q_0] \cup [q_0, +\infty)$, where its edges are glued together in such a way that $\lambda(k)$ is continuous throughout the cut. The Riemann surface Λ is the complex λ-plane consisting of the upper half-complex plane Λ^+ and the lower half complex plane Λ^- glued together along the real λ-line. The transformation $k \mapsto \lambda$ maps $\mathbb{K}^{\pm}$ onto $\Lambda^{\pm}$, the cut Σ onto the real λ-line, and the points $\pm q_0$ to zero. Also, $\{\lambda + k, \lambda - k\} \subset \Lambda^{\pm}$ for any $k \in \mathbb{K}^{\pm}$.

For later convenience, we write (1.3) in the form

$$\frac{\partial X}{\partial x}(x,k) = A_{\pm}(k)X(x,k) + (Q(x) - Q_{\pm})X(x,k), \tag{2.1}$$

where

$$A_{\pm}(k) = -ik\sigma_3 + Q_{\pm} = \begin{pmatrix} -ik & q_{\pm} \\ q_{\pm}^* & ik \end{pmatrix}, \quad Q_{\pm} = \begin{pmatrix} 0 & q_{\pm} \\ q_{\pm}^* & 0 \end{pmatrix}. \tag{2.2}$$

Then (1.3) and (2.1) can also be written in the equivalent form

$$\frac{\partial X}{\partial x}(x,k) = A(x,k)X(x,k) + (Q(x) - Q_f(x))X(x,k), \tag{2.3}$$

where,

$$A(x,k) = \theta(x)A_+(k) + \theta(-x)A_-(k), \quad Q_f(x) = \theta(x)Q_+ + \theta(-x)Q_-, \tag{2.4}$$

and $\theta(x)$ denotes the Heaviside function defined as $\theta(x) = 1$ for $x \geq 0$ and $\theta(x) = 0$ for $x < 0$. We associate to the ZS system (2.3) the Hamiltonian operator

$$H = i\sigma_3(d/dx - Q) \tag{2.5}$$

which is selfadjoint on the orthogonal direct sum of two copies of $L^2(\mathbb{R})$. It is also convenient to introduce the free Hamiltonian

$$H_f = i\sigma_3\left(\frac{d}{dx} - Q_f\right). \tag{2.6}$$

For $k \in \Sigma$, we define the *fundamental eigensolutions* $\tilde{\Psi}(x,k)$ and $\tilde{\Phi}(x,k)$ as those square matrix solutions to (1.3) satisfying

$$\tilde{\Psi}(x,k) = e^{xA_+(k)}[I_2 + o(1)], \qquad x \to +\infty, \tag{2.7a}$$

$$\tilde{\Phi}(x,k) = e^{xA_-(k)}[I_2 + o(1)], \qquad x \to -\infty. \tag{2.7b}$$

We can prove their existence, for $k \in \Sigma$, as the unique solutions of the Volterra integral equations [**6**, Prop. 1]

$$\tilde{\Psi}(x,k) = \mathcal{G}(x,0;k) - \int_x^{\infty} dy\, \mathcal{G}(x,y;k)[Q(y) - Q_f(y)]\tilde{\Psi}(y,k), \tag{2.8a}$$

$$\tilde{\Phi}(x,k) = \mathcal{G}(x,0;k) + \int_{-\infty}^{x} dy\, \mathcal{G}(x,y;k)[Q(y) - Q_f(y)]\tilde{\Phi}(y,k). \tag{2.8b}$$

Here the *fundamental matrix* $\mathcal{G}(x,y;k)$ is given by

$$\mathcal{G}(x,y;k) = \begin{cases} e^{xA_+(k)}e^{-yA_+(k)}, & x,y \geq 0, \\ e^{xA_+(k)}e^{-yA_-(k)}, & x \geq 0 \geq y, \\ e^{xA_-(k)}e^{-yA_+(k)}, & y \geq 0 \geq x, \\ e^{xA_-(k)}e^{-yA_-(k)}, & x,y \leq 0. \end{cases} \tag{2.9}$$

In fact, Eqs. (2.8) are uniquely solvable under the condition $q - q_{\pm} \in L^1(\mathbb{R}^{\pm})$ if $\pm q_0 \neq k \in \Sigma$; they are uniquely solvable under the condition $(1 + |x|^2)[q - q_{\pm}] \in L^1(\mathbb{R}^{\pm})$ if $k = \pm q_0$.

Let us now introduce the Jost solutions (as column vector solutions to (2.1) in terms of the fundamental eigensolutions) and the "transition scattering" matrix $S(k)$. Hence, we define the eigenvector matrices:

$$W_{\pm}(k) = \begin{pmatrix} 1 & \frac{-iq_{\pm}}{\lambda+k} \\ \frac{iq_{\pm}^{*}}{\lambda+k} & 1 \end{pmatrix}, \tag{2.10}$$

where $\det W_{\pm}(k) = 2\lambda/(\lambda+k)$ and $A_{\pm}(k)W_{\pm}(k) = W_{\pm}(k)\mathrm{diag}(-i\lambda, i\lambda)$. The *Jost solutions* from the right and the left, respectively, are defined as the columns of

$$\tilde{\Psi}(x,k)W_{+}(k) = \begin{pmatrix} \bar{\psi}(x,k) & \psi(x,k) \end{pmatrix}, \quad \tilde{\Phi}(x,k)W_{-}(k) = \begin{pmatrix} \phi(x,k) & \bar{\phi}(x,k) \end{pmatrix}, \tag{2.11}$$

and a detailed study of their analyticity properties can be found in [**6**, Prop. 3]. Since $\tilde{\Psi}(x,k)$ and $\tilde{\Phi}(x,k)$ are square matrix solutions of the homogeneous first order system (1.3), we have

$$\tilde{\Psi}(x,k) = \tilde{\Phi}(x,k)\mathbb{A}_{l}(k), \qquad \tilde{\Phi}(x,k) = \tilde{\Psi}(x,k)\mathbb{A}_{r}(k), \tag{2.12}$$

where $\mathbb{A}_{l}(k)$ and $\mathbb{A}_{r}(k)$ are called the transition coefficient matrices whose expressions are given by

$$\mathbb{A}_{l}(k) = I_2 - \int_{-\infty}^{\infty} dy\, \mathcal{G}(0,y;k)[Q(y) - Q_f(y)]\tilde{\Psi}(y,k), \tag{2.13a}$$

$$\mathbb{A}_{r}(k) = I_2 + \int_{-\infty}^{\infty} dy\, \mathcal{G}(0,y;k)[Q(y) - Q_f(y)]\tilde{\Phi}(y,k). \tag{2.13b}$$

As a result of (2.11) and (2.12), we get

$$\begin{pmatrix} \phi(x,k) & \bar{\phi}(x,k) \end{pmatrix} = \begin{pmatrix} \bar{\psi}(x,k) & \psi(x,k) \end{pmatrix} S(k), \tag{2.14a}$$

$$\begin{pmatrix} \bar{\psi}(x,k) & \psi(x,k) \end{pmatrix} = \begin{pmatrix} \phi(x,k) & \bar{\phi}(x,k) \end{pmatrix} \bar{S}(k), \tag{2.14b}$$

where [**6**]

$$S(k) = W_{+}^{-1}(k)\mathbb{A}_{r}(k)W_{-}(k) = \begin{pmatrix} a(k) & \bar{b}(k) \\ b(k) & \bar{a}(k) \end{pmatrix}, \tag{2.15}$$

and $\bar{S}(k) = W_{-}^{-1}(k)\mathbb{A}_{l}(k)W_{+}(k) = S^{-1}(k)$. The analyticity and continuity properties of the scattering coefficients $a(k), \bar{b}(k)$, $b(k), \bar{a}(k)$ follow from the analyticity and continuity properties of the Jost solutions using their Wronskian representations [**6**]. It is well known that the scattering data associated to the ZS system (1.3) are (see [**1, 2, 20, 22**]): the reflection coefficient $\rho(k) = b(k)/a(k)$, the zeros of $a(k)$ for $k \in \mathbb{K} \setminus \Sigma$ (the so-called discrete eigenvalues) and a suitable set of constants associated with the discrete eigenvalues known as *norming constants*. However, it is important to remark that for Eq. (1.1) with NZBCs, the discrete eigenvalues belong to the spectral gap $(-q_0, q_0)$ and are simple (proved in [**7**]) and, under the hypothesis $q - q_{\pm} \in L^{1,4}(\mathbb{R}^{\pm})$, are finite in number (established in [**6**]).

We conclude this section analyzing an explicit example which confirms the results obtained in [**5**, Sec. 4]

EXAMPLE 2.1 (Free Hamiltonian). Let us compute the discrete eigenvalues of the free Hamiltonian introduced in (2.6) [which corresponds to assume $Q(x) = Q_f(x)$ in the ZS system (1.3)]. In that case from (2.13) we have $\mathbb{A}_{l}(k) = \mathbb{A}_{r}(k) = I_2$

[I_2 denotes the 2×2 identity matrix], and by using (2.12) we get

$$\tilde{\Psi}(x,k) = \tilde{\Phi}(x,k) = \begin{cases} e^{xA_+(k)}, & \text{for } x > 0\,, \\ e^{xA_-(k)}, & \text{for } x < 0\,. \end{cases} \tag{2.16}$$

From (2.16) and taking into account (2.15) and (2.10), we arrive at

$$S(k) = \begin{pmatrix} a(k) & \bar{b}(k) \\ b(k) & \bar{a}(k) \end{pmatrix} = W_+^{-1}(k)W_-(k) = \frac{\lambda + k}{2\lambda} \begin{pmatrix} 1 - \dfrac{q_+ q_-^*}{(\lambda+k)^2} & i\dfrac{(q_+ - q_-)}{\lambda + k} \\ i\dfrac{(q_-^* - q_+^*)}{\lambda + k} & 1 - \dfrac{q_- q_+^*}{(\lambda+k)^2} \end{pmatrix}.$$

Putting $\lambda = i\ell$ with $\ell \in (0, q_0)$, from the preceding equation we obtain:

$$a(k) = \frac{(k+i\ell)^2 - q_0^2 e^{i(\theta_+ - \theta_-)}}{2i\ell(k+i\ell)} = \frac{k^2 - \ell^2 - q_0^2\cos(\theta_+ - \theta_-) + i\left(2k\ell - q_0^2\sin(\theta_+ - \theta_-)\right)}{2i\ell(k+i\ell)}.$$

As a result $a(k) = 0$ if and only if

$$\cos(\theta_+ - \theta_-) = \frac{k^2 - \ell^2}{q_0^2}, \qquad \sin(\theta_+ - \theta_-) = \frac{2k\ell}{q_0^2}. \tag{2.17}$$

Equation (2.17) has a unique solution $k_0 \in (-q_0, q_0)$, unless $q_+ = q_-$. This eigenvalue $k_0 = 0$ iff $\theta_+ - \theta_-$ is an odd multiple of π.

From now on, we denote with k_0 the unique discrete eigenvalue of the free Hamiltonian operator associated to the scattering problem (1.3) and computed in the example above.

3. Resolvent of the free Hamiltonian

In this section we calculate the resolvent $(k - H_f)^{-1}$ of the free Hamiltonian H_f. The result obtained will be used in the next section to determine the location of the discrete eigenvalues.

Let us compute the resolvent $(k - H_f)^{-1}$ of the free Hamiltonian for $k \notin \sigma(H_f) = \Sigma \cup \{k_0\}$ if $q_+ \neq q_-$ [or $k \notin \sigma(H_f) = \Sigma$ if $q_+ = q_-$]. Letting $F(x)$ be a column vector function in $L^2(\mathbb{R})^2 = L^2(\mathbb{R}) \oplus L^2(\mathbb{R})$, we seek the unique vector function $\Psi(k,x)$ in $L^2(\mathbb{R})^2$ such that $H_f\Psi = k\Psi - F$. Writing $\Psi(x,k) = W(x,k)\psi(x,k)$, where $W(x,k) = W_+(k)\theta(x) + W_-(k)\theta(-x)$, we get for $0 \neq x \in \mathbb{R}$

$$\psi'(x,k) = -i\lambda\sigma_3\psi(x,k) + iW^{-1}(x,k)\sigma_3 F(x), \tag{3.1}$$

where we have used that $A(x,k)W(x,k) = -i\lambda W(x,k)\sigma_3$.

THEOREM 3.1. *Let $q_+ \neq q_-$ and $k \in \mathbb{K}^+$. Then for $F \in L^2(\mathbb{R})^2$ and $k \notin \Sigma \cup \{k_0\}$ we have*

$$[(k - H_f)^{-1}F](x) = \int_{-\infty}^{\infty} dy\, [K_f(x,y;k) + K_{f1}(x;k)K_{f2}(y;k)]\, F(y), \tag{3.2}$$

where $K_f(x,y;k)$ *is given by*

$$
\text{(3.3a)} \qquad K_f(x,y;k) = \begin{cases} -i\,e^{i\lambda(y-x)}W_+(k)\begin{pmatrix}1&0\\0&0\end{pmatrix}\sigma_3W_+^{-1}(k)\sigma_3, & y>x>0,\\ -i\,e^{i\lambda(x-y)}W_+(k)\begin{pmatrix}0&0\\0&1\end{pmatrix}\sigma_3W_+^{-1}(k)\sigma_3, & x>y>0,\\ -i\,e^{i\lambda(y-x)}W_-(k)\begin{pmatrix}1&0\\0&0\end{pmatrix}\sigma_3W_-^{-1}(k)\sigma_3, & 0>y>x,\\ -i\,e^{i\lambda(x-y)}W_-(k)\begin{pmatrix}0&0\\0&1\end{pmatrix}\sigma_3W_-^{-1}(k)\sigma_3, & 0>x>y,\\ 0, & xy<0, \end{cases}
$$

$$
\text{(3.3b)} \qquad K_{f1}(x;k) = \begin{cases} e^{i\lambda x}\begin{pmatrix}0&0\\0&1\end{pmatrix}W_+^{-1}(k), & x>0,\\ e^{-i\lambda x}\begin{pmatrix}1&0\\0&0\end{pmatrix}W_-^{-1}(k), & x<0, \end{cases}
$$

$$
\text{(3.3c)} \qquad K_{f2}(y;k) = \begin{cases} -i\,e^{i\lambda y}Z^{-1}(k)\begin{pmatrix}1&0\\0&0\end{pmatrix}W_+^{-1}(k)\sigma_3, & y>0,\\ i\,e^{-i\lambda y}Z^{-1}(k)\begin{pmatrix}0&0\\0&1\end{pmatrix}W_-^{-1}(k)\sigma_3, & y<0, \end{cases}
$$

$$
\text{(3.3d)} \qquad Z(k) = \begin{pmatrix}1&0\\0&0\end{pmatrix}W_+^{-1}(k) + \begin{pmatrix}0&0\\0&1\end{pmatrix}W_-^{-1}(k) = \frac{1}{2\lambda}\begin{pmatrix}\lambda+k & iq_r\\ -iq_l^* & \lambda+k\end{pmatrix}.
$$

If $q_+ = q_- \neq 0$*, then for* $k \notin \Sigma$

$$
\text{(3.4)} \qquad [(k-H_f)^{-1}F](x) = -i\int_{-\infty}^{\infty} dy\, e^{i\lambda|x-y|}W_+(k)E(y-x)W_+^{-1}(k)\sigma_3F(y),
$$

where

$$
E(w) = \theta(w)\begin{pmatrix}1&0\\0&0\end{pmatrix} + \theta(-w)\begin{pmatrix}0&0\\0&-1\end{pmatrix}.
$$

A similar result can be proved also if $k \in \mathbb{K}^-$ *(or* $\lambda \in \Lambda^-$*).*

PROOF. Let us assume that $q_+ \neq q_-$. Then (3.1) implies the identity

$$
\frac{\partial}{\partial y}\left\{e^{-i\lambda(x-y)\sigma_3}\psi(y,k)\right\} = i\,e^{-i\lambda(x-y)\sigma_3}W^{-1}(y,k)\sigma_3F(y),
$$

where $0 \neq y \in \mathbb{R}$ and $k \notin \Sigma$. Therefore, for $0 \neq y \in \mathbb{R}$ and $k \notin \Sigma$ we obtain

$$
\text{(3.5a)} \qquad \frac{\partial}{\partial y}\left\{e^{i\lambda(y-x)}\begin{pmatrix}1&0\\0&0\end{pmatrix}\psi(y,k)\right\} = i\,e^{i\lambda(y-x)}\begin{pmatrix}1&0\\0&0\end{pmatrix}W^{-1}(y,k)\sigma_3F(y),
$$

$$
\text{(3.5b)} \qquad \frac{\partial}{\partial y}\left\{e^{i\lambda(x-y)}\begin{pmatrix}0&0\\0&1\end{pmatrix}\psi(y,k)\right\} = i\,e^{i\lambda(x-y)}\begin{pmatrix}0&0\\0&1\end{pmatrix}W^{-1}(y,k)\sigma_3F(y).
$$

Let k belong to $\mathbb{K}^+$ [or $\lambda \in \Lambda^+$] and let us integrate (3.5a) with respect to $y \in (x,+\infty)$ for $x \geq 0$ and with respect to $y \in (x,0)$ for $x < 0$ and (3.5b) with respect to $y \in (-\infty,x)$ for $x \leq 0$ and with respect to $y \in (0,x)$ for $x > 0$. Putting together the results obtained in this way for $x > 0$ and $x < 0$, respectively, we obtain [$\Psi(x,k)$ is continuous in $x = 0$ but $\psi(x,k)$ is not]

$$\Psi(x,k) = -iW_+(k)\int_x^\infty dy\, e^{i\lambda(y-x)}\begin{pmatrix}1 & 0\\ 0 & 0\end{pmatrix}W_+^{-1}(k)\sigma_3 F(y)$$

$$+ iW_+(k)\int_0^x dy\, e^{i\lambda(x-y)}\begin{pmatrix}0 & 0\\ 0 & 1\end{pmatrix}W_+^{-1}(k)\sigma_3 F(y)$$

$$+ e^{i\lambda x}W_+(k)\begin{pmatrix}0 & 0\\ 0 & 1\end{pmatrix}W_+^{-1}(k)\Psi(0,k) \quad \text{for } x > 0, \tag{3.6a}$$

$$= iW_-(k)\int_{-\infty}^x dy\, e^{i\lambda(x-y)}\begin{pmatrix}0 & 0\\ 0 & 1\end{pmatrix}W_-^{-1}(k)\sigma_3 F(y)$$

$$- iW_-(k)\int_x^0 dy\, e^{i\lambda(y-x)}\begin{pmatrix}1 & 0\\ 0 & 0\end{pmatrix}W_-^{-1}(k)\sigma_3 F(y)$$

$$+ e^{-i\lambda x}W_-(k)\begin{pmatrix}1 & 0\\ 0 & 0\end{pmatrix}W_-^{-1}(k)\Psi(0,k) \quad \text{for } x < 0, \tag{3.6b}$$

$$= Z^{-1}(k)\left[-i\int_0^\infty dy\, e^{i\lambda y}\begin{pmatrix}1 & 0\\ 0 & 0\end{pmatrix}W_+^{-1}(k)\sigma_3 F(y)\right.$$

$$\left. + i\int_{-\infty}^0 dy\, e^{-i\lambda y}\begin{pmatrix}0 & 0\\ 0 & 1\end{pmatrix}W_-^{-1}(k)\sigma_3 F(y)\right] \quad \text{for } x = 0, \tag{3.6c}$$

provided the matrix $Z(k)$ is nonsingular. Putting $k \in (-q_0, q_0)$ and $\lambda = i\ell$, we get

$$\det Z(k) = \frac{\ell^2 - k^2 + q_0^2\cos(\theta_+ - \theta_-) + i[-2k\ell + q_0^2\sin(\theta_+ - \theta_-)]}{4\ell^2}, \tag{3.7}$$

which can only vanish if (2.17) is true. In other words, in the case $q_+ \neq q_-$ we have to exclude the discrete eigenvalue $k = k_0$.

Let us now consider the case $q_+ = q_-$. Writing $\Psi(x,k) = W_+(k)\psi(x,k)$ for the left-hand side of (3.4), we verify that it satisfies

$$\Psi'(x,k) = A_+(k)\Psi(x,k) + i\sigma_3 F(x).$$

and formula (3.4) follows from (3.3a)-(3.3d) if one takes into account that $q_+ = q_-$ implies $W_+ = W_-$ and this completes the proof. □

We underline that:

a. If $q_+ \neq q_-$, the resolvent operator $(k - H_f)^{-1}$ is the sum of the integral operator with integral kernel $K_f(x,y;k)$ for $k \notin \Sigma$ and a rank two operator for $k \notin \Sigma \cup \{k_0\}$;
b. If $q_+ = q_-$ the resolvent operator $(k - H_f)^{-1}$ is the integral operator with integral kernel $-i\, e^{i\lambda|x-y|}W_+(k)E(y-x)W_+^{-1}(k)\sigma_3$.

4. Location of the discrete eigenvalues

The aim of this section is to characterize the location of the discrete eigenvalues in $(-q_0, q_0)$. In fact, Example 2.1 shows that if $q_- \neq q_+$ at least one such eigenvalue exists. We arrive at our main results mimicking the proofs given by Klaus et al. in [**14–16**] for focusing NLS with decaying potential. We need the following technical results, i.e., Theorems 4.1 and 4.2 below.

THEOREM 4.1. *Let $W^{(1)}(x)$ and $W^{(2)}(x)$ be two 2×2 matrices whose entries belong to $L^2(\mathbb{R})$, and let $k\notin\Sigma$. Then the integral operator on $L^2(\mathbb{R})^2$ with integral kernel*

$$W^{(1)}(x)K_f(x,y;k)W^{(2)}(y)$$

is Hilbert-Schmidt, also if we take the limit as k approaches an interior point of Σ.

PROOF. The squared Hilbert-Schmidt norm of $W^{(1)}(k-H_f)^{-1}W^{(2)}$ minus the rank two contribution is given by

$$\int_{-\infty}^{\infty}dx\int_{-\infty}^{\infty}dy\,\|W^{(1)}(x)\|_{\text{HS}}^2\|W^{(2)}(y)\|_{\text{HS}}^2|K_f(x,y;k)|^2.$$

As the exponential factor $e^{i\lambda|x-y|}$ in the expression for $K_f(x,y;k)$ is bounded above by 1 in absolute value, we get the following upper bound for the squared Hilbert-Schmidt norm of $W^{(1)}(k-H_f)^{-1}W^{(2)}$ minus the rank two contribution:

$$\frac{k^2}{\lambda^2}\left(\int_{-\infty}^{\infty}dx\,\|W^{(1)}(x)\|_{\text{HS}}^2\right)\left(\int_{-\infty}^{\infty}dy\,\|W^{(2)}(y)\|_{\text{HS}}^2\right), \tag{4.1}$$

which, for each $\varepsilon>0$, is uniformly bounded in k satisfying $\sqrt{|k^2-q_0^2|}\geq\varepsilon$. □

The following theorem is immediate as a consequence of the definition of the Hilbert-Schmidt norm. In the norm estimate it does not matter which unitarily equivalent matrix norm is taken, because $K_{f1}(x;k)$ and $K_{f2}(y;k)$ both have rank one.

THEOREM 4.2. *Let $W^{(1)}(x)$ and $W^{(2)}(x)$ be two 2×2 matrices whose entries belong to $L^2(\mathbb{R})$, and let $k\notin\Sigma$. Then the integral operator on $L^2(\mathbb{R})^2$ with integral kernel*

$$W^{(1)}(x)K_{f1}(x;k)K_{f2}(y;k)W^{(2)}(y)$$

is rank two and its norm coincides with

$$\left[\int_{-\infty}^{\infty}dx\,\left\|W^{(1)}(x)K_{f1}(x;k)\right\|^2\right]^{1/2}\left[\int_{-\infty}^{\infty}dy\,\left\|K_{f2}(y;k)W^{(2)}(y)\right\|^2\right]^{1/2}. \tag{4.2}$$

It follows from [**14**, Eq. (4.10)] and [**16**, proof of Thm. 4.2] that the norm of the integral operator on $L^2(\mathbb{R})$ with kernel $e^{-\tau(y-x)}\theta(y-x)\sqrt{m(x)m(y)}$ is less than $(2/\pi)\|m\|_1$. To generalize this result to the nonvanishing case, we start with the polar decomposition

$$\begin{aligned}Q(x)-Q_f(x)&=\begin{pmatrix}0&\sigma(x)\\ \sigma(x)^*&0\end{pmatrix}\sqrt{|q(x)-q_f(x)|I_2}\cdot\sqrt{|q(x)-q_f(x)|I_2}\\ &=U_\delta(x)|\Delta(x)|^{1/2}|\Delta(x)|^{1/2},\end{aligned}\tag{4.3}$$

where

$$q_f(x)=(q(x)-q_-)\theta(-x)+(q(x)-q_+)\theta(x),\qquad \Delta(x)=|q(x)-q_f(x)|I_2, \tag{4.4}$$

$|\sigma(x)|=1$, and hence $U_\delta(x)=\begin{pmatrix}0&\sigma(x)\\ \sigma(x)^*&0\end{pmatrix}$ is a unitary matrix. For $k\notin\Sigma$ we now define

$$\mathcal{W}(k)=I+|\Delta|^{1/2}(k-H_f)^{-1}U_\delta|\Delta|^{1/2}. \tag{4.5}$$

Then, for $k \notin \Sigma$ [and for $k \neq k_0$ if $q_+ \neq q_-$], $\mathcal{W}(k) - I$ is a Hilbert-Schmidt operator on $L^2(\mathbb{R})^2$ with integral kernel

$$\begin{cases} |\Delta(x)|^{1/2}[K_f(x,y;k) + K_{f1}(x;k)K_{f2}(y;k)]U_\delta(y)|\Delta(y)|^{1/2}, & q_+ \neq q_-, \\ -i\, e^{i\lambda|x-y|}W_+(k)E(y-x)W_+^{-1}(k)\sigma_3, & q_+ = q_-. \end{cases} \tag{4.6}$$

It is now verified that[1]

$$(k-H)^{-1} - (k-H_f)^{-1} = -(k-H_f)^{-1}U_\delta|\Delta|^{1/2}\mathcal{W}^{-1}(k)|\Delta|^{1/2}(k-H_f)^{-1} \tag{4.7}$$

provided $\mathcal{W}(k)$ is invertible. The right-hand side of this identity has a finite limit as k approaches the interior points of Σ, provided $\mathcal{W}(k)$ is invertible.[2] *For $k_0 \neq k \notin \Sigma$, the points of noninvertibility of $\mathcal{W}(k)$ are exactly the discrete eigenvalues.*

Our first result is as follows:

THEOREM 4.3. *Let us consider the case $q_+ \neq q_-$. Then, for $k \in (-q_0, q_0)$, $\mathcal{W}(k)$ is invertible if the right-hand side of equation (4.8)*

$$\|\mathcal{W}(k) - I\| \leq \|q - q_f\|_1 \left|\frac{k(k+\lambda)}{2\lambda^2 \det Z(k)}\right| \left\{\frac{2}{\pi} + \sqrt{\left|\frac{k(k+\lambda)}{2\lambda^2}\right|}\right\}. \tag{4.8}$$

is strictly less than one. As a consequence, the discrete eigenvalues belong to these neighborhoods of q_0, $-q_0$, and k_0 within $(-q_0, q_0)$ for which the right-hand side of (4.8) is at least one.

PROOF. As already noted at the end of Section 3, when $q_+ \neq q_-$ the resolvent operator $(k - H_f)^{-1}$ is the sum of the integral operator with integral kernel $K_f(x,y;k)$ for $k \notin \Sigma$ and a rank two operator for $k \notin \Sigma \cup \{k_0\}$. The idea is to estimate the integral kernel $K_f(x,y;k)$ by using (4.1) and the rank two contribution through (4.2). To use equations (4.1) and (4.2), we introduce $W^{(1)}(x)$ and $W^{(2)}(y)$ as

$$W^{(1)}(x) = |\Delta(x)|^{1/2} = \sqrt{|q(x) - q_f(x)|}\, I_2, \tag{4.9a}$$

$$W^{(2)}(y) = U_\delta(y)|\Delta(y)|^{1/2} = \sqrt{|q(y) - q_f(y)|} \begin{pmatrix} 0 & \sigma(y) \\ \sigma(y)^* & 0 \end{pmatrix}. \tag{4.9b}$$

First of all, it is easily verified that $W^{(1)}(x)K_{f1}(x;k)$ has as its (spectral) norm

$$e^{-|x|\operatorname{Im}\lambda}\sqrt{|q(x) - q_f(x)|}\sqrt{\left|\frac{k(k+\lambda)}{2\lambda^2}\right|}. \tag{4.10}$$

By straightforward calculations, we also get the (spectral) norm of $K_{f2}(y;k)W^{(2)}(y)$

$$e^{-|y|\operatorname{Im}\lambda}\sqrt{|q(y) - q_f(y)|}\left|\frac{k(k+\lambda)}{2\lambda^2 \det Z(k)}\right|, \tag{4.11}$$

and, finally,

$$\left|\frac{K_f(x,y;k)}{e^{i\lambda|x-y|}}\right| = \left|\frac{k}{\lambda}\right|. \tag{4.12}$$

---

[1]By using eqs. (4.3) and (4.5) and that $kI - H = kI - H_f + Q - Q_f$, long but straightforward calculations show that the right and the left hand side of the equation obtained by multiplying both member of (4.7) by $kI - H$ coincide and then identity (4.7) holds.

[2]For k an interior point of Σ, the points of noninvertibility of $\mathcal{W}(k)$ would be the spectral singularities, but they are known not to exist [**6**].

Applying (4.10), (4.11) and (4.12) to (4.1), it is easily verified that the integral operator on $L^2(\mathbb{R})$ has the integral kernel in the form $e^{-\tau(y-x)}\theta(y-x)\sqrt{m(x)m(y)}$ which has, for $\tau \geq 0$, norm at most $(2/\pi)\|m\|_1$ (as stated above). Thus, $\mathcal{W}(k) - I$ minus the rank two contribution has norm at most

$$\frac{2}{\pi}\|q - q_f\|_1 \left|\frac{k(k+\lambda)}{2\lambda^2 \det Z(k)}\right|.$$

By using (4.2), we also verify that the rank two contribution to $\mathcal{W}(k) - I$ has norm at most

$$\|q - q_f\|_1 \sqrt{\left|\frac{k(k+\lambda)}{2\lambda^2}\right|}\left|\frac{k(k+\lambda)}{2\lambda^2 \det Z(k)}\right|.$$

As a result,

$$\|\mathcal{W}(k) - I\| \leq \|q - q_f\|_1 \left|\frac{k(k+\lambda)}{2\lambda^2 \det Z(k)}\right| \left\{\frac{2}{\pi} + \sqrt{\left|\frac{k(k+\lambda)}{2\lambda^2}\right|}\right\},$$

which completes the proof. □

The next theorem sheds light on the case $q_+ = q_-$.

THEOREM 4.4. *Let $q_+ = q_-$. Then, for $k \in (-q_0, q_0)$, $\mathcal{W}(k)$ is invertible if the right-hand side of equation* (4.13)

$$\|\mathcal{W}(k) - I\| \leq \frac{2}{\pi}\|q - q_f\|_1 \left|\frac{k}{\lambda}\right|. \tag{4.13}$$

is strictly less than one. As a consequence, the discrete eigenvalues belong to these neighborhoods of q_0, $-q_0$ for which the right-hand side of (4.13) *is at least one.*

PROOF. Equation (4.13) immediately follows from the second of (4.6) taking into account (4.10), (4.11) and (4.12). □

We remark that these results agree with those found in [**5**, Sec. 4].

Finally, for $q_+ \neq q_-$ we prove the existence of a discrete eigenvalue in the spectral gap if $\|q - q_f\|_1$ is sufficiently small.

THEOREM 4.5. *Let $q_+ \neq q_-$ and let $(k_0 - \epsilon,\ k_0 + \epsilon) \subseteq (-q_0,\ q_0)$. Put $C_\epsilon = \max_{|k-k_0|=\epsilon}\left(\|(k - H_f)^{-1}\|^2 \|\mathcal{W}^{-1}(k)\|\right)$. Then for*

$$\|q - q_f\|_1 < \frac{1}{\epsilon C_\epsilon}$$

there exists a simple discrete eigenvalue $k \in (k_0 - \epsilon,\ k_0 + \epsilon)$.

PROOF. Let $\Gamma(\epsilon)$ be the positively oriented circle with center k_0 and radius ϵ. Using (4.7) and $\||\Delta|^{\frac{1}{2}}\|^2 = \|q - q_f\|_1$ (where $|\Delta|$ is defined by (4.4)), we estimate

$$\left\|\frac{1}{2\pi i}\int_{\Gamma(\epsilon)} dk\left\{(k - H)^{-1} - (k - H_f)^{-1}\right\}\right\| \leq \epsilon\|q - q_f\|_1 C_\epsilon < 1$$

whenever $\|q - q_f\|_1 < \dfrac{1}{\epsilon C_\epsilon}$. Now recall that $P = \frac{1}{2\pi i}\int_{\Gamma(\epsilon)} dk(k - H)^{-1}$ and $P_f = \frac{1}{2\pi i}\int_{\Gamma(\epsilon)} dk(k - H_f)^{-1}$ are the orthogonal projections onto the combined eigenvector subspaces of H and H_f corresponding to the eigenvalues in $(k_0 - \epsilon,\ k_0 + \epsilon)$. We have proved above that

$$\|P - P_f\| < 1.$$

Since $I + P - P_f$ and $I - P + P_f$ are nonsingular and

$$(I + P - P_f)\, P_f = P P_f = P\,(I - P + P_f)\,,$$

the projections P and P_f have the same rank, which equals +1. This completes the proof. □

Acknowledgements

The research leading to this article was supported in part by INdAM-GNFM, INFN IS-CSN4, and MIUR under the PRIN 2011 grant *Teorie geometriche e analitiche dei sistemi Hamiltoniani in dimensioni finite e infinite.*

References

[1] M. J. Ablowitz, B. Prinari, and A. D. Trubatch, *Discrete and continuous nonlinear Schrödinger systems*, London Mathematical Society Lecture Note Series, vol. 302, Cambridge University Press, Cambridge, 2004. MR2040621 (2005c:37117)

[2] Mark J. Ablowitz and Harvey Segur, *Solitons and the inverse scattering transform*, SIAM Studies in Applied Mathematics, vol. 4, Society for Industrial and Applied Mathematics (SIAM), Philadelphia, Pa., 1981. MR642018 (84a:35251)

[3] Naruyosi Asano and Yusuke Kato, *Nonselfadjoint Zakharov-Shabat operator with a potential of the finite asymptotic values. I. Direct spectral and scattering problems*, J. Math. Phys. **22** (1981), no. 12, 2780–2793, DOI 10.1063/1.525185. MR638083 (83a:35086)

[4] Naruyoshi Asano and Yusuke Kato, *Nonselfadjoint Zakharov-Shabat operator with a potential of the finite asymptotic values. II. Inverse problem*, J. Math. Phys. **25** (1984), no. 3, 570–588, DOI 10.1063/1.526181. MR737304 (86c:35128)

[5] Gino Biondini and Barbara Prinari, *On the spectrum of the Dirac operator and the existence of discrete eigenvalues for the defocusing nonlinear Schrödinger equation*, Stud. Appl. Math. **132** (2014), no. 2, 138–159, DOI 10.1111/sapm.12024. MR3167091

[6] F. Demontis, B. Prinari, C. van der Mee, and F. Vitale, *The inverse scattering transform for the defocusing nonlinear Schrödinger equations with nonzero boundary conditions*, Stud. Appl. Math. **131** (2013), no. 1, 1–40, DOI 10.1111/j.1467-9590.2012.00572.x. MR3081238

[7] L. D. Faddeev and L. A. Takhtajan, *Hamiltonian methods in the theory of solitons*, Springer Series in Soviet Mathematics, Springer-Verlag, Berlin, 1987. Translated from the Russian by A. G. Reyman [A. G. Reĭman]. MR905674 (89m:58103)

[8] C. Hamner, J.J. Chang, P. Engels, and M. Hoefer, *Generation of dark-bright soliton trains in superfluid-superfluid counterflow*, Phys. Rev. Lett. 106:065302 (2011).

[9] D. Yan, J.J. Chang, C. Hamner, P.G. Kevrekidis, P. Engels, V. Achilleos, D.J. Frantzeskakis, R. Carretero-Gonzáles, and P. Schmelcher, *Multiple dark-bright solitons in atomic Bose-Einstein condensates*, Phys. Rev. A 84:053630 (2011).

[10] A. Hasegawa and F. Tappert, *Transmission of stationary nonlinear optical pulses in dispersive dielectric fibers. I. Anomalous dispersion*, App. Phys. Lett. **23**(3) 142-144 (1973).

[11] A. Hasegawa and F. Tappert, *Transmission of stationary nonlinear optical pulses in dispersive dielectric fibers. II. Normal dispersion*, App. Phys. Lett. **23**(3) 171-172 (1973).

[12] Tutomu Kawata and Hiroshi Inoue, *Eigen value problem with nonvanishing potentials*, J. Phys. Soc. Japan **43** (1977), no. 1, 361–362. MR0450809 (56 #9102)

[13] Tutomu Kawata and Hiroshi Inoue, *Inverse scattering method for the nonlinear evolution equations under nonvanishing conditions*, J. Phys. Soc. Japan **44** (1978), no. 5, 1722–1729. MR0499839 (58 #17600)

[14] Martin Klaus, *On the Zakharov-Shabat eigenvalue problem*, Mathematical studies in nonlinear wave propagation, Contemp. Math., vol. 379, Amer. Math. Soc., Providence, RI, 2005, pp. 21–45, DOI 10.1090/conm/379/07023. MR2149044 (2006k:34232)

[15] Martin Klaus, *On the eigenvalues of the Lax operator for the matrix-valued AKNS system*, Topics in operator theory. Volume 2. Systems and mathematical physics, Oper. Theory Adv. Appl., vol. 203, Birkhäuser Verlag, Basel, 2010, pp. 289–323, DOI 10.1007/978-3-0346-0161-0_12. MR2683245 (2011m:34259)

[16] M. Klaus and J. K. Shaw, *On the eigenvalues of Zakharov-Shabat systems*, SIAM J. Math. Anal. **34** (2003), no. 4, 759–773, DOI 10.1137/S0036141002403067. MR1969601 (2004c:34260)
[17] Jérôme Jp. Leon, *The Dirac inverse spectral transform: kinks and boomerons*, J. Math. Phys. **21** (1980), no. 10, 2572–2578, DOI 10.1063/1.524362. MR586340 (82d:81022)
[18] C.J. Pethick and H. Smith, *Bose-Einstein Condensation in Dilute Gases*, Cambridge University Press, Cambridge (2002).
[19] A. Fratalocchi, C. Conti, G.Ruocco, S. Trillo *Free-energy transition in a gas of noninteracting nonlinear wave particles*, Phys. Rev. Lett. **101**: 044101, (2008).
[20] C. van der Mee, *Nonlinear Evolution Models of Integrable Type*, 100 pages, SIMAI e-Lecture Notes **11**, 2013.
[21] V. E. Zakharov and A. B. Shabat, *Exact theory of two-dimensional self-focusing and one-dimensional self-modulation of waves in nonlinear media* (Russian, with English summary), Ž. Èksper. Teoret. Fiz. **61** (1971), no. 1, 118–134; English transl., Soviet Physics JETP **34** (1972), no. 1, 62–69. MR0406174 (53 #9966)
[22] V.E. Zakharov and A.B. Shabat, *Interaction between solitons in a stable medium*, Sov. Phys. JETP **37**, 823–828 (1973).

Università di Cagliari, Dip. Matematica e Informatica, viale Merello 92, Cagliari
E-mail address: fdemontis@unica.it

Università di Cagliari, Dip. Matematica e Informatica, viale Merello 92, Cagliari
E-mail address: cornelis@krein.unica.it

Università del Salento, Dip. Matematica e Fisica "E. De Giorgi", via per Arnesano, Lecce and Sezione INFN
E-mail address: federica.vitale@le.infn.it

Contemporary Mathematics
Volume **635**, 2015
http://dx.doi.org/10.1090/conm/635/12718

The Novikov-Veselov Equation: Theory and Computation

R. Croke, J. L. Mueller, M. Music, P. Perry, S. Siltanen, and A. Stahel

ABSTRACT. We review recent progress in theory and computation for the Novikov-Veselov (NV) equation with potentials decaying at infinity, focusing mainly on the zero-energy case. The inverse scattering method for the zero-energy NV equation is presented in the context of Manakov triples, treating initial data of conductivity type rigorously. Special closed-form solutions are presented, including multisolitons, ring solitons, and breathers. The computational inverse scattering method is used to study zero-energy exceptional points and the relationship between supercritical, critical, and subcritical potentials.

CONTENTS

1. Introduction

The Novikov-Veselov (NV) equation is the completely integrable, nonlinear dispersive equation

$$
\begin{aligned}
q_t &= 4\,\mathrm{Re}\left(4\partial^3 q + \partial\left(qw\right) - E\partial w\right) \\
\overline{\partial} w &= -3\partial q
\end{aligned}
\tag{1.1}
$$

2010 *Mathematics Subject Classification.* Primary 37K15; Secondary 35Q53, 65M32, 47J35, 70H06.

Michael Music supported in part by NSF Grant DMS-1208778.
Peter Perry supported in part by NSF Grant DMS-1208778.
Samuli Siltanen supported in part by the Finnish Centre of Excellence in Inverse Problems Research 2012-2017 (Academy of Finland CoE-project 250215).

Here E is a real parameter, the unknown function q is a real-valued function of two space variables and time, and the operators ∂ and $\overline{\partial}$ are given by

$$\begin{aligned}\partial &= \frac{1}{2}\left(\partial_{x_1} - i\partial_{x_2}\right)\\ \overline{\partial} &= \frac{1}{2}\left(\partial_{x_1} + i\partial_{x_2}\right).\end{aligned}$$

At zero energy ($E = 0$) it can also be written (after trivial rescalings) as

$$q_t = -\partial_z^3 q - \overline{\partial}^3 q + 3\partial_z(q\nu) + 3\overline{\partial}_z(q\bar{\nu}), \quad \text{where } \overline{\partial}_z \nu = \partial_z q. \tag{1.2}$$

The NV equation (1.2) generalizes the celebrated Korteweg-de Vries (KdV) equation

$$q_t = -6qq_x - q_{xxx}$$

in the sense that, if $q(x_1, t)$ solves KdV and $\nu_{x_1}(x_1, t) = -3q_{x_1}(x_1, t)$, then $q(x_1, t)$ solves NV.

The NV equation was introduced by Novikov and Veselov in [**64**, **65**] as one of a hierarchy of completely integrable equations that generate isospectral flows for the two-dimensional Schrödinger operator at fixed energy E. Indeed, the Novikov-Veselov equation (1.1) admits the Manakov Triple Representation [**48**]

$$\dot{L} = [A, L] - BL \tag{1.3}$$

where

$$\begin{aligned}L &= -\Delta + q - E,\\ A &= 8\left(\partial^3 + \overline{\partial}^3\right) + 2\left(w\partial + \overline{w}\overline{\partial}\right),\\ B &= -2\left(\partial w + \overline{\partial}\overline{w}\right).\end{aligned}$$

Here B is the operator of multiplication by the function $2\left(\partial w + \overline{\partial}\overline{w}\right)$. That is, a pair (q, w) solves the NV equation if and only if the operator equation (1.3) holds.

The Manakov triple representation implies that the NV equation is, formally at least, a completely integrable equation. Thus one expects that, for a suitable notion of "scattering data for L at fixed energy E," the associated scattering transform will linearize the flow.

For nonzero energy E and potentials q which vanish at infinity, the scattering transform and inverse scattering method was developed by P. Grinevich, R. G. Novikov, and S.-P. Novikov (see Kazeykina's thesis [**36**] for an excellent survey and see [**23**, **26**–**28**] for the original papers). Roughly and informally, there is a scattering transform $\mathcal{T}$ which maps the potential q to scattering data that obey a linear equation if q obeys the NV equation, and an inverse scattering transform $\mathcal{Q}$ which inverts $\mathcal{T}$, so that the function

$$q(x, \tau) = \mathcal{Q}\left[e^{i(\diamond^3 + (\overline{\diamond})^3)\tau}\mathcal{T}(q_0)\right](x) \tag{1.4}$$

solves the NV equation with initial data q_0. The inverse scattering method may be visualized by the following commutative diagram:

$$
\begin{array}{ccc}
\mathbf{t}_0(k) & \xrightarrow{\exp(i\tau(k^3+\overline{k}^3))(\,\cdot\,)} & \mathbf{t}_\tau(k) \\
\mathcal{T}\Big\uparrow & & \Big\downarrow\mathcal{Q} \\
q_0(z) & \xrightarrow{\text{Novikov-Veselov evolution}} & q_\tau^{\mathrm{NV}}(z),
\end{array}
\tag{1.5}
$$

where $\mathcal{T}$ and $\mathcal{Q}$ stand for the direct and inverse nonlinear Fourier transform, respectively, and the function $\mathbf{t}_\tau : \mathbb{C} \to \mathbb{C}$ is called the *scattering transform.* In the case $E = 0$, the inverse scattering method was studied by Boiti et. al. [**9**], Tsai [**89**], Nachman [**58**], Lassas-Mueller-Siltanen [**44**], Lassas-Mueller-Siltanen-Stahel [**45, 46**], Music [**54**], Music-Perry [**55**], and Perry [**68**]. Recently, Angelopoulos [**1**] proved that the Novikov-Veselov equation at $E = 0$ is locally well-posed in the Sobolev spaces $H^s(\mathbb{R}^2)$ for $s > 1/2$, placing the local existence theory for this equation on a sound footing. The potential utility of the inverse scattering method is to elucidate *global* behavior of the solutions.

The case $E = 0$ is somewhat special and is intimately connected with the following trichotomy of behaviors for the two-dimensional Schrödinger operator L at zero energy.

Definition 1.1. *The operator $L = -\Delta + q$ is said to be:*
(i) subcritical *if the operator L has a positive Green's function and the equation $L\psi = 0$ has a strictly positive distributional solution,*
(ii) critical *if $L\psi = 0$ has a bounded strictly positive solution but no positive Green's function, and*
(iii) supercritical *otherwise.*

This distinction first arose in the study of Schrödinger semigroups, i.e., the operators e^{-tL} where $L = -\Delta + q$. Simon [**79, 80**] (see also [**81**]) studied L^p-mapping properties of e^{-tL} and asymptotics of $\left\|e^{-tL}\right\|_{p,p}$, where $\|\cdot\|_{p,p}$ denotes the operator norm as maps from L^p to itself. Simon shows that

$$\alpha_p(q) = \lim_{t\to\infty} t^{-1} \ln \left\|e^{-tL}\right\|_{p,p}$$

is independent of $p \in [1, \infty]$. In the language of Schrödinger semigroups, a potential q is:

(i) subcritical if $\alpha_\infty\left((1+\varepsilon)\,q\right) = 0$ for some $\varepsilon > 0$,
(ii) critical if $\alpha_\infty\left(q\right) = 0$ but $\alpha_\infty\left((1+\varepsilon)\,q\right) > 0$ for all $\varepsilon > 0$, and
(iii) supercritical if $\alpha_\infty(q) > 0$.

Clearly, a sufficient condition for q to be supercritical is that L have a negative eigenvalue.

In [**52**], Murata showed that, for two-dimensional Schrödinger operators with potentials q with $q(x)$ uniformly Hölder continuous and $q(x) = \mathcal{O}\left(|x|^{-4-\epsilon}\right)$ for some $\epsilon > 0$, the trichotomy of behaviors for Schrödinger semigroups is equivalent to Definition 1.1. Murata further studied the existence and properties of positive solutions of the Schrödinger equation in [**52**], and showed that for his class of potentials, the trichotomy could be characterized as follows: a potential is

(i) subcritical if and only if $L\psi = 0$ has a strictly positive solution of the form $c\log(|x|) + d + \mathcal{O}\left(|x|^{-1}\right)$ with $c \neq 0$,

(ii) critical if $L\psi = 0$ if and only if $L\psi = 0$ has a strictly positive bounded solution, and

(iii) supercritical if $L\psi = 0$ has no strictly positive solutions.

Extending Murata's result, Gesztesy and Zhao [**20**] used Brownian motion techniques to prove the following optimal result for critical potentials. Suppose that q is a real-valued measurable function with

$$\lim_{\alpha\downarrow 0}\left\{\sup_{x\in\mathbb{R}^2}\int_{|x-y|\leq\alpha}\ln\left(|x-y|)^{-1}\right)|q(y)|\,dy\right\} = 0$$

and

$$\int_{|y|\geq 1}\ln(|y|)|q(y)|\,dy < \infty.$$

Then q is critical if and only if there exists a positive, bounded distributional solution ψ of $H\psi = 0$. These two conditions mean essentially that

$$q(x) = \mathcal{O}\left(|x|^{-2}\left(\ln(|x|)\right)^{-2-\epsilon}\right)$$

for some $\epsilon > 0$. We refer the reader to [**20**] for further references and history.

As we will see, corresponding to the trichotomy in Definition 1.1, the scattering transform of q is either mildly singular, nonsingular, or highly singular. This is illustrated dramatically in the examples studied by Music, Perry, and Siltanen [**56**], described in Section 5 below. One would expect the singularities of the scattering transform to be mirrored in the behavior of solutions to the NV equation. We will discuss the following conjecture, and some partial results toward its resolution, in the last section of this paper:

CONJECTURE 1.2. *The Novikov-Veselov equation* (1.1) *has a global solution for critical and subcritical initial data, but its solution may blow up in finite time for supercritical initial data.*

To elucidate this conjecture, it is helpful to recall how the scattering transform for Schrödinger's equation is connected with Calderón's inverse conductivity problem [**15**]. Critical potentials are also known in the literature as "potentials of conductivity type" because of their connection with the Calderón inverse conductivity problem: suppose one wishes to determine the conductivity σ of a bounded plane region Ω by boundary measurements. The potential u of Ω with voltage f on the boundary is determined by the equation

$$\begin{aligned}\nabla\cdot(\sigma\nabla u) &= 0\\ u|_{\partial\Omega} &= f\end{aligned}$$

Calderón's problem is to reconstruct σ from knowledge of the *Dirichlet-to-Neumann* map, defined as follows. If Ω has smooth boundary then the above boundary value problem has a unique solution u for given $f \in H^{1/2}(\partial\Omega)$, so that

$$\Lambda_\sigma f = \sigma\left.\frac{\partial u}{\partial\nu}\right|_{\partial\Omega}$$

is uniquely determined. The map $\Lambda_\sigma : H^{1/2}(\partial\Omega) \to H^{-1/2}(\partial\Omega)$ is the *Dirichlet-to-Neumann map.*

This boundary value problem is equivalent, under the change of variables $u = \sigma^{-1/2}\psi$, to the Schrödinger problem

$$\Delta\psi - q\psi = 0$$

where $q = \sigma^{-1/2}\Delta\left(\sigma^{1/2}\right)$. A potential of this form for strictly positive $\sigma \in L^\infty$ (and some additional regularity) is called a *potential of conductivity type*. More precisely, the class of potentials originally studied by Nachman [**58**] is as follows. We denote by $L^p_\rho(\mathbb{R}^2)$ the space of measurable functions with norm $\|f\|_{L^p_\rho} = \|\langle x\rangle^\rho f\|_p$.

DEFINITION 1.3. *Let $p \in (1,2)$ and $\rho > 1$. A real-valued measurable function $q \in L^p_\rho(\mathbb{R}^2)$ is a* potential of conductivity type *if there is a function $\sigma \in L^\infty(\mathbb{R}^2)$ with $\sigma(x) \geq c_0 > 0$ so that $q = (\Delta(\sigma^{1/2}))/(\sigma^{1/2})$ in the sense of distribution derivatives.*

A real-valued potential in $L^p_\rho(\mathbb{R}^2)$ is of conductivity type if and only if it is critical: the bounded, positive solution to $\Delta\psi - q\psi = 0$ is exactly $\psi = \sigma^{1/2}$.

As shown by Murata [**52,53**], critical potentials are very unstable: if $w \in \mathcal{C}_0^\infty(\mathbb{C})$ is a nonnegative bump function and q_0 is a critical potential, the potential $q_\lambda = q_0 + \lambda w$ is supercritical for any $\lambda < 0$. This means that the set of critical potentials is nowhere dense in any reasonable function space! Music, Siltanen, and Perry [**56**] studied the behavior of the scattering transform for families of this type when q_0 and w are both smooth, compactly supported, and radial. The corresponding scattering transforms are mildly singular for subcritical potentials, regular for critical potentials, and have circles of singularities for supercritical potentials.

The NV equation may be solved by inverse scattering for subcritical and critical potentials, but it is not yet clear how to construct a solution by inverse scattering for supercritical potentials.

In this article, we will focus primarily on the Novikov-Veselov equation at zero energy. We will report on recent progress on both the theoretical and the numerical analysis of this equation, and pose a number of open problems. In section 2 we review the history of the inverse scattering method, the dispersion relation, symmetries and scaling properties, and conservation laws for the NV equation. In section 3, we give an exposition of the inverse scattering method for the NV equation at zero energy from the point of view of the Manakov triple representation, treating with full mathematical rigor the case of "smooth potentials of conductivity type" (see Definition 3.2). We discuss the numerical implementation of the maps $\mathcal{T}$ and $\mathcal{Q}$ in sections 3.2 and 3.3, respectively. In section 4, we discuss special closed-form solutions of the NV equation including ring solitons and breathers. In section 5, the computational inverse scattering method is used to study zero-energy exceptional points and the relationship between supercritical, critical, and subcritical potentials. Finally, in section 6, we discuss open problems. In an appendix, we collect some useful tools for the mathematical analysis of the direct and inverse scattering maps.

Notation. In what follows, we use the variable t to denote time except when discussing the solution of NV via the inverse scattering method. In this case, τ denotes time in order to distinguish t from $\mathbf{t}$, the scattering transform.

2. Background for the Zero-Energy NV Equation

First, we summarize the historical progress that led to the completion of the diagram (1.5) for the NV equation at zero energy. In 1987, Boiti, Leon, Manna and Pempinelli [**9**] studied the evolution under the assumption that q_0 is such that the solution q_τ^{NV} to (1.1) exists and does not have exceptional points and established that the scattering data evolves as

$$\mathcal{T}(q_\tau^{\mathrm{NV}}) = e^{i\tau(k^3+\bar{k}^3)}\mathcal{T}(q_0).$$

In 1994, Tsai [**89**] considered a certain class of small and rapidly decaying initial data (which excludes conductivity-type potentials) and assumed that q_0 has no exceptional points and that q_τ is well-defined. Under such assumptions, he then showed that q_τ is a solution of the Novikov-Veselov equation (1.1). In 1996, Nachman [**58**] established that initial data of conductivity type does not have exceptional points and the scattering data $\mathcal{T}(q_0)$ is well-defined. Nachman's work paved the way for rigorous results: all studies about diagram (1.5) published before [**58**] were formal as they had to assume the absence of exceptional points without specifying acceptable initial data. In 2007, Lassas, Mueller and Siltanen [**44**] established for smooth, compactly supported conductivity-type initial data with $\sigma \equiv 1$ outside $\mathrm{supp}(q_0)$ that there is a well-defined continuous function $q_\tau : \mathbb{R}^2 \to \mathbb{C}$ from the inverse scattering method satisfying the estimate $|q_\tau(z)| \leq C(1+|z|)^{-2}$ for all $\tau > 0$. In [**45**] it was shown that an initially radially-symmetric conductivity-type potential evolved under the ISM does not have exceptional points and is itself of conductivity-type. Note that in [**56**] the set of conductivity type potentials is shown to be unstable under C_0^∞ perturbations. In [**46**] evolutions computed from a numerical implementation of the inverse scattering method of rotationally symmetric, compactly supported conductivity type initial data are compared to evolutions of the NV equation computed from a semi-implicit finite-difference discretization of NV and are found to agree with high precision. This supported the integrability conjecture that was then established in [**67**] for a larger class of initial data using the inverse scattering map for the Davey-Stewartson equation and Bogdanov's Miura transform.

In Section 4 of this paper, we present several closed-form solutions of the NV equation. We briefly review earlier constructions of solutions for the NV equation without presenting an exhaustive list. Grinevich, Manakov and R. G. Novikov constructed solition solutions using nonlocal Riemann problem techniques in [**22–24, 24, 27**] for nonzero energy and with small initial data. Also, solitons are constructed by Grinevich using rational potentials in [**22**], by Tagami using the Hirota method in [**82**], by Athorne and Nimmo using Moutard transformation in [**2**], by Hu and Willox using a nonlinear superposition formula in [**32**], by Xia, Li and Zhang using hyperbola function method and Wu-elimination method in [**95**], by Ruan and Chen using separation of variables in [**71–73, 98**], and by J.-L. Zhang, Wang, Wang and Fang using the homogeneous balance principle and Bäcklund transformation in [**32**]. Lump solutions are constructed by Dubrovsky and Formusatik using the $\bar{\partial}$-dressing method in [**16**]. Dromion solutions are constructed by Ohta and Ünal using Pfaffians in [**66, 90**]. The Darboux transformation is used by Hu, Lou, Liu, Rogers, Konopelchenko, Stallybrass and Schief to construct solutions in [**33, 70**]. Taimanov and Tsarëv [**84–88**] use the Moutard transformation to construct examples of Schrödinger operators L with L^2 eigenvalues at zero energy, and solutions of

NV which blow up in finite time. In [**96, 97**] Zheng, J.-F. Zhang and Sheng explore chaotic and fractal properties of solutions to the NV equation.

2.1. Dispersion, group velocity and phase velocity. Solitons form when there is a balance between nonlinearity and dispersion. The dispersion relation is the relation that gives the frequency as a function of the wave vector (k_1, k_2). To find the dispersion relation for the NV equation, consider the linear part of the equation

$$q_t = -\frac{1}{4}q_{xxx} + \frac{3}{4}q_{xyy} \tag{2.1}$$

The plane wave functions $q(x, y, t) = e^{i(k_1 x + k_2 y - \omega t)}$ are solutions to (2.1) provided

$$\omega = -\frac{1}{4}k_1^3 + \frac{3}{4}k_1 k_2^2. \tag{2.2}$$

Equation (2.2) defines the dispersion relation for the NV equation.

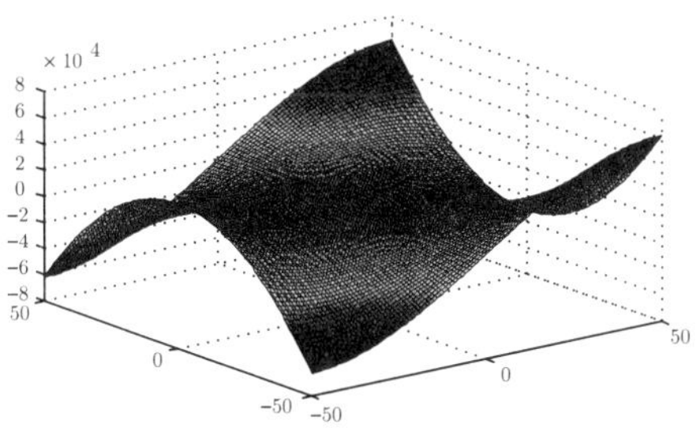

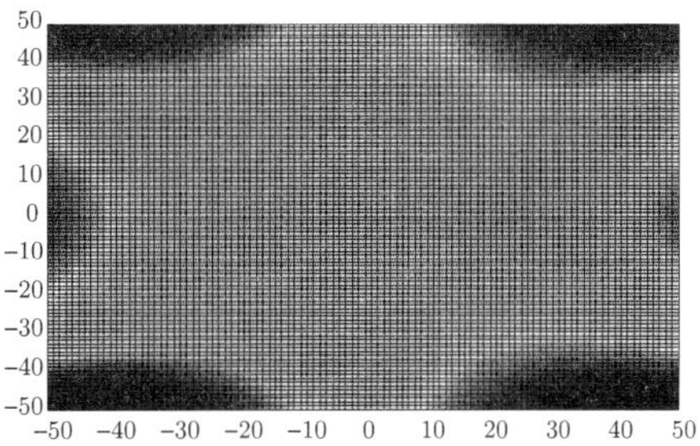

FIGURE 1. Surface plot (left) and heat map (right) of the NV dispersion relation $\omega(k) = k_1^2/4 + 3k_1k_2^2/4$

The phase velocity, $\mathbf{c}_p$, which gives the velocity of the wavefronts, is defined by $\mathbf{c}_p = \frac{\omega(\mathbf{k})}{|\mathbf{k}|^2}(k_1,\, k_2)^T$ and for the NV equation is

$$\mathbf{c}_p = \frac{k_1^3 - 3k_1k_2^2}{4(k_1^2 + k_2^2)}(k_1,\, k_2)^T. \tag{2.3}$$

The group velocity, which gives the velocity of the wave packet, is

$$\mathbf{c}_g \equiv \nabla\omega = \frac{3}{4}\left(-k_1^2 + k_2^2,\, 2k_1k_2\right)^T.$$

2.2. Symmetries and Scaling. To understand how scaling of the dependent and independent variables change the NV equation, let us first consider the auxilliary $\overline{\partial}$ equation in equation (1.2) in the form

$$\overline{\partial}\nu = \partial q, \quad \nu = v + iw.$$

Note that under the transformation $r(x, y, t) \equiv \gamma\nu(\alpha t, \beta x, \beta y)$ and $s(x, y, t) \equiv \gamma q(\alpha t, \beta x, \beta y)$, the $\overline{\partial}$ equation remains unchanged, i.e $\overline{\partial}\nu = \partial q$ if and only if $\overline{\partial}r = \partial s$. Then $r_x(x, y, t) = \beta\gamma\nu_x(\alpha t, \beta x, \beta y)$ and $r_y(x, y, t) = \beta\gamma\nu_y(\alpha t, \beta x, \beta y)$.

Now, we examine the main equation as presented in equation (1.2). Note that $s_t(x, y, t) = \alpha\gamma q_t(\alpha t, \beta x, \beta y)$, $s_{xxx}(x, y, t) = \beta^3\gamma q_{xxx}(\alpha t, \beta x, \beta y)$, $s_{xyy}(x, y, t) =$

$\beta^3\gamma q_{xyy}(\alpha t, \beta x, \beta y)$, and $(qv)_x + (qw)_y = \gamma^2\beta((s\Re(r))_x + (s\Im(r)_y)$. Assuming q is a solution to the NV equation (1.1), we find

$$\begin{aligned} 4q_t &= -u_{xxx} + 3q_{xyy} + 3(qv)_x + 3(qw)_y \\ \implies \frac{4}{\alpha\gamma}s_t &= -\frac{1}{\beta^3\gamma}s_{xxx} + \frac{3}{\beta^3\gamma}s_{xyy} + \frac{3}{\beta\gamma^2}(s\Re(r))_x + \frac{3}{\beta\gamma^2}(s\Im(r))_y. \end{aligned}$$

Multiplying by $\alpha\gamma$ leads to

$$4s_t = -\frac{\alpha}{\beta^3}s_{xxx} + \frac{3\alpha}{\beta^3}s_{xyy} + \frac{3\alpha}{\beta\gamma}((sr_1)_x + (sr_2)_y). \tag{2.4}$$

The table below shows the possible sign conventions possible for each term of the right hand side of equation (2.4).

α	β	γ	Signs in (2.4)		
+	+	+	-	+	+
-	-	+	-	+	+
-	+	+	+	-	-
+	-	+	+	-	-
+	+	-	-	+	-
-	-	-	-	+	-
-	+	-	+	-	+
+	-	-	+	-	+

Thus, there is a fixed ratio of -3 of the coefficients of the linear spatial terms, and any other coefficient is possible by proper rescaling of independent and dependent variables.

We also consider under what rotations the Novikov-Veselov equation is invariant. Writing the NV equation as

$$q_t = -\partial^3 q - \overline{\partial}^3 q + 3\partial(q\nu) + 3\overline{\partial}(q\overline{\nu}), \tag{2.5}$$

$$\overline{\partial}\nu = \partial q, \tag{2.6}$$

it is easy to see by conjugating (2.5), that if q is real at time t_0, then q stays real.

If the initial value $q_0(z)$ is invariant under rotations by $\pm\frac{2\pi}{3}$, then the Novikov-Veselov evolution preserves this symmetry, see [**46**]. In particular, all radially symmetric initial values will lead to solutions with this three-fold symmetry.

Under such rotations, equation (2.5) becomes

$$q_t = e^{i\theta}\partial_{z'}(\nu' q) + e^{i\theta}\overline{\partial}_{z'}(\overline{\nu' }q) - e^{3i\theta}\overline{\partial}_{z'}^3 q - e^{-3i\theta}\partial_{z'}^3 q \tag{2.7}$$

where $\nu' = e^{-i\theta}\nu$. The auxiliary equation becomes

$$e^{i\theta}\overline{\partial}_{z'}\nu = e^{-i\theta}\partial_{z'}q$$

or

$$\overline{\partial}_{z'}\nu' = e^{-3i\theta}\partial_{z'}q,$$

and so we have invariant solutions under rotations of $2\pi/3$ and $4\pi/3$. This shows that if a solution to the NV equation has this symmetry, it must be preserved under the evolution. It does not mean that all solutions will display this type of symmetry.

2.3. Conservation Laws for the NV equation. In order to present the conservation laws for (1.1), we need to recall some ideas from the inverse scattering method. A rigorous derivation of the conservation laws for smooth potentials of conductivity type is given below in section 3.

Suppose that $q \in L^p(\mathbb{R}^2)$ for some $p \in (1,2)$. The scattering data, or *scattering transform* $\mathbf{t} : \mathbb{C} \to \mathbb{C}$ of q is defined via Faddeev's [**18**] complex geometric optics (CGO) solutions, which we now recall. Let $z = x + iy$ and $k = k_1 + ik_2$. For $k \in \mathbb{C}$ with $k \neq 0$, the function $\psi(z,k)$ is the exponentially growing solution of the Schrödinger equation

$$(-\Delta + q)\psi(\cdot,k) = 0 \tag{2.8}$$

with asymptotic behavior $\psi(z,k) \sim e^{ikz}$ in the following sense: for $\tilde{p}$ defined by $\frac{1}{\tilde{p}} = \frac{1}{p} - \frac{1}{2}$, we have

$$e^{-ikz}\psi(z,k) - 1 \in L^{\tilde{p}}(\mathbb{R}^2) \cap L^\infty(\mathbb{R}^2). \tag{2.9}$$

It is more convenient to work with the *normalized complex geometric optics solutions* (NCGO) $\mu(z,k)$ defined by

$$\mu(z,k) = \psi(z,k)e^{-ikz}. \tag{2.10}$$

A straightforward computation shows that μ obeys the equation

$$\overline{\partial}\left(\partial + ik\right)\mu = \frac{1}{4}q\mu, \qquad \mu(\cdot,k) - 1 \in L^{\tilde{p}} \cap L^\infty. \tag{2.11}$$

One can reduce the problem (2.11) to an integral equation of Fredholm type (see the discussion in section 3). Faddeev's Green's function g_k is the fundamental solution for the equation

$$-4\overline{\partial}(\partial + ik)u = f$$

(the factor of -4 is chosen so that, if $k = 0$, the equation reduces to $-\Delta u = f$ whose fundamental solution is the logarithmic potential; see Appendix A.1 for details). One has

$$\mu = 1 - g_k * (q\mu) \tag{2.12}$$

(where $*$ denotes convolution) and it can be shown that the operator $\varphi \mapsto g_k * (q\varphi)$ is compact on $L^{\tilde{p}}$. Thus, for given k, the solution $\mu(z,k)$ exists if and only if it is unique.

It is known that such solutions exist for any $q \in L^p$ provided that $|k|$ is sufficiently large. In general, however, the equation (2.11) need not have a unique solution for every k.

Points for which uniqueness fails, i.e., points for which the homogeneous problem

$$\overline{\partial}\left(\partial + ik\right)\mu = \frac{1}{4}q\mu,$$
$$\mu(\cdot,k) \in L^{\tilde{p}} \cap L^\infty.$$

has a nontrivial solution are called *exceptional points*. It is known that the exceptional points form a bounded closed set in $\mathbb{C}$. Nachman [**58**] proved that the exceptional set is empty for potentials of conductivity type; more recently, Music [**54**] has shown that the same is true for subcritical potentials. We discuss this further in section 5.

Let

$$e_k(z) = e^{i\left(kz + \overline{k}\overline{z}\right)}. \tag{2.13}$$

Then, if q decays rapidly at infinity (say $q \in \mathcal{S}(\mathbb{R}^n)$), the function $\mu(z,k)$ obeys the large-z asymptotic formula

$$\mu(z,k) \sim 1 + \frac{1}{4\pi i k z} s(k) - \frac{e_{-k}(z)}{4\pi i \overline{k}\overline{z}} \mathbf{t}(k) + \mathcal{O}\left(|z|^{-2}\right) \tag{2.14}$$

where

$$\mathbf{t}(k) = \int e_k(z) q(z) \mu(z,k)\, dz, \tag{2.15}$$

$$s(k) = \int q(z)\mu(z,k)\, dz. \tag{2.16}$$

Note that $s(k)$ and $\mathbf{t}(k)$ are *always* defined for large $|k|$, whether the potential q is subcritical, critical, or supercritical.

The asymptotic formula (2.14) is a consequence of the following simple lemma.

LEMMA 2.1. *Suppose $k \in \mathbb{C}$ and $k \neq 0$, that $p > 2$ and suppose that $u \in C^2(\mathbb{C}) \cap L^p(\mathbb{C})$ satisfies*

$$-4\overline{\partial}(\partial + ik) u = f$$

for $f \in \mathcal{S}(\mathbb{C})$. Then

$$u(z) \underset{|z|\to\infty}{\sim} \sum_{\ell \geq 0} \frac{a_\ell}{z^{\ell+1}} + e_{-k}(z) \frac{b_\ell}{\overline{z}^{\ell+1}} \tag{2.17}$$

where

$$a_0 = \frac{1}{4\pi i k} \int f(z)\, dz,$$
$$b_0 = \frac{1}{4\pi i \overline{k}} \int e_k(z) f(z)\, dz.$$

PROOF. The conditions on u imply that

$$u = g_k * f$$

so using the asymptotic expansion (A.3) for g_k we obtain

(2.18)

$$u(z) = -\frac{1}{4\pi} \sum_{j=0}^{N} \int \left[\frac{j!}{(ik(z-z'))^{j+1}} + e^{-i(kz + \overline{k}\overline{z})} \frac{j!}{\left(-i\overline{k}\left(\overline{z - z'}\right)\right)^{j+1}} \right] f(z')\, dz'$$
$$+ \mathcal{O}\left(|z|^{-N-2}\right)$$

It is not difficult to see that for $f \in \mathcal{S}(\mathbb{C})$ and any positive integer N, the expansion

$$\int \frac{1}{(z-z')^j} f(z')\, dz' = \sum_{\ell=0}^{N} c_{\ell,j} z^{-j-\ell} + \mathcal{O}\left(|z|^{-N-1-j}\right)$$

holds, with an analogous expansion for the terms involving $\overline{z} - \overline{z'}$. The existence of the expansion (2.17) is immediate. The leading terms come from the $j = 0$ term of (2.18). □

If $q(z,\tau)$ solves the NV equation at zero energy, it can be shown that $\mathbf{t}(\cdot,\tau)$, the scattering transform of $q(\cdot,\tau)$, is then given by

$$\mathbf{t}(k,\tau) = \mathbf{m}(k,\tau)\mathbf{t}(k,0), \tag{2.19}$$

where

$$\mathbf{m}(k,\tau) = \exp(i\tau(k^3 + \overline{k}^3)).$$

On the other hand,

$$s(k,\tau) = s(k,0). \tag{2.20}$$

Here $\mathbf{t}(k,\tau)$ and $s(k,\tau)$ are computed from the solution $\mu(z,k,\tau)$ of

$$\overline{\partial}_z\left(\partial_z + ik\right)\mu(z,k,\tau) = \frac{1}{4}q(z,\tau)\mu(z,k,\tau). \tag{2.21}$$

A rigorous proof of (2.19) and (2.20) for smooth potentials of conductivity type is given in section 3 below.

If q is smooth, rapidly decreasing, and either critical or subcritical, the Schrödinger potential q can be recovered using the $\overline{\partial}$-method of Beals and Coifman [**5**] (see [**44**] for the critical case, and [**54**] for the subcritical case). Both of these papers use techniques developed by Nachman [**58**] in the context of the inverse conductivity problem.

We can now derive a set of conservation laws for the NV equation by using the large-k asymptotic expansion of $s(k)$. Since $s(k,\tau)$ is conserved we set $\tau = 0$ and suppress τ-dependence henceforward. We will give a formal derivation of the conserved quantities by using a large-k asymptotic expansion of $\mu(z,k)$ and inserting this expansion into the formula (2.16). Since $s(k)$ is conserved, the coefficients of that large-k expansion are also conserved quantities. For the moment, we assume that $\mu(z,k)$ admits a large-k expansion of the form

$$\mu(z,k) \sim 1 + \sum_{j=1}^{\infty} \frac{a_j(z)}{k^j}. \tag{2.22}$$

We will derive such an expansion for smooth potentials of conductivity type in the next section (see Lemma 3.7). It is expected to hold in general.

Substituting the series (2.22) into (2.21), we may solve the resulting system for the coefficients a_j

$$-\sum_{j=1}^{\infty} \frac{\Delta a_j(z)}{k^j} - \sum_{j=1}^{\infty} \frac{4i\overline{\partial} a_j(z)}{k^{j-1}} + q\sum_{j=1}^{\infty} \frac{a_j(z)}{k^j} = -q. \tag{2.23}$$

We find

$$a_1 = \frac{1}{4i}\overline{\partial}^{-1} q.$$

A recursion formula can then be derived,

$$a_{j+1} = \frac{1}{4i}\overline{\partial}^{-1}(-4\overline{\partial}\partial a_j + q a_j) = i\partial a_j + \frac{1}{4i}\overline{\partial}^{-1}(q a_j). \tag{2.24}$$

From this, it is clear that

$$s(k) \sim \sum_{j=0}^{\infty} \frac{s_j}{k^j}, \quad s_j = \int_{\mathbb{R}^2} q(z) a_j(z)\, dz.$$

Thus, the first three conserved quantities are

$$
\begin{aligned}
s_0 &= \int_{\mathbb{R}^2} q(z)dz, \\
s_1 &= \int_{\mathbb{R}^2} \frac{1}{4i} q(z)(\overline{\partial}^{-1} q)(z)dz, \\
s_2 &= \int_{\mathbb{R}^2} \left(\frac{1}{4} q(z)v(z) - \frac{1}{16} q\overline{\partial}^{-1}(q\overline{\partial}^{-1} q)(z) \right) dz,
\end{aligned}
$$

where with $z = x + iy$ and $\zeta = \zeta_1 + i\zeta_2$,

$$(\overline{\partial}^{-1} q)(z) = \frac{1}{\pi} \int_{\mathbb{R}^2} \frac{q(\zeta)d\zeta}{z - \zeta}.$$

3. Inverse Scattering via Manakov Triples

In this section we develop the inverse scattering method to solve the Cauchy problem for the Novikov-Veselov (NV) equation at zero energy

$$
\begin{aligned}
q_\tau &= 2\,\mathrm{Re}\left(\partial^3 q - \frac{3}{4} \partial(uq) \right) \\
\overline{\partial} u &= \partial q \\
q\big|_{\tau=0} &= q_0
\end{aligned}
\tag{3.1}
$$

for smooth Cauchy data q_0 of *conductivity type* (see Definition 3.2). Note that our convention for the NV equation differs slightly from (1.1); the form used here is more convenient for the zero-energy inverse scattering formalism. This section should be regarded as expository and the material here is undoubtedly "well known to the experts" (see the original paper of Manakov [**48**] and see e.g. Boiti, Leon, Manna, and Pempenelli [**9**] for the NV equation), although we give an essentially self-contained and mathematically rigorous presentation. An extension of these ideas to broader classes of potentials will appear in [**54**], [**55**], and [**69**].

In this section, we draw on previous work of Lassas, Mueller, and Siltanen [**44**], Lassas, Mueller, Siltanen, and Stahel [**45**], and Perry [**67**], particularly for mapping properties of the scattering transform and its inverse on the space of smooth functions of conductivity type as defined below. The main ingredient in our analysis (as contrasted to [**44, 45, 67**]) is the systematic use of the Manakov triple representation for the NV equation.

To describe the Manakov triple representation, suppose that q is a smooth function of z and $t\tau$. Suppose that there is a smooth function $u(z, \tau)$ with the property that $\overline{\partial} u = \partial q$ (existence of such a function for suitable classes of q can be deduced from properties of the Beurling transform; see the Appendix). Define[1]

$$L = \partial\overline{\partial} - q/4, \tag{3.2}$$

$$A = \partial^3 + \overline{\partial}^3 - \frac{3}{4}\left(u\partial + \overline{u}\overline{\partial} \right), \tag{3.3}$$

$$B = \frac{3}{4}\left(\partial u + \overline{\partial} u \right), \tag{3.4}$$

[1] This Manakov triple differs from that of the introduction by numerical factors since we use, for convenience, the version (3.1) of the NV equation.

where B is a multiplication operator. The (L, A, B) representation means the following:

PROPOSITION 3.1. *Let $q \in \mathcal{C}^1([0,T]; \mathcal{C}^\infty(\mathbb{C}))$ and suppose that there is a smooth function u with $\overline{\partial} u = \partial q$. Then q is a classical solution to the Cauchy problem* (3.1) *if and only if the operator identity*

$$\dot{L} = [A, L] - BL \tag{3.5}$$

holds.

The proof is a straightforward calculation. The significance of the Manakov triple is that it defines a scattering problem at zero energy for the operator L, and a law of evolution of scattering data through the operator A. We will fully describe the inverse scattering method for smooth initial data of conductivity type, defined as follows:

DEFINITION 3.2. *A function $q_0 \in \mathcal{C}^\infty(\mathbb{C})$ is a smooth function of conductivity type if*

$$q_0 = 2\partial u_0 + |u_0|^2$$

for some $u_0 \in \mathcal{S}(\mathbb{C})$ with $\partial u_0 = \overline{\partial} \overline{u_0}$.

The regularity requirements can be considerably relaxed but we make them here to ease the exposition. To compare this definition with Nachman's definition (see Definition 1.3 in the introduction), one should think of u_0 as $2\overline{\partial} \log \sigma$.

We develop in turn the direct scattering transform, the inverse scattering transform, and the solution formula for NV. We also comment on numerical methods for implementing the direct and inverse scattering transforms.

3.1. The Direct Scattering Map. To compute the scattering transform of q_0, one first constructs the complex geometric optics (CGO) solutions to (2.8). Analytically, it is more convenient to study the normalized complex geometric optics (NCGO) solutions μ defined by (2.10). As shown by Nachman [**58**], there exists a unique solution of (2.11) for every nonzero k, so that $\mathbf{t}(k)$ is defined for every nonzero k. Nachman also shows that $\mathbf{t}(k)$ is $\mathcal{O}(|k|^\varepsilon)$ as $k \to 0$ for conductivity-type potentials.

DEFINITION 3.3. *The map $\mathcal{T} : q \mapsto \mathbf{t}$ defined by the problem* (2.11) *and the representation formula* (2.15) *is called the* direct scattering map.

We will use without proof the following result of Lassas, Mueller, Siltanen, Stahel [**45**] (see also [**68**] for a different proof)

LEMMA 3.4. *Suppose that q is a smooth function of conductivity type, and let $\mathbf{t}$ be the scattering transform of q. Then $\mathbf{t}(k)/\overline{k} \in \mathcal{S}(\mathbb{C})$.*

3.2. Computation of Scattering Transforms. We describe two approaches for the computation of $\mathbf{t} = \mathcal{T} q$ for a given compactly supported and continuous q. The *LS method* is most accurate for k away from zero, and the *DN method* is more effective for k near zero. Matlab codes for both approaches are available at the webpage of the book [**51**].

Without loss of generality we can assume that $\operatorname{supp}(q) \subset \Omega$ where $\Omega \subset \mathbb{R}^2$ is the open unit disc. The LS method is based on the definition

$$\mathbf{t}(k) = \int_{\mathbb{R}^2} e^{i\overline{k}\overline{z}} q(z) \psi(z,k)\, dz, \tag{3.6}$$

and the DN method uses integration by parts to transform (3.6) into

$$\mathbf{t}(k) = \int_{\partial\Omega} e^{i\bar{k}\bar{z}}(\Lambda_q - \Lambda_0)\psi(\,\cdot\,, k)\, dS(z), \tag{3.7}$$

where Λ_q is the *Dirichlet-to-Neumann map* defined below. A rigorous derivation of formulas (3.6) and (3.7) was given by Nachman in [**58**].

The LS method requires numerical evaluation of the *complex geometrical optics solutions* $\psi(z, k)$. Numerically, it is better to solve the Lippmann-Schwinger equation for μ

$$\mu(z, k) = 1 - \int_{\Omega} g_k(z - w)q(w)\mu(w, k)\, dw. \tag{3.8}$$

A rigorous solvability analysis for equation (3.8) can be found in [**56**–**58**]. Here g_k is the fundamental solution satisfying $(-\Delta - 4ik\bar{\partial})g_k(z) = \delta(z)$. The origin of g_k is Faddeev's 1965 article [**18**]. Computationally, $g_1(z)$ can be evaluated using the Matlab expression "`exp(-1i*z).*real(expint(-1i*z))/(2*pi);`" The symmetry relation $g_k(z) = g_1(kz)$ extends this to all values $k \neq 0$. Note that g_k has a $\log k$ singularity when $k \to 0$, causing numerical difficulties for k near zero.

Equation (3.8) is defined in the whole plane $z \in \mathbb{R}^2$, so some kind of truncation is needed for practical computation. The first numerical computation of complex geometrical optics solutions was reported in [**75**] in the context of (3.8). That computation was used as a part of the first numerical implementation [**76**] of the $\bar{\partial}$ method for electrical impedance tomography. A more effective approach for computing μ is based on the periodization technique introduced by Gennadi Vainikko in [**91**]; see also [**74**, Section 10.5]. The adaptation of Vainikko's method to equation (3.8) was first introduced in [**50**]. For more details see [**51**, Section 14.3].

Now let us turn to the DN method. This method has practical use in the $\bar{\partial}$ reconstruction method for electrical impedance tomography. (See [**51**] and the relevant references therein.) We first define the Dirichlet-to-Neumann map Λ_q for the Dirichlet problem

$$\begin{aligned} (-\Delta + q)\, u &= 0 \text{ in } \Omega \\ u|_{\partial\Omega} &= f. \end{aligned} \tag{3.9}$$

If zero is not a Dirichlet eigenvalue of $-\Delta + q$ in Ω, the problem (3.9) has a unique solution u for given $f \in H^{1/2}(S^1)$, and we set

$$\Lambda_q f = \left.\frac{\partial u}{\partial \nu}\right|_{\partial\Omega} \tag{3.10}$$

where $\partial/\partial\nu$ denotes differentiation with respect to the outward normal on $\partial\Omega$.

Formula (3.7) requires the trace $\psi(\,\cdot\,, k)|_{\partial\Omega}$. According to [**58**], the traces can be solved from the boundary integral equation

$$\psi(z, k)|_{\partial\Omega} = e^{ikz} - \int_{\partial\Omega} G_k(z - w)\,(\Lambda_q - \Lambda_0)\psi(w, k)|_{\partial\Omega}\, dS(w), \tag{3.11}$$

if k is not an exceptional point of q. Here $G_k(z) := e^{ikz}g_k(z)$ is Faddeev's Green function for the Laplace operator. For details of the numerical solution of (3.11) see [**51**, Section 15.3].

3.3. The Inverse Scattering Map. It turns out the the NCGO solutions $\mu(z,k)$ also solve a $\overline{\partial}$-problem in the k variable determined by $\mathbf{t}(k)$. Letting

$$\mathbf{t}^{\sharp}(k) = \frac{\mathbf{t}(k)}{4\pi\overline{k}}, \tag{3.12}$$

we have, for any $p > 2$:

LEMMA 3.5. *Suppose that q is a smooth potential of conductivity type, and let $\mu(z,k)$ be the corresponding NCGO solutions. Then:*

$$\begin{aligned} \overline{\partial}_k \mu(z,k) &= e_{-k}\mathbf{t}^{\sharp}(k)\overline{\mu(z,k)} \\ \mu(z,\ \cdot\ ) - 1 &\in L^p(\mathbb{C}). \end{aligned} \tag{3.13}$$

Moreover, $q(z)$ is recovered from $\mathbf{t}$ and $\mu(z,k)$ by the formula

$$q(z) = \frac{4i}{\pi}\overline{\partial}_z\left(\int e_{-k}(z)\mathbf{t}^{\sharp}(k)\overline{\mu(z,k)}\,dk\right) \tag{3.14}$$

REMARK 3.6. This equation has at most one solution for each z, provided only that $\mathbf{t}^{\sharp} \in L^2$ by a standard uniqueness theorem for the $\overline{\partial}$-problem (see Brown-Uhlmann [**11**], Corollary 3.11).

The fact that $\mu(z,k)$ obeys a $\overline{\partial}$-equation in the k-variable also implies a large-k asymptotic expansion for $\mu(z,k)$.

LEMMA 3.7. *Suppose that q is a smooth potential of conductivity type, and let $\mu(z,k)$ be the corresponding NCGO solution. Then*

$$\mu(z,k) \sim 1 + \sum_{\ell \geq 0} \frac{c_\ell(z)}{k^{\ell+1}} \tag{3.15}$$

where

$$c_0 = -\frac{i}{4}\overline{\partial}^{-1} q \tag{3.16}$$

$$c_1 = -\frac{1}{4}\partial\overline{\partial}^{-1} q + \frac{1}{16}\overline{\partial}^{-1}\left(q\overline{\partial}^{-1} q\right). \tag{3.17}$$

and the remaining c_j are determined by the recurrence

$$i\overline{\partial}c_{j+1} = \left(\frac{q}{4} - \partial\overline{\partial}\right)c_j \tag{3.18}$$

PROOF. The coefficients in the asymptotic expansion may be computed recursively from the equation $\overline{\partial}\left(\partial + ik\right)\mu = (q/4)\mu$ once the existence of the asymptotic expansion is established. To do so, note that

$$\mu(z,k) = 1 + \frac{1}{\pi}\int_{\mathbb{C}} \frac{1}{k-\kappa} e_{-k}\mathbf{t}^{\sharp}(\kappa)\overline{\mu(z,\kappa)}\,dm(\kappa).$$

Writing

$$(k-\kappa)^{-1} = k^{-1}\sum_{j=0}^{N}\left(\frac{\kappa}{k}\right)^j + \left(\frac{\kappa}{k}\right)^{N+1}\frac{1}{k-\kappa}$$

we obtain an expansion of the desired form with

$$c_j(z) = \frac{1}{\pi}\int \kappa^j e_{-k}\mathbf{t}^{\sharp}(\kappa)\overline{\mu(z,\kappa)}\,dm(\kappa)$$

and remainder

$$\frac{1}{\pi} k^{-(N+1)} \int \frac{1}{k-\kappa} \kappa^{N+1} e_k \mathbf{t}^\sharp(\kappa) \overline{\mu(z,\kappa)}\, dm(\kappa)$$

Finally to obtain explicit formulae for the c_j, we substitute the expansion (3.15) into (2.11) to obtain

$$i\overline{\partial} c_0 = \frac{1}{4} q$$

and the recurrence (3.18) from which (3.17) follows. □

Motivated by these results, suppose given $\mathbf{t}^\sharp \in \mathcal{S}(\mathbb{C})$ and let $\mu(z,k)$ be the unique solution to (3.13). *Define* q by the reconstruction formula (3.14). Then the solution μ of (3.13) obeys the partial differential equation

$$\overline{\partial}\left(\partial + ik\right)\mu = \frac{q}{4}\mu, \qquad \lim_{|z|\to\infty} \mu(z,\ \cdot\ ) - 1 = 0.$$

DEFINITION 3.8. *The map* $\mathcal{Q} : \mathbf{t} \to q$ *defined by* (3.13) *and* (3.14) *is called the* inverse scattering map.

We conclude this subsection by obtaining a full asymptotic expansion for $\mu(z,k)$ which encodes relations between s and $\mathbf{t}$.

LEMMA 3.9. *Suppose that* q *is a smooth potential of conductivity type, and let* $\mathbf{t}^\sharp$ *be given by* (3.12)*. Then, the expansion*

$$\mu(z,k) \underset{|z|\to\infty}{\sim} 1 + \sum_{\ell\geq 0} \left(\frac{a_\ell}{z^{\ell+1}} + e_{-k} \frac{b_\ell}{\overline{z}^{\ell+1}} \right) \tag{3.19}$$

holds, where:

$$a_0 = -i\overline{\partial}_k^{-1}\left(\left|\mathbf{t}^\sharp\right|^2\right),$$
$$b_0 = i\mathbf{t}^\sharp$$

and the subsequent a_ℓ, b_ℓ *are determined by the recurrence relations*

$$\overline{\partial}_k a_\ell = \mathbf{t}^\sharp \overline{b_\ell},$$
$$b_{\ell+1} = i\overline{a_\ell}\mathbf{t}^\sharp - i\overline{\partial}_k b_\ell.$$

PROOF. The existence of an expansion of the form (3.19) was already established in Lemma 2.1. To compute the coefficients, we substitute the asymptotic series into (3.13). □

REMARK 3.10. Comparing Lemmas 2.1 and 3.9, we see that

$$s(k) = 4\pi k \overline{\partial}_k^{-1}\left(\left|\mathbf{t}^\#\right|^2\right)$$

3.4. Computation of Inverse Scattering Transforms. The first step in the computation of the inverse scattering map is to solve the $\overline{\partial}$ equation

$$\frac{\partial}{\partial \overline{k}} \mu_\tau(z,k) = \frac{\mathbf{t}_\tau(k)}{4\pi \overline{k}} e_{-k}(z) \overline{\mu_\tau(z,k)} \tag{3.20}$$

with a fixed parameter $z \in \mathbb{R}^2$ and requiring large $|k|$ asymptotics $\mu_\tau(z,\cdot) - 1 \in L^\infty \cap L^r(\mathbb{C})$ for some $2 < r < \infty$. Since $q_0(z)$ is compactly supported and of conductivity type, by [**44**] the scattering transform $\mathbf{t}_\tau(k)$ is in the Schwartz class, and the solution μ_τ to equation (3.20) can be computed by numerically solving the integral equation

$$\mu_\tau(z,k) = 1 + \frac{1}{4\pi^2}\int_{\mathbb{R}^2} \frac{\mathbf{t}_\tau(k')}{\bar{k}'(k-k')} e_{-k'}(z)\overline{\mu_\tau(z,k')}dk'. \tag{3.21}$$

Computational solution of the $\overline{\partial}$ equation (3.20) is based on truncating the scattering transform $\mathbf{t}_\tau(k)$ to a large disc of radius R, generally chosen by inspection of the scattering transform. The truncated integral equation is solved numerically by the method described in [**41**] for each point z at which the evolved potential is to be computed. The method in [**41**] is based on Vainikko's periodic method [**91**]; see also [**74**, Section 10.5]. Note that since the $\bar{\partial}$ equation (3.20) is real-linear and not complex-linear due to the complex conjugate on the right-hand side of (3.20), one must write the real and imaginary parts of the unknown function μ separately in the vector of function values at the grid points. It is proven in [**40**] that the error decreases as R tends to infinity. The first computational solutions of equation (3.20) can be found in [**76**], and the first computations based on [**91**] are found in [**41**]; for more details see [**51**, Section 15.4].

The inverse scattering transform is defined by

$$(\mathcal{Q}\mathbf{t}_\tau)(z) := \frac{i}{\pi^2}\overline{\partial}_z \int_{\mathbb{C}} \frac{\mathbf{t}_\tau(k)}{\overline{k}} e^{-ikz}\,\overline{\psi_\tau(z,k)}dk, \tag{3.22}$$

where $\psi_\tau(z,k) := e^{ikz}\mu_\tau(z,k)$. The inverse transform (3.22) first appeared in [**9**, formula (4.10)]. See [**44**] and the references therein for an analysis of the solvability of (3.20) and the domain of definition for (3.22). Under the assumption that real-valued, smooth initial data of conductivity type remain of conductivity type under evolution by the ISM, the conductivity γ_τ associated with the potential q_τ is given by

$$\gamma_\tau^{1/2}(z) = \mu_\tau(z,0).$$

Then q_τ is computed by numerical differentiation of γ_τ by the formula

$$q_\tau(z) = \gamma_\tau^{-1/2}\Delta\gamma_\tau^{1/2}.$$

The reader is referred to [**46**] for numerical examples of the computation of the time evolution of conductivity-type potentials by the ISM.

3.5. Time Evolution of NCGO Solutions. In order to prove the solution formula (1.4), we first study the time evolution of NCGO solutions using the Manakov triple representation. First we note the following important uniqueness theorem which is actually a special case of results of Nachman.

THEOREM 3.11. *Let $k \in \mathbb{C}$, $k \neq 0$. Suppose that q is a smooth potential of conductivity type and that ψ is a solution of $L\psi = 0$ with $\lim_{|z|\to\infty}\left(e^{-ikz}\psi(z)\right) = 0$. Then $\psi(z) = 0$.*

Now suppose that $q(z,\tau)$ solves the NV equation and that $t \mapsto q(z,\tau)$ is a C^1 map from $[0,T]$ into $\mathcal{S}(\mathbb{C})$. Suppose that, for each τ, $\varphi(\tau)$solves $L(\tau)\varphi(\tau) = 0$. Differentiating the equation $L(\tau)\varphi(\tau) = 0$ and using the Manakov triple representation, we find

$$[A(\tau), L(\tau)]\,\varphi(\tau) + L(\tau)\dot{\varphi}(\tau) = 0$$

or

$$L(\tau)\left[\dot{\varphi}(\tau) - A(\tau)\varphi(\tau)\right] = 0$$

From this simple computation and Theorem 3.11, we can derive an equation of motion for the NCGO solutions, and recover (2.19) and (2.20) from a careful calculation of asymptotics. Later, we will show by explicit construction that, if q_0 is a smooth function of conductivity type, then there is a solution $q(z,\tau)$ of the NV equation so that $q(z,\tau)$ is smooth and of conductivity type for each τ.

LEMMA 3.12. *Suppose that $q(z,\tau)$ is a solution of the NV equation where, for each τ, $q(z,\tau)$ is a smooth function of conductivity type. Let $u = \overline{\partial}^{-1}\partial q$. Then*

$$\dot{\mu} = ik^3\mu + (\partial + ik)^3 \mu + \overline{\partial}^3 \mu - \frac{3}{4}u(\partial + ik)\mu - \frac{3}{4}\overline{u}\overline{\partial}\mu \tag{3.23}$$

PROOF. Before giving the proof we make several remarks. Since $\psi = e^{ikz}\mu$, the evolution equation (3.23) is equivalent to

$$\psi_\tau = ik^3\psi + \partial^3\psi + \overline{\partial}^3\psi - \frac{3}{4}\left(u\partial + \overline{u}\overline{\partial}\right)\psi. \tag{3.24}$$

Next, let $\varphi(z,k,\tau) = e^{iS}\mu(z,k,\tau)$ with $S(z,k,\tau) = kz - k^3 t$. From the argument above we have

$$L(\tau)\left[\dot{\varphi}(\tau) - A(\tau)\varphi(\tau)\right] = 0.$$

To conclude that $\dot{\varphi}(\tau) = A(\tau)\varphi(\tau)$, we must show that

$$\lim_{|z|\to\infty}\left(e^{-ikz}\left[\dot{\varphi}(\tau) - A(\tau)\varphi(\tau)\right]\right) = 0.$$

Write

$$f \sim_k g$$

if

$$\lim_{|z|\to\infty}\left[e^{-ikz}(f-g)\right] = 0$$

Noting that $\mu - 1$ and its derivatives in z and $\overline{z}$ vanish as $|z| \to \infty$, a simple calculation shows that

$$\begin{aligned}\dot{\varphi} - A\varphi &\sim_k e^{iS}\left(-ik^3\mu - (\partial + ik)^3\mu\right)\\ &\sim_k 0\end{aligned}$$

Hence $\dot{\varphi} = A\varphi$ from which (3.24) follows. □

Hence:

LEMMA 3.13. *Suppose that $q(z,\tau)$ is a solution of the NV equation where, for each τ, $q(z,\tau)$ is a smooth function of conductivity type. Let $\mu(z,k,\tau)$ be the corresponding NCGO solution with*

$$\mu(z,k,\tau) \sim 1 + \frac{1}{4\pi ikz}s(k,t) - \frac{e_{-k}(z)}{4\pi i\overline{k}\overline{z}}\mathbf{t}(k,t) + \mathcal{O}\left(|z|^{-2}\right). \tag{3.25}$$

Then

$$\dot{s}(k,t) = 0, \tag{3.26}$$

$$\dot{\mathbf{t}}(k,t) = i\left(k^3 + \overline{k}^3\right)\mathbf{t}(k,t). \tag{3.27}$$

PROOF. Substituting the asymptotic relation (3.25) into (3.23), we may compute, modulo terms of order z^{-2},

$$-\frac{1}{\pi i k z}\dot{\mathbf{t}}+\frac{e_{-k}(z)}{\pi i \overline{k}\overline{z}}\dot{s}=-\frac{e_{-k}(z)}{\pi i \overline{k}\overline{z}}\left(ik^3+i\overline{k}^3\right)\mathbf{t}.$$

The computation uses the following facts. If

$$\mu(z,k,\tau)=1+\frac{a_0}{z}+e_{-k}\frac{b_0}{\overline{z}}+\mathcal{O}\left(|z|^{-2}\right)$$

then ("$\sim$" means "is asymptotic as $|z|\to\infty$ to"modulo $\mathcal{O}\left(|z|^{-2}\right)$)

$$\begin{aligned}\partial^3\mu &\sim e_{-k}\left(-ik\right)^3\frac{b_0}{\overline{z}},\\ \overline{\partial}^3\mu &\sim e_{-k}\left(-i\overline{k}\right)^3\frac{b_0}{\overline{z}},\\ 3ik\partial^2\mu-3k^2\partial\mu &\sim 0\end{aligned}$$

together with the fact that u defined by $\overline{\partial}u=\partial q$ satisfies $u=\mathcal{O}\left(|z|^{-2}\right)$. The identities (3.26) and (3.27) are immediate. □

3.6. Solution by Inverse Scattering. Motivated by the computations of the preceding subsection, we now consider the problem

$$\overline{\partial}_k\mu=e^{i\tau S}\mathbf{t}^\sharp\mu \tag{3.28}$$

$$\mu(z,\,\cdot\,,t)-1\in L^p\left(\mathbb{C}\right)$$

for a function $\mu(z,k,\tau)$ and the putative reconstruction

$$q(z,\tau)=\frac{4i}{\pi}\overline{\partial}_z\left(\int_{\mathbb{C}}e^{i\tau S}\mathbf{t}^\sharp(k)\overline{\mu(z,k,\tau)}\,dk\right). \tag{3.29}$$

Here

$$S(z,k,\tau)=-\frac{1}{\tau}\left(kz+\overline{k}\overline{z}\right)+\left(k^3+\overline{k}^3\right) \tag{3.30}$$

and $\mathbf{t}^\sharp$ is obtained from the Cauchy data q_0. We will show that $q(z,\tau)$ solves the NV equation by deriving an equation of motion for $\mu(z,k,\tau)$ and using this equation to compute q_τ if q is given by (3.29) and μ is the unique solution of (3.28)

First, we establish an equation of motion for the solution μ of (3.28). Although this equation is the same equation as (3.23) for the solution of the direct problem, our starting point here is (3.28).

LEMMA 3.14. *Suppose that $\mathbf{t}^\sharp\in\mathcal{S}\left(\mathbb{C}\right)$ and μ solves (3.28). For each τ, define $q(z,\tau)$ by (3.29) and define $u(z,\tau)$ by $u=\overline{\partial}^{-1}\partial q$. Then*

$$\dot{\mu}=ik^3\mu+\left(\partial+ik\right)^3\mu+\overline{\partial}^3\mu-\frac{3}{4}u\left(\partial+ik\right)\mu-\frac{3}{4}\overline{u}\overline{\partial}\mu$$

where ∂ and $\overline{\partial}$ denote differentiation with respect to the z and $\overline{z}$ variables.

PROOF. Let

$$w=\dot{\mu}-\left(ik^3\mu+\left(\partial+ik\right)^3\mu+\overline{\partial}^3\mu-\frac{3}{4}u\left(\partial+ik\right)\mu-\frac{3}{4}\overline{u}\overline{\partial}\mu\right).$$

We will show that $w=0$ in two steps. First, we show that

$$\overline{\partial}_k w=e^{i\tau S}\mathbf{t}^\sharp\overline{w}.$$

This is an easy consequence of the formulas

$$\begin{aligned}\overline{\partial}_k\left(\partial+ik\right)\mu &= e^{itS}\mathbf{t}^\sharp\overline{\overline{\partial}\mu},\\ \overline{\partial}_k\left(\overline{\partial}\mu\right) &= e^{i\tau S}\mathbf{t}^\sharp\overline{\left(\partial+ik\right)\mu}\end{aligned}$$

and holds for any smooth function u.

Next, we show that for any fixed values of the parameters τ and z,

$$\lim_{|k|\to\infty} w(z,k,\tau)=0.$$

Here we must choose $u=\overline{\partial}^{-1}\partial q$ in order for the assertion to be correct. Owing to Lemma 3.7 and the formula

$$w=\dot{\mu}-\left(\partial^3\mu+\overline{\partial}^3\mu+3ik\partial^2\mu-3k^2\partial\mu-\frac{3}{4}u\left(\partial+ik\right)\mu-\frac{3}{4}\overline{u}\overline{\partial}\mu\right),$$

we have

$$\begin{aligned}w &= -3ik\partial^2\mu+3k^2\partial\mu+\frac{3}{4}u\left(\partial+ik\right)\mu+\mathcal{O}\left(k^{-1}\right)\\ &= A_{-1}k+A_0+\mathcal{O}\left(k^{-1}\right)\end{aligned}$$

where

$$A_{-1}=3\left(\partial a_0+\frac{i}{4}u\right)$$

and

$$A_0=3\left[-i\partial^2 a_0+\partial a_1+\frac{i}{4}ua_0\right].$$

The condition $A_{-1}=0$ forces the choice $u=\overline{\partial}^{-1}\partial q$. We may then compute

$$A_0=\frac{3}{16}\left[\partial\overline{\partial}^{-1}\left(q\overline{\partial}^{-1}q\right)-\left(\overline{\partial}^{-1}q\right)\cdot\left(\partial\overline{\partial}^{-1}q\right)\right].$$

One the one hand, A_0 vanishes as $|z|\to\infty$ for each fixed τ by the decay of q. On the other hand, a straightforward computation shows that $\overline{\partial}A_0=0$. It now follows from Liouville's Theorem that $A_0=0$, and hence $w=O\left(k^{-1}\right)$. We now used the generalized Liouville Theorem to conclude that $w=0$. ☐

Finally, we prove:

PROPOSITION 3.15. *Suppose that* $\mathbf{t}^\sharp\in\mathcal{S}\left(\mathbb{R}^2\right)$. *Then, the formula*

$$q(z,\tau)=\frac{4i}{\pi}\overline{\partial}_z\left(\int e^{itS(z,k,\tau)}\mathbf{t}^\sharp(k)\overline{\mu(z,k,\tau)}\,dm(k)\right)$$

yields a classical solution of the NV equation.

PROOF. In what follows, we will freely use the commutation relations

$$\partial e^{i\tau S}=e^{i\tau S}\left(\partial-ik\right),\tag{3.31}$$

$$\overline{\partial}e^{i\tau S}=e^{i\tau S}\left(\overline{\partial}-i\overline{k}\right)\tag{3.32}$$

and the equation

$$\overline{\partial}\left(\partial+ik\right)\mu=\frac{1}{4}q\mu.\tag{3.33}$$

For notational brevity we'll write $c = 4i/\pi$. We compute

$$\begin{aligned}\dot{q} &= c\overline{\partial}\left(\int e^{i\tau S}\mathbf{t}^{\sharp}\left\{ik^3 + i\overline{k}^3\right\}\mu\right)\\ &\quad + c\overline{\partial}\left(\int e^{i\tau S}\mathbf{t}^{\sharp}\left\{-i\overline{k}^3 + \left(\overline{\partial} - i\overline{k}\right)^3 + \partial^3 - \frac{3}{4}\overline{u}\left(\overline{\partial} - i\overline{k}\right) - \frac{3}{4}u\partial\right\}\overline{\mu}\right)\end{aligned}$$

where in the second term we used Lemma 3.14. Using the commutation relations above to move differential operators to the left of $\exp(itS)$, we conclude that

$$\dot{q} = \partial^3 q + \overline{\partial}^3 q - \frac{3}{4}\overline{\partial}\left(u\partial\overline{\partial}^{-1}q\right) - \frac{3}{4}\overline{\partial}\left(\overline{u}q\right) + I$$

where

$$I = c\overline{\partial}\left(\int\left\{3ik\partial^2 - 3k^2\partial - \frac{3}{4}iku\right\}e^{i\tau S}\mathbf{t}^{\sharp}\overline{\mu}\right).$$

We claim that

$$I = \frac{3}{4}\left\{\overline{\partial}\left(u\partial\overline{\partial}^{-1}q\right) - \partial\left(uq\right)\right\}. \tag{3.34}$$

If so, then q solves the NV equation as claimed. To compute I, write $I = I_1 - I_2$ where

$$\begin{aligned}I_1 &= c\overline{\partial}\left(\int\left\{3ik\partial^2 - 3k^2\partial\right\}e^{i\tau S}\mathbf{t}^{\sharp}\overline{\mu}\right),\\ I_2 &= c\overline{\partial}\left(\int e^{i\tau S}\mathbf{t}^{\sharp}\left\{\frac{3}{4}iku\right\}\overline{\mu}\right).\end{aligned}$$

Using (3.33) and (3.32) we may write

$$\begin{aligned}I_1 &= \frac{3}{4}c\partial\left(\int e^{i\tau S}\left(ik\right)\mathbf{t}^{\sharp}q\overline{\mu}\right)\\ &= \frac{3}{4}c\partial\left(q\int\left(-\partial e^{i\tau S}\right)\mathbf{t}^{\sharp}\overline{\mu}\right)\\ &= -\frac{3}{4}c\partial\left(q\partial\left(\int e^{i\tau S}\mathbf{t}^{\sharp}\overline{\mu}\right) - q\int e^{i\tau S}\mathbf{t}^{\sharp}\partial\overline{\mu}\right)\\ &= -\frac{3}{4}\partial\left(q\partial\overline{\partial}^{-1}q\right) + \frac{3}{4}c\partial\left(q\overline{\partial}^{-1}\overline{\partial}\left(\int e^{i\tau S}\mathbf{t}^{\sharp}\partial\overline{\mu}\right)\right)\end{aligned}$$

where in the third line we used $u\partial v = \partial\left(uv\right) - v\partial u$. In the second term on the fourth line, we may use

$$\overline{\partial}e^{i\tau S}\partial\overline{\mu} = e^{i\tau S}\partial\left(\overline{\partial} - i\overline{k}\right)\overline{\mu} \tag{3.35}$$

and (3.33) to conclude that

$$\begin{aligned}\frac{3}{4}c\partial\left(q\overline{\partial}^{-1}\overline{\partial}\left(\int e^{i\tau S}\mathbf{t}^{\sharp}\partial\overline{\mu}\right)\right) &= \frac{3}{16}c\partial\left(q\overline{\partial}^{-1}q\left(\int e^{i\tau S}\mathbf{t}^{\sharp}\overline{\mu}\right)\right)\\ &= \frac{3}{16}\partial\left(q\overline{\partial}^{-1}\left(q\overline{\partial}^{-1}q\right)\right)\end{aligned}$$

so that

$$\begin{aligned}I_1 &= -\frac{3}{4}\partial\left(q\partial\overline{\partial}^{-1}q\right) + \frac{3}{16}\partial\left(q\overline{\partial}^{-1}\left(q\overline{\partial}^{-1}q\right)\right)\\ &= -\frac{3}{4}\partial\left(qu\right) + \frac{3}{16}\partial\left(q\overline{\partial}^{-1}\left(q\overline{\partial}^{-1}q\right)\right).\end{aligned}$$

Similarly, we may compute

$$\begin{aligned}
I_2 &= \frac{3}{4} c \overline{\partial} \left(u \int ik \; e^{itS} \mathbf{t}^\sharp \overline{\mu} \right) \\
&= \frac{3}{4} c \overline{\partial} \left(u \int \left(-\partial e^{i\tau S}\right) \; \mathbf{t}^\sharp \overline{\mu} \right) \\
&= -\frac{3}{4} c \overline{\partial} \left(u \partial \left(\int e^{i\tau S} \mathbf{t}^\sharp \overline{\mu} \right) - u \int e^{i\tau S} \mathbf{t}^\sharp \partial \overline{\mu} \right) \\
&= -\frac{3}{4} \overline{\partial} \left(u \overline{\partial}^{-1} \partial q \right) + \frac{3}{4} c \overline{\partial} \left(u \overline{\partial}^{-1} \left(\int \mathbf{t}^\sharp \overline{\partial} \left(e^{i\tau S} \partial \overline{\mu} \right) \right) \right).
\end{aligned}$$

Using (3.35) again we find

$$\begin{aligned}
I_2 &= -\frac{3}{4} \overline{\partial} \left(u \overline{\partial}^{-1} \partial q \right) + \frac{3}{16} \overline{\partial} \left(u \overline{\partial}^{-1} \left(q \overline{\partial}^{-1} q \right) \right) \\
&= -\frac{3}{4} \overline{\partial} \left(u^2 \right) + \frac{3}{16} \overline{\partial} \left(u \overline{\partial}^{1} \left(q \overline{\partial}^{-1} q \right) \right)
\end{aligned}$$

where we used (3.33) in the first line, and in the second line we used $u = \overline{\partial}^{-1} \partial q$. Hence

$$\begin{aligned}
I_1 - I_2 = & -\frac{3}{4} \partial (qu) + \frac{3}{4} \overline{\partial} \left(u^2 \right) \\
& + \frac{3}{16} \partial \left(q \overline{\partial}^{-1} \left(q \overline{\partial}^{-1} q \right) \right) - \frac{3}{16} \overline{\partial} \left(u \overline{\partial}^{1} \left(q \overline{\partial}^{-1} q \right) \right).
\end{aligned}$$

Since $\partial q = \overline{\partial} u$ and $\overline{\partial}^{-1} \left(q \overline{\partial}^{-1} q \right) = \frac{1}{2} \left(\overline{\partial}^{-1} q \right)^2$ we can conclude that the second line is zero and (3.34) holds. The conclusion now follows. □

4. Special Solutions

There are various powerful methods to find solutions of nonlinear evolution equations, most notably the inverse scattering method. However, the inverse scattering method is not readily useful for finding closed-form solutions to the NV equation, and so techniques including Hirota's method and the extended mapping approach (EMA) are presented here to construct closed-form solutions of several types of solitons. We begin by explaining the close connection between plane-wave solutions to NV and solutions to the KdV equation and present evolutions of KdV ring-type solutions. Although the KdV ring-type solition is not of conductivity type, the scattering transform is computed in Section 4.3, and the numerical results provide evidence of the presence of an exceptional circle.

4.1. KdV-type Solutions. Consider the NV equation (3.1) in the form

$$\dot{q} = -\frac{1}{4} q_{xxx} + \frac{3}{4} q_{xyy} + \frac{3}{4} \operatorname{div}((q - E)\mathbf{u}), \tag{4.1}$$

where $u = u_1 + iu_2$ and $\mathbf{u} = (u_1, u_2)$, and the auxiliary equation $\overline{\partial} u = \partial q$ as

$$\left\{ \begin{array}{rcl} (u_1)_x - (u_2)_y & = & +q_x \\ (u_2)_x + (u_1)_y & = & -q_y \end{array} \right. \tag{4.2}$$

As in [**13**] we use a FFT-based method to solve the equations on the square $-L \leq x, y \leq +L$ with periodic boundary conditions.

To examine the linear contributions we introduce a function

$$q(t,x,y) = \exp(i\,(\xi\,x + \eta\,y)).$$

Then the $\overline{\partial}$ equation (4.2) is solved by

$$\begin{aligned} u_1(t,x,y) &= \frac{\xi^2-\eta^2}{\xi^2+\eta^2}\,\exp(i\,(\xi\,x+\eta\,y)) \\ u_2(t,x,y) &= \frac{-2\,\xi\,\eta}{\xi^2+\eta^2}\,\exp(i\,(\xi\,x+\eta\,y)), \end{aligned}$$

and thus the linear part of the NV equation (4.1)

$$\dot{q} = -\frac{1}{4}\frac{\partial^3}{\partial x^3}\,q + \frac{3}{4}\frac{\partial^3}{\partial x \partial y^2}\,q + E\,\frac{3}{4}\nabla\cdot\mathbf{u}$$

is transformed into a elementary linear ODE for the Fourier coefficient $c(t)$

$$4\,\frac{d}{dt}\,c(t) = i\,\left(\xi^3 - 3\,\xi\,\eta^2\right)\left(1 - \frac{3\,E}{\xi^2+\eta^2}\right)\,c(t).$$

Assuming a Fourier approximation of the solutions

$$q(t,x,y) = \sum_{j,k=0}^{N-1} c_{j,k}(t)\,\exp(ipi\,(k\,x + j\,y)/L)$$

this leads to a coupled system of ODEs for the Fourier coefficients $c_{j,k}(t)$. We use a Crank–Nicolson scheme for the linear part of NV and an explicit method for the nonlinear contribution $\mathrm{div}(q\,\mathbf{u})$. For details see [**13, 14**].

There is a close connection between plane wave solutions to NV and solutions to KdV (see [**13**]):

REMARK 4.1. Assume the solutions to NV are planar waves

$$\begin{aligned} q(t,s) &= q(t,x,y) = q(t,n_1 s, n_2 s) \\ u_i(t,s) &= u_i(t,x,y) = u_i(t,n_1 s, n_2 s) \end{aligned}$$

for some direction vector $\vec{n} = (n_1,n_2) = (\cos(\alpha),\sin(\alpha))$. Then the bounded solutions to the $\overline{\partial}$ equation (4.2) are given by

$$\begin{aligned} u_1(t,s) &= +(n_1^2 - n_2^2)\,q(t,s) + c_1 \\ u_2(t,s) &= -(2\,n_1\,n_2)\,q(t,s) + c_2 \end{aligned}$$

for arbitrary constants c_1, c_2, and the NV equation (4.1) reduces to an equation similar to the KdV equation,

$$\frac{4}{\kappa}\,q_t = -q''' + 6\,q\,q' + \frac{3\,\beta}{\kappa}\,q' \tag{4.3}$$

with $\kappa = \cos(3\,\alpha)$ and $\beta = -\kappa\,E + c_1\,n_1 + c_2\,n_2$. If $v(t,s)$ denotes a solution to the standard KdV equation

$$\dot{v}(t,x) = -v'''(t,x) + 6\,v(t,x)\,v'(t,x), \tag{4.4}$$

we obtain explicit solutions to NV by

$$q(t,s) = v\left(\frac{\kappa}{4}\,t\,,\,s + k_1\,t\right) - k_2 = v\left(\frac{\kappa}{4}\,t\,,\,s + \frac{3}{4}\,(c_1\,n_1 + c_2\,n_2)\,t\right) + \frac{E}{2}.$$

For the special case $c_1 = c_2 = 0$ we find

$$q(t, n_1 s, n_2 s) = q(t, s) = v\left(\frac{\kappa}{4}\, t\,,\, s\right) + \frac{E}{2}$$

Thus we can relate all solutions to KdV as planar solutions to NV, with different speeds of propagation depending on the direction of the plane wave.

EXAMPLE 4.2. Based on the above remark we have an exact solution of the NV equation at zero energy given by

$$q(t, x, y) = -2\,c\,\cosh^{-2}(\sqrt{c}(x - c\,t)).$$

This solution is unstable with respect to perturbations periodic in y direction with period $\frac{2\,\pi}{k\,\sqrt{c}}$ for $0.363 < k < 1$, see [**13**]. This is confirmed by numerical evolution of the NV equation using the above spectral method.

EXAMPLE 4.3. A KdV ring initial condition.
Using $r = \sqrt{x^2 + y^2}$ we choose a radially symmetric initial value

$$q_0(x, y) = q(0, x, y) = f(r) = -\frac{1}{2}\cosh^{-2}\left(\frac{1}{2}\,(r - 20)\right) < 0\,.$$

This corresponds to a solution of equation (4.3) in the radial variable r, a KdV ring with radius 20.

Using the argument in [**56**, Appendix B] (based on [**53**]) we may consider q_0 as a non-positive deviation from 0, which is a potential of conductivity type. Thus q_0 is subcritical and consequently not of conductivity type. In [**53**] Murata classifies general solutions of the Schrödinger equation $\Delta u = q\,u$, For our special case we have an elementary proof for the required result. For $q(0, x, y)$ to be of conductivity type we would need a positive, radially symmetric function u such that $\Delta u = q\,u$ or in radial coordinates $(r\,u'(r))' = r\,f(r)\,u(r)$. Since the function is radially symmetric we use $u'(0) = 0$ and an integration leads to

$$r\,u'(r) = 0 + \int_0^r s\,f(s)\,u(s)\,ds < 0$$

Thus $u'(r)$ is negative and $r\,u'(r)$ is decreasing. Consequently we have a constant $C = -r_0\,u'(r_0) > 0$ and for all $r \geq r_0$ we conclude $u'(r) \leq -C/r$. This implies

$$u(r) \;=\; u(r_0) + \int_{r_0}^r u'(s)\,ds \leq u(r_0) - \int_{r_0}^r \frac{C}{s}\,ds = u(r_0) - C\,(\ln(r) - \ln(r_0))$$

and for r large enough this is in contradition to $u(r) > 0$ and thus $q_0(x, y)$ is not of conductivity type.

We solve the NV equation at zero energy $E = 0$ with the above initial condition. Based on the speed profile with the angularly dependent speed factor $\kappa = \cos(3\,\alpha)$ from Remark 4.1, one expects that the initially circular shape will be deformed and its shape will be more triangular at later times. This is confirmed by Figure 2, which shows graphs of $-q(t, x, y)$ at different times t.

To examine the possible blowup of the solution at a finite time we ran the algorithm based on the spectral method on a domain $-50 \leq x, y \leq +50$ with Fourier grids of sizes 1024×1024, 2048×2048 and 4096×4096. With time steps $dt = 0.01$ and $dt = 0.001$ we examined the solution and its L_2 norm. In all cases the solution either blew up at times just beyond $t = 38$ or displayed a sudden occurrence of sizable noise. For a final decision of a blow up time the exact shape and size of the spikes in Figure 2 have to be examined carefully.

Letting the KdV ring initial condition evolve for negative times, we observed the same solutions as for positive times, but rotated by in the spatial plane by 60°. We observed blowup at times smaller that $t = -38$.

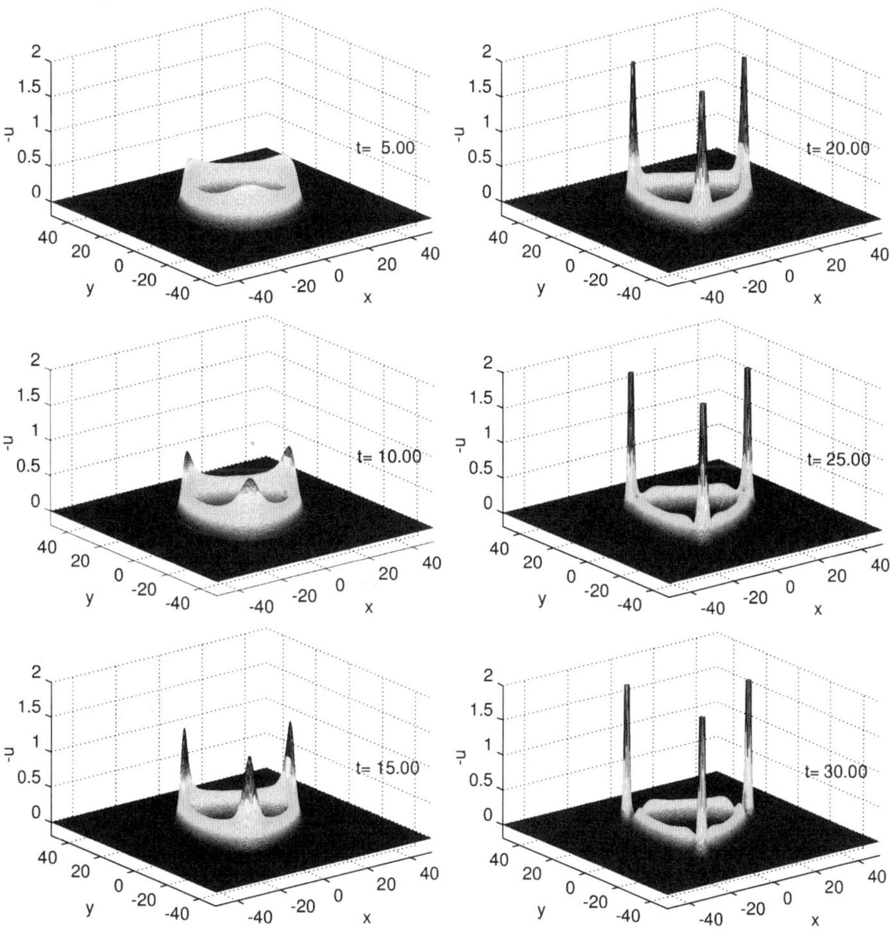

FIGURE 2. Evolution of a KdV ring by NV at zero energy

EXAMPLE 4.4. With the initial value of the KdV ring in Example 4.3, we evolve the solution by the NV equation (4.1) and (4.2) with a positive energy $E = 1/8$. Time snapshots of the evolution are plotted in Figure 3. The initial dynamics are comparable to the previous example at zero energy: three spikes appear and grow rapidly in size, but as time progresses, these spikes decay in amplitude, and separate from the previous KdV ring, and a new triple of spikes appears. The process is repeated. Observe that the solution exists at least until time $t = 100$, also confirmed by the graph of the L_2 norm of the solution as function of time in Figure 4.

4.2. Closed-form Solutions. Most, if not all, soliton equations admit traveling waves solutions that involve the hyperbolic secant function, which can be

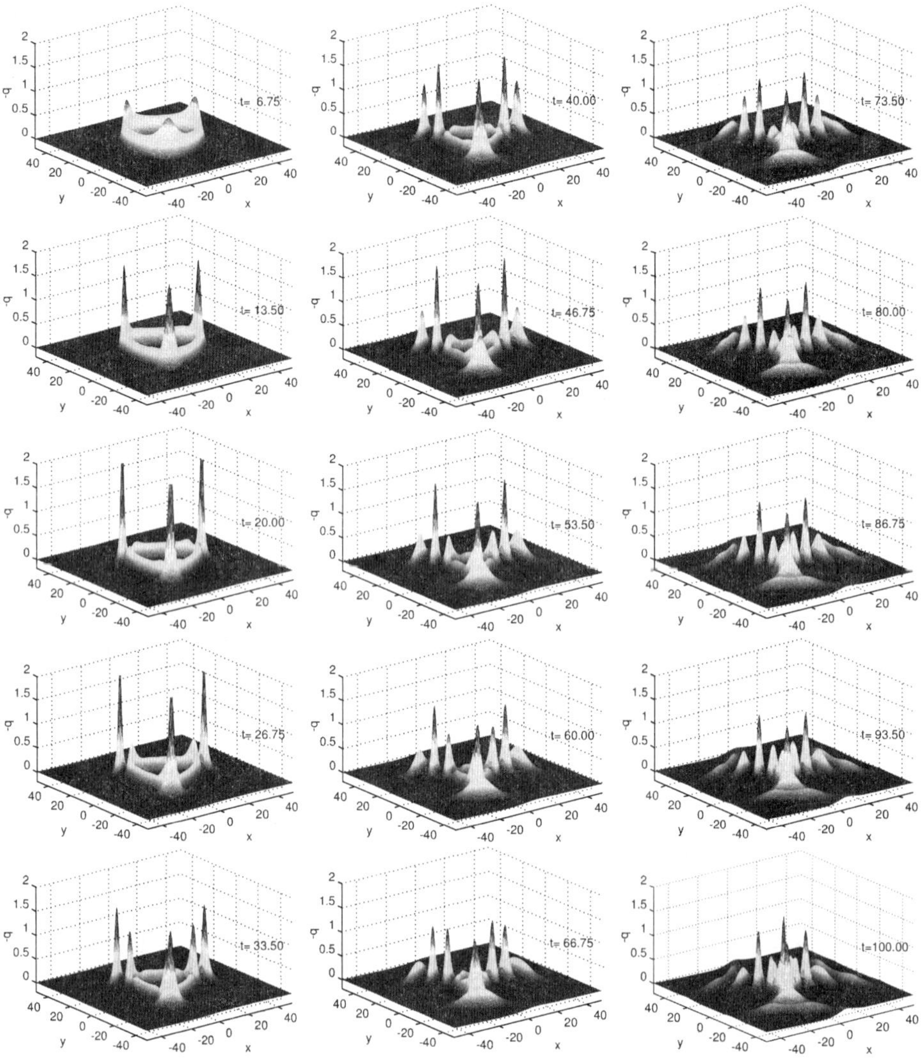

FIGURE 3. Evolution of a KdV ring by NV at positive energy

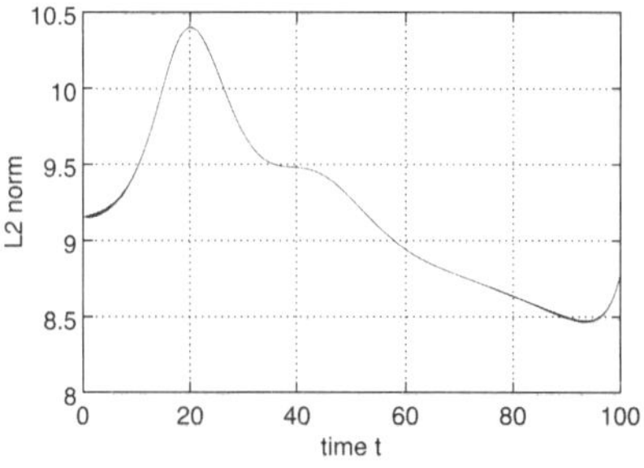

FIGURE 4. L_2 norm of the NV solution at positive energy

written in terms of the hyperbolic tangent function. Moreover, the hyperbolic tangent function is a solution to the Riccati equation, $\phi' = l_0 + \phi^2$, for $l_0 < 0$ for certain initial conditions. The ubiquity of the hyperbolic functions as traveling wave solutions naturally leads to the idea of expansion methods for solving soliton equations.

Solutions in the literature to the NV equation include the solutions from the inverse scattering transform [**44**–**46**, **64**], the classic hyperbolic secant and cnoidal solutions [**60**], and rational solutions derived using Darboux transformations that lead to finite time blow–up (see [**88**] and references therein). In this section we present new solutions to the NV equation using Hirota's bilinear method and the Extended Mapping Approach (EMA). New multiple traveling wave solutions using the Modified Extended Tanh-Function Method can be found in [**14**]. This approach results in closed-form solutions, most of which contain singularities. We note that the solutions found by Hirota's method are plane-wave solutions, that is, KdV-type solutions, while the EMA-derived solutions are not.

4.2.1. *Hirota's Method.* Following the pioneering work of Hirota [**31**], multi-soliton solutions can be derived using *Hirota's bilinear method.* This method yields soliton solutions as a sum of polynomials of exponentials and was used in [**94**] to find multisoliton solutions to the Nizhnik-Novikov-Veselov equation

$$\begin{aligned} \dot{q} &= -aq_{xxx} + bq_{yyy} - 3a(qu_1)_x - 3b(qu_2)_y \\ q_x &= (u_1)_y \\ q_y &= (u_2)_x \end{aligned}$$

The main idea is to reduce the nonlinear equation to a bilinear form through a transformation involving the logarithmic function. To express the wave velocity c in terms of the dispersive coefficients, assume u is a plane wave solution with $k_1 = k_2 = k$, $u = e^{kx+ky-ct}$, and substitute u into (2.1). This results in $c = -k^3/2$. Under the transformations $q = R(\ln(f))_{xx}$, $v = R(\ln(f))_{xy}$, and $w = R(\ln(f))_{yy}$ where $f(x,y,t) = 1 + Ce^{kx+ky+\frac{k^3}{2}t}$ and C is an arbitrary constant, one can algebraically solve for R to find a bilinear form (one finds $R = 2$). This method results in the soliton solution

$$q(x,y,t) = u_1(x,y,t) = u_2(x,y,t) = \frac{2\,C\,k^2 e^{k\,(2\,x+2\,y+k^2t)/2}}{\left(1+e^{k(2\,x+2\,y+k^2t)/2}\right)^2}. \tag{4.5}$$

Choosing

$$f(x,y,t) = 1 + e^{\theta_1} + e^{\theta_2} + a_{12}e^{\theta_1+\theta_2},$$

where $\theta_i = k_i x + k_i y + \frac{1}{2}k_i^3 t$, $i = 1, 2$, in the logarithmic transformations above results in the two-soliton solution with a_{12} given in terms of k_1 and k_2 by $a_{12} = (k_1 - k_2)^2/(k_1 + k_2)^2$

$$\begin{aligned} q(x,y,t) = {} & \frac{2\left(k_1^2 e^{\theta_1} + k_2^2 e^{\theta_2} + (k_1-k_2)^2 e^{\theta_1+\theta_2}\right)}{1 + e^{\theta_1} + e^{\theta_2} + \frac{(k_1-k_2)^2}{(k_1+k_2)^2}e^{\theta_1+\theta_2}} \\ & - \frac{2\left(k_1 e^{\theta_1} + k_2 e^{\theta_2} + \frac{(k_1-k_2)^2}{k_1+k_2}e^{\theta_1+\theta_2}\right)^2}{\left(1 + e^{\theta_1} + e^{\theta_2} + \frac{(k_1-k_2)^2}{(k_1+k_2)^2}e^{\theta_1+\theta_2}\right)^2}. \end{aligned} \tag{4.6}$$

The evolution of the two-soliton solution is plotted in Figure 5. Further details and a three-soliton solution are found in [**14**].

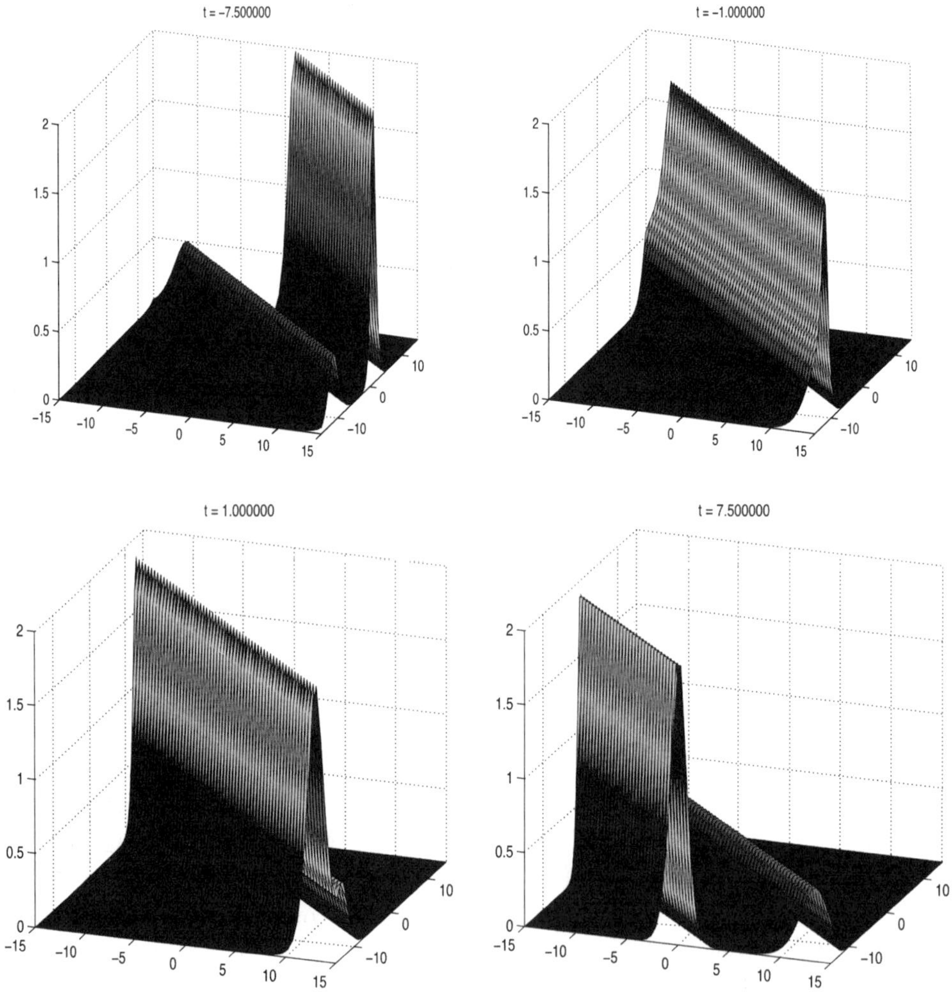

FIGURE 5. Time snapshots of the evolution of a 2-soliton solution derived Hirota's bilinear method.

4.2.2. *Extended Mapping Approach.* The extended mapping approach (EMA) was presented formally by Zheng [**97**] and extends results by Lou and Ni [**47**]. The method is designed to find mappings between nonlinear PDE's. In this approach, q, v, w are expanded in terms of a function ϕ_i that satisfies the Riccati equation

$$\frac{d\phi}{dR} = \ell_0 + \phi^2,$$

where $R = R(x, y, t)$. Thus, $q(x, y, t) = \sum_{i=0}^{n} a_i \phi^i$, $v(x, y, t) = \sum_{i=0}^{m} b_i \phi^i$, and $w(x, y, t) = \sum_{i=0}^{k} c_i \phi^i$, where the values of n, m and k are determined by balancing the highest order derivative terms with the nonlinear terms of the PDE. The method is described nicely in [**78**]. The balancing method results in $n = m = k = 2$. Substituting these expansions into the NV equation and equating coefficients of the resulting polynomial in ϕ results in a system of thirteen PDE's from which we need to solve for the coefficients $a_i(x,y,t), b_i(x,y,t)$, and $c_i(x,y,t)$, $i = 1,2$. Using a separation technique for R, namely, $R(x,y,t) = p(x,t) + q(y,t)$ results in $sech^2$ solutions, static (time-independent) solutions, and breather-type solutions of the NV equation. Further details, including the choices of ϕ are found in [**14**]. A time-independent solution is given by

$$q(x,y,t) = \frac{-1728\,y^6 + (-96+1728\,C)y^4 + (-40+288\,C)\,y^2 - 36\,C + 5}{432\,y^4 - 36\,y^2} - 4\,\tanh(x+y^2) + (2+8\,y)\,\tanh^2(x+y^2)$$

$$v(x,y,t) = \frac{144\,y^4 + (-12+432\,C)y^2 - 36\,C + 5}{36\,y^2} + 4\,\tanh(x+y^2) + (2-8\,y)\,\tanh^2(x+y^2)$$

$$w(x,y,t) = 8\,y - 8\,y\,\tanh^2(x+y^2),$$

where C is an arbitrary constant. See Figure 6 for a plot of the solution q with $C = 1$.

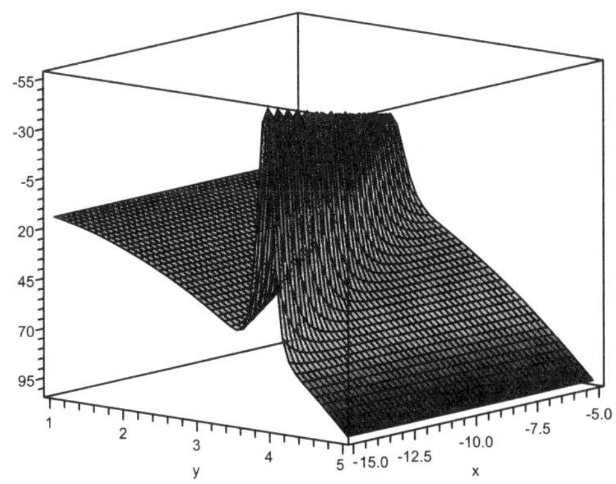

FIGURE 6. A static solution to the NV equation

Breather solutions are solutions with a type of periodic back–and–forth motion in time. One particular breather solution from [**14**] is

$$\begin{aligned} q(x,y,t) &= \frac{-1728\,y^6 + (-96 + 1728\,C)\,y^4 + (-40 + 288\,C)\,y^2 - 36\,C + 5}{432\,y^4 - 36\,y^2} \\ &\quad - 4\tanh(1 + x + y^2 + 4\cos t) \\ &\quad + (2 + 8\,y^2)\tanh^2(1 + x + y^2 + 4\cos t) \end{aligned}$$

$$\begin{aligned} v(x,y,t) &= \frac{-192\sin(t)\,y^2 + 144\,y^4 + (-12 + 432\,C)\,y^2 - 36\,C + 5}{36\,y^2} \\ &\quad + 4\tanh(1 + x + y^2 + 4\cos t) \\ &\quad + (2 - 8\,y^2)\tanh^2(1 + x + y^2 + 4\cos t) \end{aligned}$$

$$w(x,y,t) = 8\,y - 8\,y\tanh^2(1 + x + y^2 + 4\cos t).$$

Several time snapshots are shown in Figure (7). For multisoliton solutions the reader is referred to [**14**].

For multisoliton solutions the reader is referred to [**14**]

4.3. Scattering transform of the ring soliton at time zero. In this section we compute numerically the scattering transform of the KdV ring soliton discussed in example 4.3 and illustrated in Figure 2. Since the initial potential is supercritical, we expect the scattering transform to have a singularity. Since the intial potential is real-valued and rotationally symmetric in the z-plane, also the scattering transform is real-valued and rotationally symmetric in the k-plane. See [**56**, Appendix A] for a proof. Therefore it is enough to compute $\mathbf{t}(k)$ only for real and positive k.

We use both the LS and the DN methods described in Section 3.2 and compare the results to verify accuracy. The values in the range $0.1 \leq |k| \leq 4$ are reliable as the results of both NV and LS methods closely agree there. However, the DN method does not give reliable results for $|k| > 4$.

To assure accuracy for $|k| > 4$, we compare the results of the LS method with two different grids in the z-domain. The coarser grid has 4096×4096 points, and the finer grid has 8192×8192 points. The coarser grid is not a subset of the finer grid. We remark that both of these grids are significantly finer than those we typically use for computing scattering transforms for conductivity-type potentials. Due to high memory requirements, we used a liquid-cooled HP Z800 Workstation with 192 GB of memory. Even with that powerful machine, the evaluation of one point value using the finer grid takes more than 11 hours. The LS method gives closely matching results in the regions $3 \leq |k| \leq 5$ and $|k| > 9$.

Figure 8 shows the profile of the scattering transform. In the interval $5 < |k| < 9$ the numerical computation does not converge, resulting either in inaccurate evaluation of the point values of the scattering transform or in complete failure of the algorithm due to using up all the memory. We suspect that the observed numerical divergence arises from the existence of at least one exceptional circle in the interval $5 < |k| < 9$.

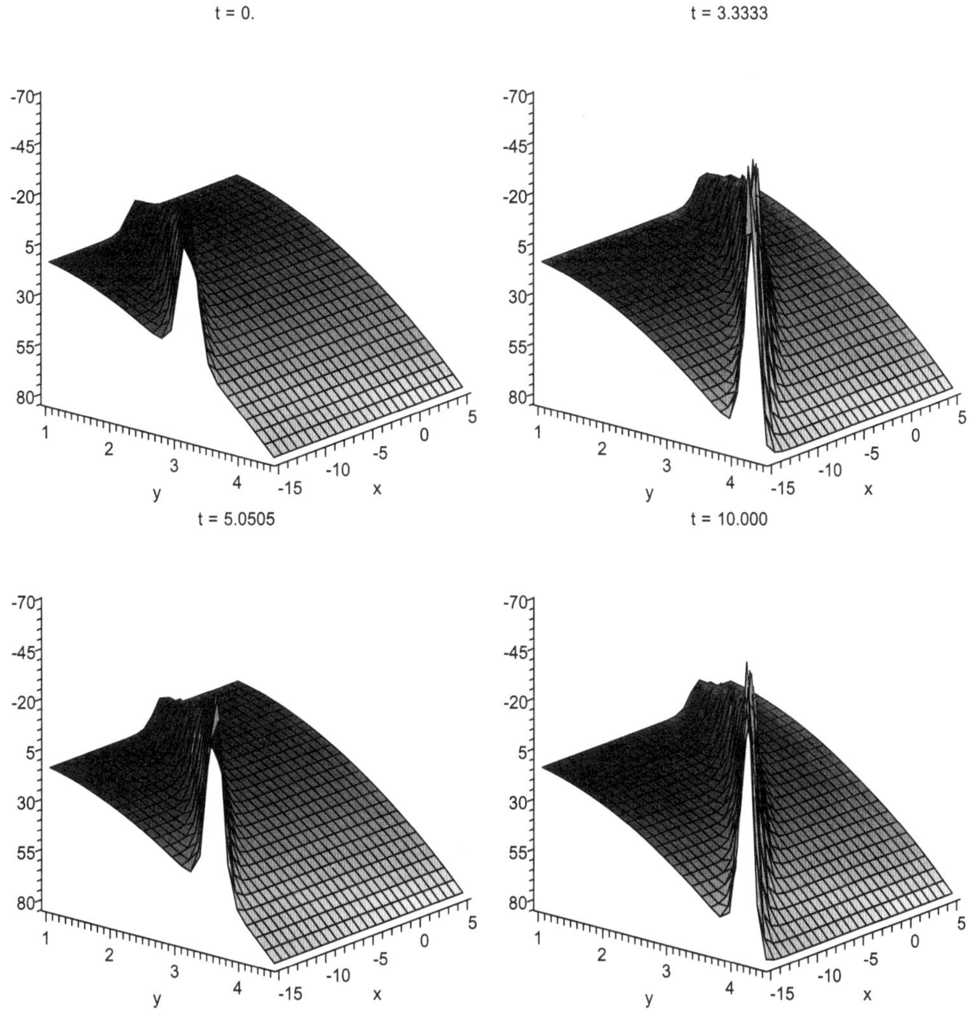

FIGURE 7. Time snapshots of a breather solution derived using the EMA.

5. Zero-energy exceptional points

The inverse scattering method for the solution of the Novikov-Veselov equation is based on the complex geometrical optics (CGO) solutions ψ of the Schrödinger equation (2.8). The function $\psi(z,k)$ is asymptotically close to the exponential function e^{ikz} in the sense of formula (2.9); the point is that ψ can be used to define a nonlinear Fourier transform $\mathbf{t}(k)$ specially designed for linearizing the NV equation. See diagram (1.5) above.

However, there is a possible difficulty in using ψ and $\mathbf{t}$. Even in the case of a smooth and compactly supported potential $q \in C_0^\infty$, there may exist complex numbers $k \neq 0$ for which equation (2.8) does not have a unique solution satisfying the asymptotic condition (2.9). Such k are called *exceptional points* of q. It is shown in [**56**] that that exceptional points of rotationally symmetric potentials come in circles centered at the origin and that the scattering transform has a strong

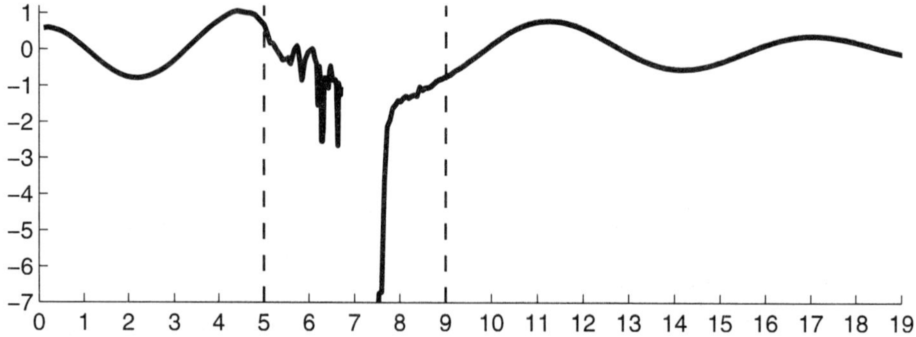

FIGURE 8. Profile of the scattering transform $\mathbf{t}(k)$ of the KdV ring soliton for k ranging in the positive real axis. The computation was unstable in the region $5 < |k| < 9$, suggesting the presence of an exceptional circle.

singularity at the circles. The singularity prevents any currently understood use of the inverse nonlinear Fourier transform in the diagram (1.5). It seems safe to assume that the situation becomes only worse for more general potentials.

What is the connection between exceptional points and dynamics of solutions of the Novikov-Veselov equation? For example, does the absence of exceptional points in the initial data ensure smooth NV evolution? Do exceptional points perhaps correspond to lumps or solitons or finite-time blow-ups? Such a conjecture was presented already in [**9**, page 27], but the question is still open.

This section is devoted to a computational experiment illustrating exceptional points of a parametric family of rotationally symmetric potentials. The example clarifies the relationship between exceptional points and the trichotomy supercritical/critical/subcritical presented in Definition 1.1.

Take a radial C_0^2 function $w(z) = w(|z|)$ as shown in Figure 9. A detailed definition of w is given in [**55**, Section 5.1]. Define a family of potentials by $q_\lambda = \lambda w$, parameterized by $\lambda \in \mathbb{R}$. Now the case $\lambda = 0$ gives $q_0 \equiv 0$, which is a critical potential since it arises as $q_0 = \sigma^{-1}\Delta\sigma$ with the positive function $\sigma \equiv 1$. From Murata [**52**] we see that $\lambda < 0$ gives a supercritical potential and $\lambda > 0$ gives a subritical potential. See [**56**, Appendix B] for details.

We use the DN method described in Section 3.2 to compute the scattering transforms of the potentials q_λ for the parameter λ ranging in the interval $[-25, 5]$. Since each potential $q_\lambda(z)$ is real-valued and rotationally symmetric in the z-plane, also the scattering transform is real-valued and rotationally symmetric in the k-plane. See [**56**, Appendix A] for details. Therefore it is enough to compute $\mathbf{t}(k)$ only for k ranging along the positive real axis. In Figure 11 we show the result of the computation as a two-dimensional grayscale image.

It is known that critical potentials do not have nonzero exceptional points; see [**56, 58**]. Thus there are no singularities in Figure 11 for $\lambda = 0$ (actually in this simple example we have $\mathcal{T}q_0 \equiv 0$). Furthermore, a Neumann series argument shows that for a fixed λ there exists such a positive constant $K = K(\lambda)$ that there are no exceptional points for q_λ satisfying $|k| > K$. See the analysis in [**58**, above formula (1.12)].

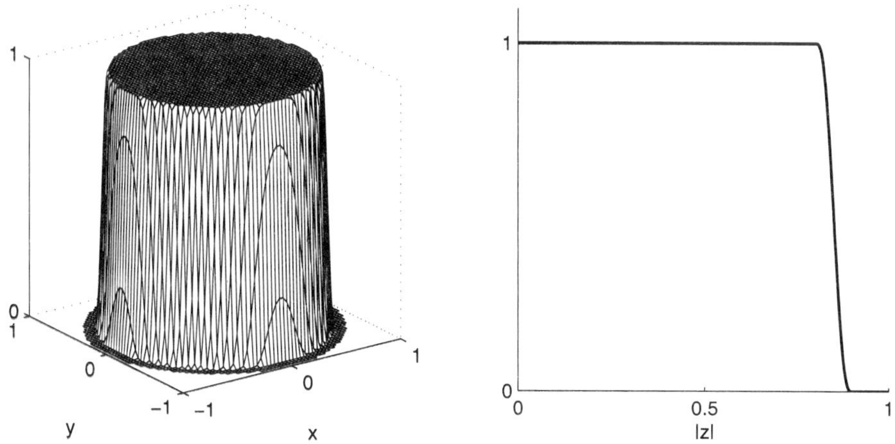

FIGURE 9. Left: rotationally symmetric test function $w(z)$. Right: the profile $w(|z|)$.

According to [**54**], subcritical potentials do not have nonzero exceptional points. Thus there are no singularities in Figure 11 for parameter values $\lambda \geq 0$. (We remark that the seemingly exceptional curves in the upper right corner of Figures 3 and 9 in [**56**] are due to deteriorating numerical accuracy for large positive values of λ. Those figures are trustworthy only for λ close to zero.)

For negative λ close to zero it is known from [**56**] that there is exactly one circle of exceptional points. The asymptotic form of that radius as a function of λ is calculated explicitely in [**56**].

For $\lambda << 0$ there is no precise understanding of exceptional points as of now; numerical evidence such as Figure 11 suggests that there may be several exceptional circles. Also, something curious seem to happen around $\lambda \approx -8$ and $\lambda \approx -20.5$; at present there is no explanation available.

See [**56**] for more zero-energy examples and [**77**] for analogous evidence of exceptional points at positive energies.

6. Open Problems

6.1. Applications of the NV Equation. The stationary NV equation [**19**] and the modified NV equation [**83**] have applications in differential geometry. Although the NV equation (1.1) is not known to be a mathematical model for any physical dynamical system, there has been some research in this direction for the dispersionless Novikov-Veselov (dNV) equation

$$\begin{aligned} q_\xi &= (uq)_z + (\overline{u}q)_{\bar{z}} \\ u_{\bar{z}} &= -3q_z. \end{aligned} \tag{6.1}$$

Equation (6.1) was derived in [**43**] as the geometrical optics limit of Maxwell's equations in an anisotropic medium. The model governs the propagation of monochromatic electromagnetic (EM) waves of high frequency ω. In particular, they consider nonlinear media with Cole-Cole dependence [**12**] of the dielectric function and magnetic permeability on the frequency. Assuming slow variation of all quantities along the z axis, it is shown that Maxwell's equations reduce to (6.1) where $q = n^2$, the

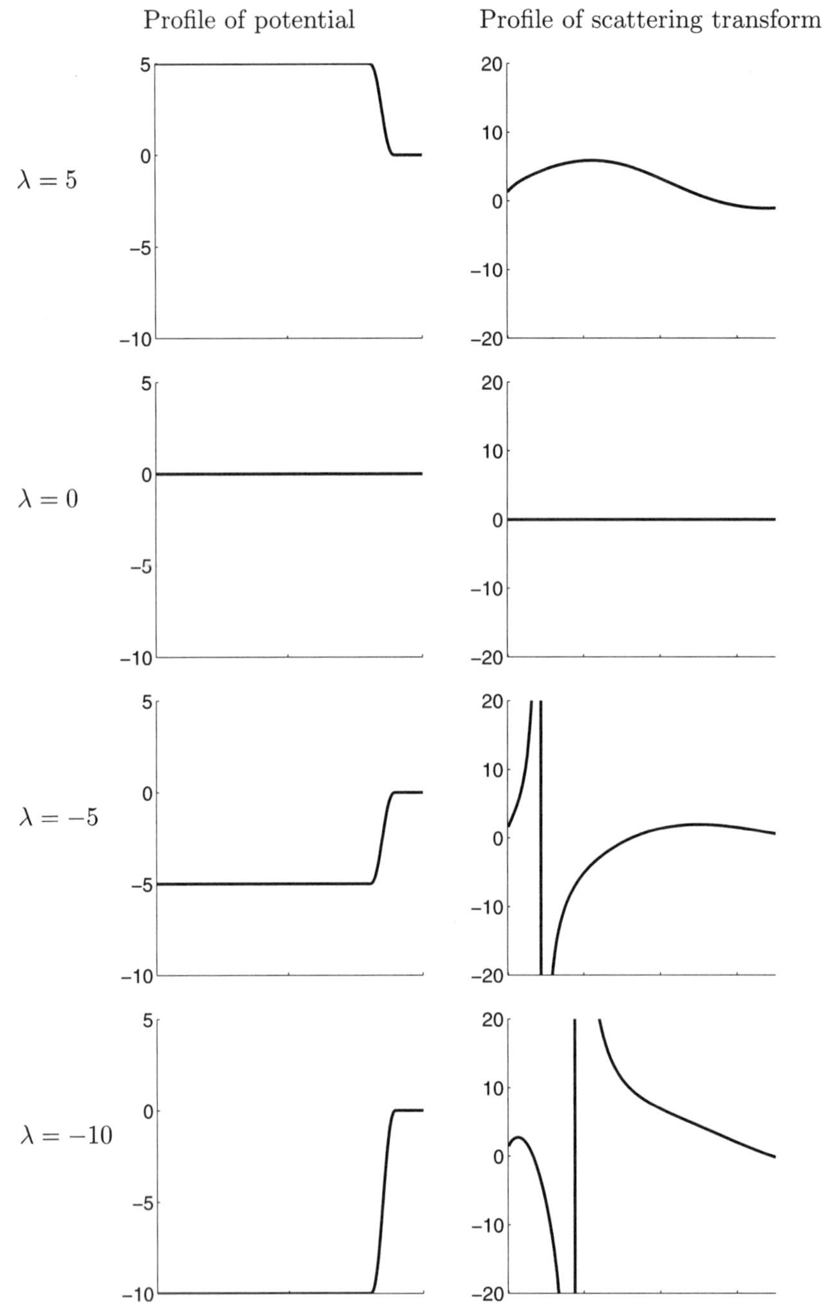

FIGURE 10. Left column: profiles of rotationally symmetric potentials $q_\lambda(|z|)$ resulting from different values of λ. Right column: profiles of corresponding scattering transforms $t_\lambda(|k|)$. Note that negative values of λ lead to exceptional circles where the scattering transform is singular. See also Figure 11 which shows scattering transform profiles corresponding to more values of λ.

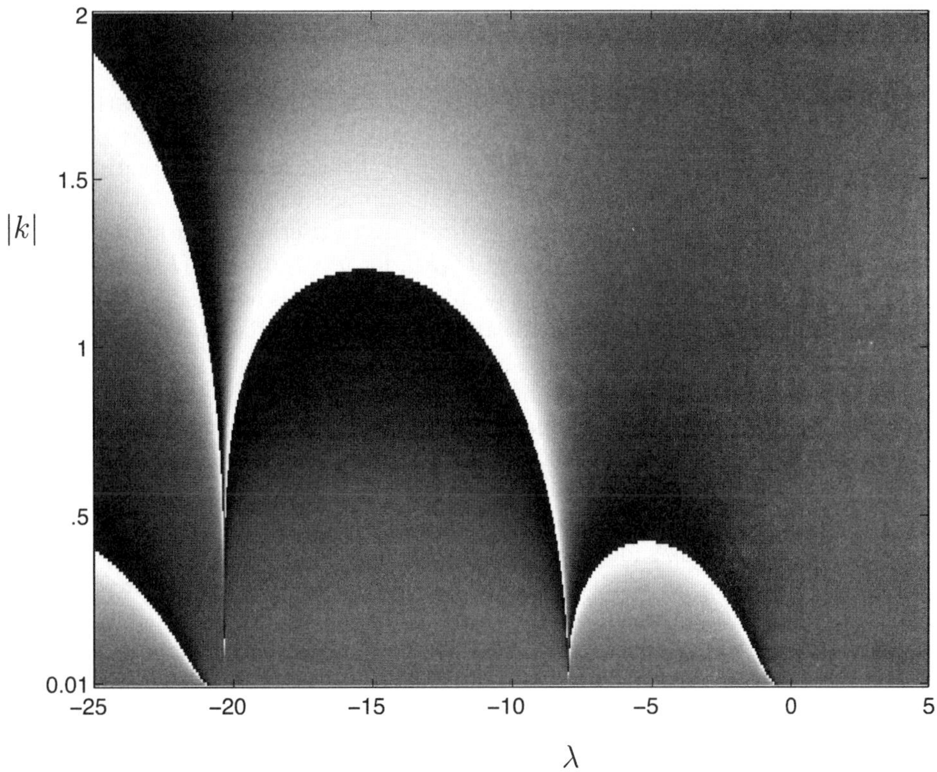

FIGURE 11. Profiles of rotationally symmetric scattering transforms. Lighter shade indicates larger values and darker shade smaller values. The horizontal axis is the parameter λ in the definition $q_\lambda(z) = \lambda w(z)$ of the potential. The vertical axis is $|k|$. There are curves along which a singular jump "from $-\infty$ to $+\infty$" appears, indicated by an abrupt change from black to white. The k values at those curves are exceptional points. Note that the profiles corresponding to $\lambda = -10, -5, 0, 5$ are shown also in Figure 10.

refractive index, and ξ is a "slow" variable defined by $z = \omega^\nu \xi$. The phase of the electric field is governed by [**43**]

$$
\begin{aligned}
S_x^2 + S_y^2 &= n^2(x, y, \xi) \\
S_\xi &= \phi(x, y, \xi; S_x, S_y)
\end{aligned}
\tag{6.2}
$$

for a real-valued function ϕ, and the dNV heirarchy characterizes both the phase and refractive index.

In [**8**] hydrodynamic-type reductions of the dNV equation are presented, but the physical interpretation of these reductions are left for future work. Interesting open problems are whether the inclusion of the dispersion terms in the NV equation models EM waves in a manner related to those derived for the dNV equation, and whether either NV or dNV serves as a physical model for any kind of hydrodynamic phenomenon.

6.2. Exceptional Sets and Large-Time Behavior for the NV Equation. Recall that the *exceptional set* of a potential q_0 is the set of all those $k \in \mathbb{C}$ for which there is not a unique NCGO solution $\mu(z,k)$. Here we discuss the relationship between: (1) the "size" of the exceptional set," (2) whether the potential q_0 is subcritical, critical, or supercritical (recall this trichotomy, Definition 1.1 from the introduction) and (3) whether the NV equation with initial data q_0 has a global solution. In these remarks, we will usually restriction attention to real-valued potentials q_0 belonging to the space $L^p_\rho(\mathbb{R}^2)$ for $\rho > 1$ and $p \in (1,2)$ (see the remarks preceding Definition 1.3); one may think of potentials of order $\mathcal{O}\left(|z|^{-2-\varepsilon}\right)$ as $|z| \to \infty$. Recall that, in the inverse scattering literature, critical potentials are usually referred to as "potentials of conductivity type."

To date, the only rigorous results on the size of exceptional sets for the zero-energy NV equation are due to Nachman [**58**] and Music [**54**]. As explained above, Nachman showed that a potential is of conductivity type (or, equivalently, a critical potential as defined in the introduction, Definition 1.1) if and only if the exceptional set is empty and the scattering transform $\mathbf{t}(k)$ is $\mathcal{O}(|k|^\varepsilon)$ as $|k| \to 0$ for some $\varepsilon > 0$. Music, extending Nachman's ideas and techniques, showed that a subcritical potential with sufficient decay at infinity has an empty exceptional set and characterized the singularity of the potential as $|k| \to 0$. Perry [**67**] showed that, if q_0 is a sufficiently smooth critical potential, the NV equation with initial data q_0 has a solution global in time. There is strong evidence to suggest that a similar result can be proved for the NV equation with subcritical initial data, based on the work of Music [**54**].

Thus, it remains to understand the singularities of the scattering transform for supercritical potentials. Examples due to Grinevich and Novikov [**30**] and Music, Perry and Siltanen [**56**] show that supercritical potentials may have circles of exceptional points. It is not known whether supercritical potentials *must* or *may* have exceptional points, nor is it known how to extend the inverse scattering formalism to potentials with nonempty exceptional sets. The following result due to Brown, Music, and Perry [**10**] gives an initial constraint on the size of exceptional sets for particularly nice potentials.

THEOREM 6.1. [**10**] *Suppose that q is a real-valued measurable valued function with the property that $|q(z)| \leq C_1 \exp(-C_2|z|)$ for some constants C_1 and C_2. Then the exceptional set of q consists at most of isolated points together with at most finitely many smooth curves with at most finitely many intersections.*

To analyze the exceptional set, the authors define a renormalized determinant whose zero set is exactly the exceptional set. To describe it, let T_k is the integral operator

$$T_k\psi = \frac{1}{4} g_k * (q\psi).$$

The differential equation for $\mu(z,k)$, the NCGO solution, may be rewritten $\mu = 1 + T_k\mu$. Hence, uniqueness of solutions is equivalent to invertibility of $(I - T_k)$, and the exceptional set is exactly the set of points k for which $(I - T_k)$ fails to be invertible. It can be shown that T_k is a compact linear operator from L^p to itself for any $p > 2$, and that T_k belongs to the so-called Mikhlin-Itskovich algebra of integral operators on L^p. For this reason we can apply the theory renormalized determinants due to Gohberg, Goldberg, and Krupnik [**21**] and define

$$\Delta(k) = \det_2(I - T_k)$$

where the determinant $\det_2$ is the renormalized determinant. Brown, Music, and Perry show that this determinant is a real-analytic function of k for exponentially decaying potentials. It now follows from the Weierstrass preparation theorem that the zero set of $\Delta(k)$ is locally the zero set of a polynomial. Since the exceptional set is known to be closed and bounded, one can completely analyze the behavior of $\Delta(k)$ near the exceptional set using finitely many such local representations. It can be shown that $\Delta(k)$ is also real-valued, from which it follows that the zero set has the claimed form.

Theorem 6.1 opens up several areas for further investigation.

First, it would be of considerable interest to determine what additional data is needed to reconstruct a potential from $\mathbf{t}(k)$ when $\mathbf{t}(k)$ has point or line singularities.

Second, it would be very interesting to know whether singularities are always present for supercritical potentials, or whether, on the other hand, singularities are generically absent.

Third, our understanding of the NV equation and its dynamics would be greatly improved by connecting 'spectral' properties of the scattering transform (i.e., the nature of its singularities) to long-term behavior of solutions. The form of the time evolution for $\mathbf{t}(k)$ suggests that the 'trichotomy' of subcritical, critical, and supercritical potentials is invariant under the NV flow. It is known that critical initial data give rise to global solutions (see [**68**]), and there is strong evidence that the same is true of subcritical initial data. On the other hand, numerical experiments such as the ring soliton, Example 4.3, and analytical solutions such as those produced by Taimanov and Tsarev [**85**–**88**] strongly suggest that supercritical initial data lead to solutions of NV that blow up in finite time. It would be very interesting to obtain a rigorous proof that this is the case, and to analyze the nature of the blow-ups by inverse scattering methods.

Appendix A. Some Useful Analysis

In the direct scattering problem at zero energy, Faddeev's Green's function plays a critical role in elucidating properties of the CGO solutions that define the scattering transform. Recall that the normalized CGO solutions solve the equation

$$\overline{\partial}\,(\partial + ik)\,\mu = (1/4)q\mu$$

and that Faddeev's Green's function is Green's function for the operator $\overline{\partial}\,(\partial + ik)$. On the other hand, the solid Cauchy transform is an inverse for the $\overline{\partial}$ operator with range in L^p functions for $p > 2$, and hence is a fundamental tool for solving the $\overline{\partial}$ problem that defines the inverse scattering transform. Finally, the Beurling transform is an integral operator which gives a meaning to the operator $\overline{\partial}^{-1}\partial$ that occurs in the definition of the nonlinearity in the NV equation. Here we collect some useful properties of these transforms and some essential estimates.

A.1. Faddeev's Green's Function at Zero Energy. We recall some key facts about Faddeev's Green's function g_k. We refer the reader to Siltanen's thesis [**75**] for details and references to the literature.

Recall that g_k is defined by the formula

$$g_k(z) = \frac{1}{(2\pi)^2} \int e^{i\xi\cdot x} \frac{1}{\overline{\xi}(\xi+k)}\, dm(\xi)$$

where $z = x_1 + ix_2$, $\xi \cdot x = \xi_1 x_1 + \xi_2 x_2$, $\xi = \xi_1 + i\xi_2$, and $\overline{\xi} = \xi_1 - i\xi_2$. By the Hausdorff-Young inequality, $g_k \in L^p$ for any $p > 2$. In fact, the estimate

$$\|g_k\|_p \leq C_p \, |k|^{-2/p}$$

(see Siltanen [**75**], Theorem 3.10) holds for any $k \neq 0$. It is important to note that

$$g_k(z) = g_1(kz).$$

The following large-x asymptotic expansion of $g_1(x)$ is proved in [**75**], Theorem 3.11.

LEMMA A.1. *Let $z = x_1 + ix_2$ with $z \neq 0$ and $x_1 > 0$. For any integer $N \geq 0$,*

$$g_1(z) = -\frac{1}{4\pi} \sum_{j=0}^{N} \left[\frac{j!}{(iz)^{j+1}} - e^{-2ix_1} \frac{j!}{(-i\overline{z})^{j+1}} \right] + E_N(z) \tag{A.1}$$

where

$$|E_N(z)| \leq \frac{(N+1)!2^{(N+1)/2}}{\pi \, |z|^{N+2}}. \tag{A.2}$$

Since $g(-x_1 + ix_2) = \overline{g_1(x_1 + ix_2)}$, similar formulas hold for $x_1 < 0$.

REMARK A.2. Since the error estimate (A.2) does not depend on the condition $x_1 > 0$, and since $g_1(z)$ is continuous, we can conclude that the expansion (A.1) remains valid for $z \neq 0$ and $\mathrm{Re}(z) = 0$.

Now consider $g_k(z) = g_1(kz)$. Since $\mathrm{Re}\,(kz) = \frac{1}{2}\left(kz + \overline{k}\overline{z}\right)$, we immediately obtain:

LEMMA A.3. *Let $z = x_1 + ix_2$ and $k \in \mathbb{C}$. For any integer $N \geq 0$, the expansion*

$$g_k(z) = -\frac{1}{4\pi} \sum_{j=0}^{N} \left[\frac{j!}{(ikz)^{j+1}} - e^{-i\left(kz+\overline{k}\overline{z}\right)} \frac{j!}{\left(-i\overline{k}\overline{z}\right)^{j+1}} \right] + E_N(kz) \tag{A.3}$$

holds, where

$$|E_N(kz)| \leq C_N \, |kz|^{-(N+2)}.$$

.

A.2. The Cauchy Transform and the Beurling Operator. Following [**3**], chapter 4, we study the Cauchy transform and the Beurling operator through the logarithmic potential associated with Poisson's equation in two dimensions. For $\varphi \in C_0^\infty(\mathbb{R}^2)$, we may define the logarithmic potential

$$(L\varphi)(z) = \frac{2}{\pi} \int \log|z - z'| \; \varphi(z')\, dm(z')$$

which has the property

$$\partial\overline{\partial}\,(L\varphi) = \varphi.$$

Associated to L are the *Cauchy transform,*

$$(P\varphi)(z) = \frac{\partial}{\partial z}(L\varphi)(z),$$

the transform

$$\left(\overline{P}\varphi\right)(z) = \frac{\partial}{\partial \overline{z}}(L\varphi)(z),$$

and the *Beurling transform*

$$(\mathcal{S}\varphi)(z) = \frac{\partial^2}{\partial z^2}(L\varphi)(z).$$

From these definitions and (3.2), it is easy to see that

$$P \circ \frac{\partial}{\partial \overline{z}} = \frac{\partial}{\partial \overline{z}} \circ P = I \tag{A.4}$$

where I is the identity on $\mathcal{C}_0^\infty(\mathbb{R}^2)$, and

$$\mathcal{S}\left(\frac{\partial\varphi}{\partial\overline{z}}\right) = \frac{\partial\varphi}{\partial z}. \tag{A.5}$$

We have

$$(P\varphi)(z) = \frac{1}{\pi}\int \frac{1}{z-z'}\varphi(z')\; dm(z'),$$

$$\left(\overline{P}\varphi\right)(z) = \frac{1}{\pi}\int \frac{1}{\overline{z-z'}}\varphi(z)\; dm(z'),$$

and

$$(\mathcal{S}\varphi)(z) = -\frac{1}{\pi}\lim_{\varepsilon\downarrow 0}\left(\int_{|z-z'|>\varepsilon} \frac{1}{(z-z')^2}\varphi(z')\; dm(z')\right). \tag{A.6}$$

The following estimates on P extend the Cauchy transform to L^p spaces and are standard consequences of the Hardy-Littlewood-Sobolev and Hölder inequalities (see Vekua [**92**] or [**3**], §4.3). They are used to prove existence and uniqueness of solutions to the $\overline{\partial}$ problem that defines the inverse problem.

LEMMA A.4. *(i) For any* $p \in (2,\infty)$ *and* $f \in L^{2p/(p+2)}(\mathbb{R}^2)$,

$$\|Pf\|_p \le C_p \|f\|_{2p/(p+2)}\,. \tag{A.7}$$

(ii) For any p, q *with* $1 < q < 2 < p < \infty$ *and any* $f \in L^p\left(\mathbb{R}^2\right) \cap L^q\left(\mathbb{R}^2\right)$, *the estimate*

$$\|Pf\|_\infty \le C_{p,q}\|f\|_{L^p\cap L^q} \tag{A.8}$$

holds. Moreover, P *is Hölder continuous of order* $(p-2)/p$ *with*

$$|(Pf)(z) - (Pf)(w)| \le C_p |z-w|^{(p-2)/p}\|f\|_p\,. \tag{A.9}$$

(iii) If $v \in L^s(\mathbb{R}^2)$ *and* $q > 2$ *with* $q^{-1} + 1/2 = p^{-1} + s^{-1}$, *then for any* $f \in L^p(\mathbb{R}^2)$,

$$\|P(vf)\|_q \le C_{p,q}\|v\|_s\|f\|_p\,. \tag{A.10}$$

REMARK A.5. Since $\mathcal{C}_0^\infty(\mathbb{R}^2)$ is dense in $L^p \cap L^q$ and $(Pf)(z) = \mathcal{O}\left(z^{-1}\right)$ as $|z| \to \infty$ for any $f \in \mathcal{C}_0^\infty(\mathbb{R}^2)$, it follows from (ii) that if $f \in L^p \cap L^q$ for $1 < p < 2 < q < \infty$, then $Pf \in C_0(\mathbb{R}^2)$, the continuous functions that vanish at infinity.

REMARK A.6. Note that with $s = q$ in (A.10) we have

$$\left\| P\left(vf \right) \right\|_p \leq C_p \left\| v \right\|_2 \left\| f \right\|_p .$$

For any $f \in L^{2p/(p+2)}(\mathbb{R}^2)$, Lemma A.4 together with (A.4) imply that $u = Pf$ solves $\overline{\partial} u = f$ in distribution sense. Suppose, on the other hand, that $u \in L^p(\mathbb{R}^2)$ for some $p \in [1, \infty)$ and $\overline{\partial} u = 0$ in distribution sense. It follows that $\partial \overline{\partial} u = 0$ in distribution sense, so that $u \in \mathcal{C}^\infty$ by Weyl's lemma. Thus, u is actually holomorphic, so u vanishes identically by Liouville's Theorem. From this fact and (A.4), we deduce:

LEMMA A.7. *Suppose that $p \in (2, \infty)$, that $u \in L^p(\mathbb{R}^2)$, that $f \in L^{2p/(p+2)}(\mathbb{R}^2)$, and that $\overline{\partial} u = f$ in distribution sense. Then $u = Pf$. Conversely, if $f \in L^{2p/(p+2)}(\mathbb{R}^2)$ and $u = Pf$, then $\overline{\partial} u = f$ in distribution sense.*

The following expansion for solutions of $\overline{\partial} u = f$ when f is rapidly decaying gives rise to the large-k asymptotic expansion for $\mu(z, k)$.

LEMMA A.8. *Suppose that $p \in (2, \infty)$, that $u \in L^p(\mathbb{R}^2)$, that $f \in L^{2p/(p+2),N}(\mathbb{R}^2)$, and that $\overline{\partial} u = f$. Then*

$$z^N \left[u(z) - \sum_{j=0}^{N-1} \frac{1}{z^{j+1}} \int \zeta^j f(\zeta) \, dm(\zeta) \right] \in L^p(\mathbb{R}^2).$$

PROOF. An immediate consequence of the estimate (A.7), Lemma A.7 and the formula

$$\frac{1}{z - \zeta} = \frac{1}{z} \sum_{j=0}^{N-1} \left(\frac{\zeta}{z} \right)^j + \frac{1}{z^N} \frac{\zeta^N}{z - \zeta}.$$

□

REMARK A.9. If $f \in \mathcal{S}\left(\mathbb{R}^2\right)$ and depends smoothly on parameters, then the asymptotic expansion holds pointwise and is differentiable in the parameters.

The principal value integral (A.6) identifies $\mathcal{S}$ as a Calderón-Zygmund type integral operator. We have (see, for example, [**3**], §4.5.2):

LEMMA A.10. *Suppose that $p \in (1, \infty)$. The operator $\mathcal{S}$ extends to a bounded operator from $L^p(\mathbb{R}^2)$ to itself, unitary if $p = 2$. Moreover, if $\nabla \varphi$ belongs to $L^p(\mathbb{R}^2)$ for $p \in (1, \infty)$, then $\mathcal{S}\left(\partial \varphi\right) = \overline{\partial} \varphi$.*

References

[1] Angelopoulos, Y. Well-posedness and ill-posedness results for the Novikov-Veselov equation. Preprint, `arXiv:1307.4110[Math.AP]`.

[2] C. Athorne and J. J. C. Nimmo, *On the Moutard transformation for integrable partial differential equations*, Inverse Problems **7** (1991), no. 6, 809–826. MR1140316 (92k:35011)

[3] Kari Astala, Tadeusz Iwaniec, and Gaven Martin, *Elliptic partial differential equations and quasiconformal mappings in the plane*, Princeton Mathematical Series, vol. 48, Princeton University Press, Princeton, NJ, 2009. MR2472875 (2010j:30040)

[4] Kari Astala and Lassi Päivärinta, *Calderón's inverse conductivity problem in the plane*, Ann. of Math. (2) **163** (2006), no. 1, 265–299, DOI 10.4007/annals.2006.163.265. MR2195135 (2007b:30019)

[5] Richard Beals and R. R. Coifman, *Linear spectral problems, nonlinear equations and the $\overline{\partial}$-method*, Inverse Problems **5** (1989), no. 2, 87–130. MR991913 (90f:35171)

[6] L. V. Bogdanov, *The Veselov-Novikov equation as a natural generalization of the Korteweg-de Vries equation* (Russian, with English summary), Teoret. Mat. Fiz. **70** (1987), no. 2, 309–314. MR894472 (88k:35170)

[7] L. V. Bogdanov, *On the two-dimensional Zakharov-Shabat problem* (Russian, with English summary), Teoret. Mat. Fiz. **72** (1987), no. 1, 155–159. MR910489 (89j:58040)

[8] L. V. Bogdanov, B. G. Konopel′chenko, and A. Moro, *Symmetry reductions for the real dispersionless Beselov-Novikov equation* (Russian, with English and Russian summaries), Fundam. Prikl. Mat. **10** (2004), no. 1, 5–15, DOI 10.1007/s10958-006-0234-3; English transl., J. Math. Sci. (N. Y.) **136** (2006), no. 6, 4411–4418. MR2119748 (2005k:37145)

[9] Marco Boiti, Jérôme Jp. Leon, Miguel Manna, and Flora Pempinelli, *On a spectral transform of a KdV-like equation related to the Schrödinger operator in the plane*, Inverse Problems **3** (1987), no. 1, 25–36. MR875315 (88b:35167)

[10] Brown, R. M.; Music, M. E.; Perry, P. Determinants and Inverse Scattering in Two Dimensions, I: Linear Systems for the Davey-Stewartson and Novikov-Veselov Equations. In preparation.

[11] Russell M. Brown and Gunther A. Uhlmann, *Uniqueness in the inverse conductivity problem for nonsmooth conductivities in two dimensions*, Comm. Partial Differential Equations **22** (1997), no. 5-6, 1009–1027, DOI 10.1080/03605309708821292. MR1452176 (98f:35155)

[12] Cole, K. S., Cole, R. H. Dispersion and absorption in dielectrics. *J. Chem. Phys.*, **9** (1941), 341–351.

[13] Croke, R. P., Mueller, J. L., Stahel, A. Transverse instability of plane wave soliton solutions of the Novikov-Veselov equation, Preprint (2013).

[14] Ryan P. Croke, *An investigation of the Novikov-Veselov equation: New solutions, stability and implications for the inverse scattering transform*, ProQuest LLC, Ann Arbor, MI, 2012. Thesis (Ph.D.)–Colorado State University. MR3034716

[15] Alberto-P. Calderón, *On an inverse boundary value problem*, Seminar on Numerical Analysis and its Applications to Continuum Physics (Rio de Janeiro, 1980), Soc. Brasil. Mat., Rio de Janeiro, 1980, pp. 65–73. MR590275 (81k:35160)

[16] V. G. Dubrovsky and I. B. Formusatik, *New lumps of Veselov-Novikov integrable nonlinear equation and new exact rational potentials of two-dimensional stationary Schrödinger equation via $\overline{\partial}$-dressing method*, Phys. Lett. A **313** (2003), no. 1-2, 68–76, DOI 10.1016/S0375-9601(03)00715-1. MR2012392 (2004h:37114)

[17] S. A. El-Wakil and M. A. Abdou, *Modified extended tanh-function method for solving nonlinear partial differential equations*, Chaos Solitons Fractals **31** (2007), no. 5, 1256–1264, DOI 10.1016/j.chaos.2005.10.072. MR2262131

[18] Faddeev, L. D. Increasing solutions of the Schrödinger equation, *Dokl. Akad Nauk SSSR*, **165** (1965), 514–517 (English translation: *Sov. Phys. Dokl.* **10** (1966), 1033–1035).

[19] E. V. Ferapontov, *Stationary Veselov-Novikov equation and isothermally asymptotic surfaces in projective differential geometry*, Differential Geom. Appl. **11** (1999), no. 2, 117–128, DOI 10.1016/S0926-2245(99)00028-5. MR1712135 (2000m:53009)

[20] F. Gesztesy and Z. Zhao, *On positive solutions of critical Schrödinger operators in two dimensions*, J. Funct. Anal. **127** (1995), no. 1, 235–256, DOI 10.1006/jfan.1995.1010. MR1308624 (96a:35037)

[21] Israel Gohberg, Seymour Goldberg, and Nahum Krupnik, *Traces and determinants of linear operators*, Operator Theory: Advances and Applications, vol. 116, Birkhäuser Verlag, Basel, 2000. MR1744872 (2001b:47035)

[22] P. G. Grinevich, *Rational solitons of the Veselov-Novikov equations are reflectionless two-dimensional potentials at fixed energy* (Russian, with English summary), Teoret. Mat. Fiz. **69** (1986), no. 2, 307–310. MR884498 (88b:81208)

[23] P. G. Grinevich, *The scattering transform for the two-dimensional Schrödinger operator with a potential that decreases at infinity at fixed nonzero energy* (Russian, with Russian summary), Uspekhi Mat. Nauk **55** (2000), no. 6(336), 3–70, DOI 10.1070/rm2000v055n06ABEH000333; English transl., Russian Math. Surveys **55** (2000), no. 6, 1015–1083. MR1840357 (2002e:37115)

[24] P. G. Grinevich and R. G. Novikov, *Analogues of multisoliton potentials for the two-dimensional Schrödinger operator* (Russian), Funktsional. Anal. i Prilozhen. **19** (1985), no. 4, 32–42, 95. MR820082 (88a:58090)

[25] P. G. Grinevich and R. G. Novikov, *Analogues of multisoliton potentials for the two-dimensional Schrödinger operator, and a nonlocal Riemann problem* (Russian), Dokl. Akad. Nauk SSSR **286** (1986), no. 1, 19–22. MR822090 (87h:35297)

[26] P. G. Grinevich and S. P. Novikov, *Inverse scattering problem for the two-dimensional Schrödinger operator at a fixed negative energy and generalized analytic functions*, Plasma theory and nonlinear and turbulent processes in physics, Vol. 1, 2 (Kiev, 1987), World Sci. Publishing, Singapore, 1988, pp. 58–85. MR957156 (90c:35199)

[27] P. G. Grinevich and S. V. Manakov, *Inverse problem of scattering theory for the two-dimensional Schrödinger operator, the $\bar{\partial}$-method and nonlinear equations* (Russian), Funktsional. Anal. i Prilozhen. **20** (1986), no. 2, 14–24, 96. MR847135 (88g:35197)

[28] P. G. Grinevich and S. P. Novikov, *A two-dimensional "inverse scattering problem" for negative energies, and generalized-analytic functions. I. Energies lower than the ground state* (Russian), Funktsional. Anal. i Prilozhen. **22** (1988), no. 1, 23–33, 96, DOI 10.1007/BF01077719; English transl., Funct. Anal. Appl. **22** (1988), no. 1, 19–27. MR936696 (90a:35181)

[29] Piotr G. Grinevich and Roman G. Novikov, *Transparent potentials at fixed energy in dimension two. Fixed-energy dispersion relations for the fast decaying potentials*, Comm. Math. Phys. **174** (1995), no. 2, 409–446. MR1362172 (96h:35036)

[30] P. G. Grinevich and R. G. Novikov, *Faddeev eigenfunctions for point potentials in two dimensions*, Phys. Lett. A **376** (2012), no. 12-13, 1102–1106, DOI 10.1016/j.physleta.2012.02.025. MR2899587

[31] Ryogo Hirota, *Exact N-soliton solutions of the wave equation of long waves in shallow-water and in nonlinear lattices*, J. Mathematical Phys. **14** (1973), 810–814. MR0338588 (49 #3352)

[32] Xing-Biao Hu and Ralph Willox, *Some new exact solutions of the Novikov-Veselov equation*, J. Phys. A **29** (1996), no. 15, 4589–4592, DOI 10.1088/0305-4470/29/15/026. MR1413220 (97g:35151)

[33] Hu, H.-C., Lou, S.-Y., Liu, Q.-P. Darboux transformation and variable separation approach: the Nizhnik-Novikov-Veselov equation *Chinese Physics Letters* **20** (2003) no. 9, 1413–1415

[34] A. V. Kazeĭkina, *Examples of the nonexistence of a traveling wave for the generalized Korteweg-de Vries-Burgers equation* (Russian, with English and Russian summaries), Vestnik Moskov. Univ. Ser. XV Vychisl. Mat. Kibernet. **1** (2011), 17–24, 56, DOI 10.3103/S0278641911010043; English transl., Moscow Univ. Comput. Math. Cybernet. **35** (2011), no. 1, 14–21. MR2848748 (2012f:35468)

[35] A. V. Kazeykina, *A large-time asymptotics for the solution of the Cauchy problem for the Novikov-Veselov equation at negative energy with non-singular scattering data*, Inverse Problems **28** (2012), no. 5, 055017, 21, DOI 10.1088/0266-5611/28/5/055017. MR2923202

[36] Kazeykina, A. V. Solitons and large time asymptotics for solutions of the Novikov-Veselov equation. Thèse, École Polytechnique, 2012. Available at `http://hal.archives-ouvertes.fr/docs/00/76/26/62/PDF/these.pdf`

[37] A. V. Kazeykina and R. G. Novikov, *Large time asymptotics for the Grinevich-Zakharov potentials*, Bull. Sci. Math. **135** (2011), no. 4, 374–382, DOI 10.1016/j.bulsci.2011.02.003. MR2799814 (2012m:35287)

[38] A. V. Kazeykina and R. G. Novikov, *Absence of exponentially localized solitons for the Novikov-Veselov equation at negative energy*, Nonlinearity **24** (2011), no. 6, 1821–1830, DOI 10.1088/0951-7715/24/6/007. MR2802306 (2012d:37167)

[39] A. V. Kazeykina and R. G. Novikov, *A large time asymptotics for transparent potentials for the Novikov-Veselov equation at positive energy*, J. Nonlinear Math. Phys. **18** (2011), no. 3, 377–400, DOI 10.1142/S1402925111001660. MR2846100 (2012k:35472)

[40] Kim Knudsen, Matti Lassas, Jennifer L. Mueller, and Samuli Siltanen, *Regularized D-bar method for the inverse conductivity problem*, Inverse Probl. Imaging **3** (2009), no. 4, 599–624, DOI 10.3934/ipi.2009.3.599. MR2557921 (2011d:35518)

[41] K. Knudsen, J. Mueller, and S. Siltanen, *Numerical solution method for the dbar-equation in the plane*, J. Comput. Phys. **198** (2004), no. 2, 500–517, DOI 10.1016/j.jcp.2004.01.028. MR2062911 (2005g:65185)

[42] Herbert Koch and Jean-Claude Saut, *Local smoothing and local solvability for third order dispersive equations*, SIAM J. Math. Anal. **38** (2006/07), no. 5, 1528–1541 (electronic), DOI 10.1137/050630659. MR2286018 (2008c:35307)

[43] Boris G. Konopelchenko and Antonio Moro, *Geometrical optics in nonlinear media and integrable equations*, J. Phys. A **37** (2004), no. 10, L105–L111, DOI 10.1088/0305-4470/37/10/L01. MR2039848 (2004m:78001)

[44] M. Lassas, J. L. Mueller, and S. Siltanen, *Mapping properties of the nonlinear Fourier transform in dimension two*, Comm. Partial Differential Equations **32** (2007), no. 4-6, 591–610, DOI 10.1080/03605300500530412. MR2334824 (2009b:81207)

[45] M. Lassas, J. L. Mueller, S. Siltanen, and A. Stahel, *The Novikov-Veselov equation and the inverse scattering method, Part I: Analysis*, Phys. D **241** (2012), no. 16, 1322–1335, DOI 10.1016/j.physd.2012.04.010. MR2947348

[46] M. Lassas, J. L. Mueller, S. Siltanen, and A. Stahel, *The Novikov-Veselov equation and the inverse scattering method: II. Computation*, Nonlinearity **25** (2012), no. 6, 1799–1818, DOI 10.1088/0951-7715/25/6/1799. MR2929603

[47] Sen Yue Lou and Guang Jiong Ni, *The relations among a special type of solutions in some* $(D+1)$*-dimensional nonlinear equations*, J. Math. Phys. **30** (1989), no. 7, 1614–1620, DOI 10.1063/1.528294. MR1002269 (90c:35182)

[48] S. V. Manakov, *The method of the inverse scattering problem, and two-dimensional evolution equations* (Russian), Uspehi Mat. Nauk **31** (1976), no. 5(191), 245–246. MR0467037 (57 #6906)

[49] Robert M. Miura, *Korteweg-de Vries equation and generalizations. I. A remarkable explicit nonlinear transformation*, J. Mathematical Phys. **9** (1968), 1202–1204. MR0252825 (40 #6042a)

[50] Jennifer L. Mueller and Samuli Siltanen, *Direct reconstructions of conductivities from boundary measurements*, SIAM J. Sci. Comput. **24** (2003), no. 4, 1232–1266, DOI 10.1137/S1064827501394568. MR1976215 (2004c:78036)

[51] Jennifer L. Mueller and Samuli Siltanen, *Linear and nonlinear inverse problems with practical applications*, Computational Science & Engineering, vol. 10, Society for Industrial and Applied Mathematics (SIAM), Philadelphia, PA, 2012. MR2986262

[52] Minoru Murata, *Positive solutions and large time behaviors of Schrödinger semigroups, Simon's problem*, J. Funct. Anal. **56** (1984), no. 3, 300–310, DOI 10.1016/0022-1236(84)90079-X. MR743843 (85j:35050)

[53] Minoru Murata, *Structure of positive solutions to* $(-\Delta+V)u=0$ *in* $\mathbf{R}^n$, Duke Math. J. **53** (1986), no. 4, 869–943, DOI 10.1215/S0012-7094-86-05347-0. MR874676 (88f:35039)

[54] Music, M. The nonlinear Fourier transform for two-dimensional subcritical potentials. `arXiv:1312.0567[Math.AP]`, submitted to *Inverse Problems and Imaging*, Special Issue on Inverse Problems and Complex Geometrical Optics Solutions.

[55] Music, M., Perry, P. The Novikov-Veselov Equation with subcritical initial data. *In preparation.*

[56] M. Music, P. Perry, and S. Siltanen, *Exceptional circles of radial potentials*, Inverse Problems **29** (2013), no. 4, 045004, 25, DOI 10.1088/0266-5611/29/4/045004. MR3042080

[57] Adrian I. Nachman, *Global uniqueness for a two-dimensional inverse boundary value problem*, Ann. of Math. (2) **143** (1996), no. 1, 71–96, DOI 10.2307/2118653. MR1370758 (96k:35189)

[58] Adrian I. Nachman, *Global uniqueness for a two-dimensional inverse boundary value problem*, Ann. of Math. (2) **143** (1996), no. 1, 71–96, DOI 10.2307/2118653. MR1370758 (96k:35189)

[59] Nikolai A. Kudryashov, *Seven common errors in finding exact solutions of nonlinear differential equations*, Commun. Nonlinear Sci. Numer. Simul. **14** (2009), no. 9-10, 3507–3529, DOI 10.1016/j.cnsns.2009.01.023. MR2509943 (2010m:35451)

[60] Nickel, J., Serov, V. S., Schurmann, H. W. Some elliptic traveling wave solutions to the Novikov–Veselov equation *Progress In Electromagnetics Research* (2006) **61**, 323–331.

[61] L. P. Nižnik, *Integration of multidimensional nonlinear equations by the inverse problem method* (Russian), Dokl. Akad. Nauk SSSR **254** (1980), no. 2, 332–335. MR591004 (82k:35096)

[62] R. G. Novikov, *The inverse scattering problem on a fixed energy level for the two-dimensional Schrödinger operator*, J. Funct. Anal. **103** (1992), no. 2, 409–463, DOI 10.1016/0022-1236(92)90127-5. MR1151554 (93e:35080)

[63] R. G. Novikov, *Absence of exponentially localized solitons for the Novikov-Veselov equation at positive energy*, Phys. Lett. A **375** (2011), no. 9, 1233–1235, DOI 10.1016/j.physleta.2011.01.052. MR2770407 (2011k:35208)

[64] A. P. Veselov and S. P. Novikov, *Finite-gap two-dimensional potential Schrödinger operators. Explicit formulas and evolution equations* (Russian), Dokl. Akad. Nauk SSSR **279** (1984), no. 1, 20–24. MR769198 (86d:58053)
[65] S. P. Novikov and A. P. Veselov, *Two-dimensional Schrödinger operator: inverse scattering transform and evolutional equations*, Phys. D **18** (1986), no. 1-3, 267–273, DOI 10.1016/0167-2789(86)90187-9. Solitons and coherent structures (Santa Barbara, Calif., 1985). MR838332 (87k:58114)
[66] Yasuhiro Ohta, *Pfaffian solutions for the Veselov-Novikov equation*, J. Phys. Soc. Japan **61** (1992), no. 11, 3928–3933, DOI 10.1143/JPSJ.61.3928. MR1196152 (93j:35164)
[67] Perry, P. Global well-posedness and large-time asymptotics for the defocussing Davey-Stewartson II equation in $H^{1,1}(\mathbb{R}^2)$. Preprint, `arXiv.math/1110.5589[math.AP]`, 2011. Submitted to *Journal of Spectral Theory.*
[68] Peter A. Perry, *Miura maps and inverse scattering for the Novikov-Veselov equation*, Anal. PDE **7** (2014), no. 2, 311–343, DOI 10.2140/apde.2014.7.311. MR3218811
[69] Perry, P. Novikov and Veselov meet Lippmann and Schwinger. *In preparation.*
[70] C. Rogers, B. G. Konopelchenko, M. P. Stallybrass, and W. K. Schief, *The Nizhnik-Veselov-Novikov equation: associated boundary value problems*, Internat. J. Non-Linear Mech. **31** (1996), no. 4, 441–450, DOI 10.1016/0020-7462(96)00012-1. MR1409721 (97f:35194)
[71] Hang-Yu Ruan and Yi-Xin Chen, *Localized coherent structures in $(2+1)$-dimensional generalized Nizhnik-Novikov-Veselov equations*, J. Phys. Soc. Japan **70** (2001), no. 6, 1499–1504, DOI 10.1143/JPSJ.70.1499. MR1847111 (2002e:35213)
[72] Ruan, H.-Y., Chen, Y.-X. Interaction between a Line Soliton and a Y-Periodic Soliton in the $(2+1)$-dimensional Nizhnik-Novikov-Veselov Equation. *Z. Naturforsch* **57**a (2002), 948–954
[73] Hang-yu Ruan and Yi-xin Chen, *Dromion interactions of $(2+1)$-dimensional KdV-type equations*, J. Phys. Soc. Japan **72** (2003), no. 3, 491–495, DOI 10.1143/JPSJ.72.491. MR1978363 (2004c:35366)
[74] Jukka Saranen and Gennadi Vainikko, *Periodic integral and pseudodifferential equations with numerical approximation*, Springer Monographs in Mathematics, Springer-Verlag, Berlin, 2002. MR1870713 (2003h:47088)
[75] Samuli Siltanen, *Electrical impedance tomography and Faddeev Green's functions*, Ann. Acad. Sci. Fenn. Math. Diss. **121** (1999), 56. Dissertation, Helsinki University of Technology, Espoo, 1999. MR1719277 (2001b:35301)
[76] Samuli Siltanen, Jennifer Mueller, and David Isaacson, *An implementation of the reconstruction algorithm of A. Nachman for the 2D inverse conductivity problem*, Inverse Problems **16** (2000), no. 3, 681–699, DOI 10.1088/0266-5611/16/3/310. MR1766226 (2001g:35269)
[77] Siltanen, S., Tamminen, J. P. Exceptional points of radial potentials at positive energies. `arXiv:1307.2037[Math.NA]`
[78] Shingareva, I., Lizarraga-celaya, C. Solving Nonlinear Partial Differential Equations with Maple and Mathematica. *Springer* (2011)
[79] Barry Simon, *Brownian motion, L^p properties of Schrödinger operators and the localization of binding*, J. Funct. Anal. **35** (1980), no. 2, 215–229, DOI 10.1016/0022-1236(80)90006-3. MR561987 (81c:35033)
[80] Barry Simon, *Large time behavior of the L^p norm of Schrödinger semigroups*, J. Funct. Anal. **40** (1981), no. 1, 66–83, DOI 10.1016/0022-1236(81)90073-2. MR607592 (82i:81027)
[81] Barry Simon, *Schrödinger semigroups*, Bull. Amer. Math. Soc. (N.S.) **7** (1982), no. 3, 447–526, DOI 10.1090/S0273-0979-1982-15041-8. MR670130 (86b:81001a)
[82] Yukiko Tagami, *Soliton-like solutions to a $(2+1)$-dimensional generalization of the KdV equation*, Phys. Lett. A **141** (1989), no. 3-4, 116–120, DOI 10.1016/0375-9601(89)90770-6. MR1025708 (90k:35248)
[83] Iskander A. Taimanov, *Modified Novikov-Veselov equation and differential geometry of surfaces*, Solitons, geometry, and topology: on the crossroad, Amer. Math. Soc. Transl. Ser. 2, vol. 179, Amer. Math. Soc., Providence, RI, 1997, pp. 133–151. MR1437161 (98c:53071)
[84] I. A. Taĭmanov and S. P. Tsarëv, *Two-dimensional Schrödinger operators with rapidly decaying rational potential and multidimensional L_2-kernel* (Russian), Uspekhi Mat. Nauk **62** (2007), no. 3(375), 217–218, DOI 10.1070/RM2007v062n03ABEH004423; English transl., Russian Math. Surveys **62** (2007), no. 3, 631–633. MR2355430

[85] I. A. Taĭmanov and S. P. Tsarev, *Blowing up solutions of the Veselov-Novikov equation* (Russian), Dokl. Akad. Nauk **420** (2008), no. 6, 744–745, DOI 10.1134/S1064562408030393; English transl., Dokl. Math. **77** (2008), no. 3, 467–468. MR2484029

[86] I. A. Taĭmanov and S. P. Tsarev, *Two-dimensional rational solitons constructed by means of the Moutard transformations, and their decay* (Russian, with Russian summary), Teoret. Mat. Fiz. **157** (2008), no. 2, 188–207, DOI 10.1007/s11232-008-0127-3; English transl., Theoret. and Math. Phys. **157** (2008), no. 2, 1525–1541. MR2493777 (2009k:37164)

[87] Iskander A. Taimanov and Sergey P. Tsarev, *The Moutard transformation: an algebraic formalism via pseudodifferential operators and applications*, Riemann surfaces, harmonic maps and visualization, OCAMI Stud., vol. 3, Osaka Munic. Univ. Press, Osaka, 2010, pp. 169–183. MR2605796 (2012f:37155)

[88] I. A. Taĭmanov and S. P. Tsarev, *On the Moutard transformation and its applications to spectral theory and soliton equations* (Russian, with Russian summary), Sovrem. Mat. Fundam. Napravl. **35** (2010), 101–117, DOI 10.1007/s10958-010-0092-x; English transl., J. Math. Sci. (N. Y.) **170** (2010), no. 3, 371–387. MR2752642 (2012c:37149)

[89] T.-Y. Tsai, *The Schrödinger operator in the plane*, Inverse Problems **9** (1993), no. 6, 763–787. MR1251205 (94i:35154)

[90] Üna,l M. *Applications of Pfaffians to soliton theory*, PhD thesis, University of Glasgow, 1998.

[91] Gennadi Vainikko, *Fast solvers of the Lippmann-Schwinger equation*, Direct and inverse problems of mathematical physics (Newark, DE, 1997), Int. Soc. Anal. Appl. Comput., vol. 5, Kluwer Acad. Publ., Dordrecht, 2000, pp. 423–440, DOI 10.1007/978-1-4757-3214-6_25. MR1766314 (2001b:78021)

[92] I. N. Vekua, *Generalized analytic functions*, Pergamon Press, London-Paris-Frankfurt; Addison-Wesley Publishing Co., Inc., Reading, Mass., 1962. MR0150320 (27 #321)

[93] Veselov, A. P., Novikov, S. P.Finite-zone, two-dimensional, potential Schrödinger operators, explicit formulas and evolution equations, *Sov. Math. Dokl.* **30** (1984) 558–591

[94] Abdul-Majid Wazwaz, *Multiple soliton solutions for the* $(2+1)$*-dimensional asymmetric Nizhnik-Novikov-Veselov equation*, Nonlinear Anal. **72** (2010), no. 3-4, 1314–1318, DOI 10.1016/j.na.2009.08.012. MR2577533 (2010k:35438)

[95] Tie-cheng Xia, Biao Li, and Hong-qing Zhang, *New explicit and exact solutions for the Nizhnik-Novikov-Vesselov equation*, Appl. Math. E-Notes **1** (2001), 139–142 (electronic). MR1980552 (2004c:35372)

[96] Jie-Fang Zhang, *Homogeneous balance method and chaotic and fractal solutions for the Nizhnik-Novikov-Veselov equation*, Phys. Lett. A **313** (2003), no. 5-6, 401–407, DOI 10.1016/S0375-9601(03)00803-X. MR2013995 (2004k:35342)

[97] Chun-Long Zheng, Hai-Ping Zhu, and Li-Qun Chen, *Exact solution and semifolded structures of generalized Broer-Kaup system in* $(2+1)$*-dimensions*, Chaos Solitons Fractals **26** (2005), no. 1, 187–194, DOI 10.1016/j.chaos.2004.12.017. MR2144550

[98] Chun-Long Zheng, Jian-Ping Fang, and Li-Qun Chen, *New variable separation excitations of* $(2+1)$*-dimensional dispersive long-water wave system obtained by an extended mapping approach*, Chaos Solitons Fractals **23** (2005), no. 5, 1741–1748, DOI 10.1016/j.chaos.2004.06.082. MR2102786

Department of Applied Mathematics, University of Colorado, Boulder, Colorado, 80309-0526

E-mail address: ryan.croke@colorado.edu

Department of Mathematics, Colorado State University, Fort Collins, Colorado, 80523-1874

E-mail address: mueller@math.colostate.edu

Department of Mathematics, University of Kentucky, Lexington, Kentucky 40506-0027

E-mail address: Michael.Music@uky.edu

Department of Mathematics, University of Kentucky, Lexington, Kentucky 40506-0027

E-mail address: Peter.Perry@uky.edu

Department of Mathematics and Statistics, P.O. Box 68, FI-00014 University of Helsinki

E-mail address: samuli.siltanen@helsinki.fi

Bern University of Applied Sciences, Engineering and Information Technology, Mathematics, BFH-TI Biel, Postfach CH-2501 Biel, Switzerland

E-mail address: Andreas.Stahel@bfh.ch

Contemporary Mathematics
Volume **635**, 2015
http://dx.doi.org/10.1090/conm/635/12678

Transverse instability of plane wave soliton solutions of the Novikov-Veselov equation

Ryan Croke, Jennifer L. Mueller, and Andreas Stahel

ABSTRACT. The Novikov-Veselov (NV) equation is a dispersive (2+1)-dimensional nonlinear evolution equation that generalizes the (1+1)-dimensional Korteweg-deVries (KdV) equation. This paper considers the instability of plane wave soliton solutions of the NV equation to transverse perturbations. To compute time evolutions numerically, a hybrid semi-implicit/spectral scheme was developed, applicable to other nonlinear PDE systems. Numerical simulations of the evolution of transversely perturbed plane wave solutions are presented using the spectral scheme. It is established that plane wave soliton solutions are not stable for transverse perturbations.

1. Introduction

The Novikov-Veselov (NV) equation for $u(z,t) = u(x,y,t)$ was introduced in the periodic setting by Novikov and Veselov [**28**] in the form

$$u_t = -\partial_z^3 u - \overline{\partial}^3 u + 3\partial_z(uf) + 3\overline{\partial}_z(u\bar{f}), \quad \text{where } \overline{\partial}_z f = \partial_z u, \tag{1.1}$$

with $\overline{\partial}_z = \frac{1}{2}(\partial_x + i\partial_y)$, $\partial_z = \frac{1}{2}(\partial_x - i\partial_y)$. It was derived algebraically from a Lax triple, and from this point of view is considered the most natural generalization of the KdV equation [**5**].

While physical applications of the NV equation are unknown, the dispersionless NV equation models the propagation of high frequency electromagnetic waves in certain nolinear media [**24, 25**]. It is also related to two other (2+1)-dimensional integrable systems which have been more widely studied, including the Davey-Stewartson II equation, which describes the complex amplitude of surface waves in shallow water, and was proved in [**31**] to be completely integrable.

Novikov and Veselov showed that the equation (1.1) is completely integrable [**28**]. Perry [**29**] gave an implementation of the completely integrable method for the Novikov-Veselov equation at zero energy that uses the inverse scattering map for the Davey-Stewartson equation together with Bogdonov's Miura transform.

2010 *Mathematics Subject Classification.* Primary 35C08, 35C07.

If we consider real solutions $u(x,y,t)$ and let $f(x,y,t) = v(x,y,t) + i\,w(x,y,t)$, the NV equation has an equivalent representation in (x,y)-space:

$$4\,u_t = -u_{xxx} + 3\,u_{xyy} + 3\,(uv)_x + 3\,(uw)_y, \tag{1.2}$$

$$u_x = v_x - w_y, \tag{1.3}$$

$$u_y = -w_x - v_y. \tag{1.4}$$

This formulation is used throughout the paper since it is useful both for the numerics and the stability analysis. If the functions u, v and w are not dependent on y, the NV equation reduces to a KdV-type equation

$$4\,u_t + u_{xxx} - 6\,u\,u_x = 0, \qquad u = v, \qquad w = 0 \tag{1.5}$$

and admits soliton solutions of the form

$$u(x,y,t) = -2\,c\,\mathrm{sech}^2(\sqrt{c}(x - c\,t)), \tag{1.6}$$

$$v(x,y,t) = -2\,c\,\mathrm{sech}^2(\sqrt{c}(x - c\,t)), \tag{1.7}$$

$$w(x,y,t) = 0. \tag{1.8}$$

To facilitate the investigation of the qualitative nature of solutions to the NV equation, we present a version of a semi-implicit pseudo–spectral numerical scheme introduced by Feng et al [**13**] that solves the Cauchy problem for the NV equation. The method introduced by Feng et al was developed to compute solutions of 2-D nonlinear wave equations. It was used in [**2**] to compute evolutions of the modified KP equation with initial conditions chosen to be perturbed solitons. We have adapted it to solve systems of equations. We show that this method preserves the L_2 norm for the KdV type equation (1.5) and is considerably faster and requires less computer memory allocation than the finite difference scheme introduced in [**26**] for the solution of the Cauchy problem. The spectral method can be used to compute evolutions for a wide variety of initial conditions. Here, we take traveling wave solutions with small transverse perturbations as initial conditions in the Cauchy problem and compute their numerical evolution to study the nature of the instability that is proved in Section 4. The reader is referred to [**8, 22, 23**] for other examples of the use of spectral methods in soliton PDE's.

The problem of transverse stability of traveling wave solutions has been studied for many of the classic soliton equations including the KP equation [**2, 4, 7, 19**], the Boussinesq equation [**4**], the ZK equation [**3, 14, 16, 17**], and most notably, the KdV equation [**20**]. For the NV equation, we carry out a linear stability analysis by considering sinusoidal perturbations with wavefront perpendicular to the direction of propagation. In order to draw conclusions about the instability of soliton solutions, as well as approximate the growth rate, we apply the **K**-expansion method developed by Rowlands and Infeld. See [**18**] for an exposition of the method, [**3**] for an application of the method to the linearized Z-K equation, and [**6**] for an application in a soliton system modeling small amplitude long waves traveling over the surface of thin current-carrying metal film. The **K**-expansion method is based on the assumption that the nonlinear wave experiences a long-wavelength perturbation as well as a periodic change in wavelength [**18**].

In [**4**], the authors state the following conjecture: "It is interesting to speculate that when the expansion method introduced by Allen and Rowlands [**3**] is applied to any integrable system it will reveal that self similarity of the eigenfunctions will appear at each order, enabling one to write down the form of the exact solution.

Alternatively this procedure may be seen as a necessary condition for integrability." The ZK equation and the equation in [6] are not integrable, and the ordinary perturbation analysis fails. So far this conjecture is supported by the KP and Boussinesq equations [4]. The results here support the conjecture by showing that only an ordinary perturbation analysis is needed for the NV equation.

The paper is organized as follows. In Section 2 we show that planar solutions to the NV equation must be a solution of a KdV-type equation. In Section 3 the semi-implicit pseudo-spectral method is presented, with its linear stability analysis in Section 3.1. The **K**-expansion method is used to establish the instability of traveling wave solutions of NV to transverse perturbations in Section 4. Numerical results are found in Section 5 and conclusions in Section 6.

2. From planar solutions of Novikov–Veselov to KdV

We examine planar solutions to the Novikov–Veselov equations (1.2)–(1.4), i.e. solutions that depend only on one spatial variable $s = (n_1, n_2) \cdot (x, y)$, moving in a direction given by the vector $\vec{n} = (n_1, n_2) = (\cos(\alpha), \sin(\alpha))$. We seek solutions of the form

$$\begin{aligned} u(t,s) &= u(t,x,y) = u(t, n_1\, s, n_2\, s) \\ v_i(t,s) &= v_i(t,x,y) = v_i(t, n_1\, s, n_2\, s) \quad \text{for } i = 1,2 \\ u'(t,s) &= \frac{\partial}{\partial s}\, u(t,s) = n_1 u_x(t,x,y) + n_2 u_y(t,x,y) \end{aligned}$$

The assumption that u, v_1 and v_2 are independent on $n_2\, x - n_1\, y$ is equivalent to

$$n_2 u_x - n_1 u_y = n_2\, \frac{\partial\, v_1}{\partial x} - n_1\, \frac{\partial\, v_1}{\partial y} = n_2\, \frac{\partial\, v_2}{\partial x} - n_1\, \frac{\partial\, v_2}{\partial y} = 0$$

As a consequence we obtain $u_x = n_1\, u'$ and $q_x = n_2\, u'$. The goal is to find a PDE for $u(t,s)$. For a given $u(t,s)$ the $\bar{\partial}$ equation

$$\left\{ \begin{aligned} \frac{\partial}{\partial x}\, v_1 - \frac{\partial}{\partial y}\, v_2 &= +u_x \\ \frac{\partial}{\partial x}\, v_2 + \frac{\partial}{\partial y}\, v_1 &= -u_y \end{aligned} \right.$$

translates to

$$\left\{ \begin{aligned} n_1\, v_1' - n_2\, v_2' &= +n_1\, u' \\ n_2\, v_1' + n_1\, v_2' &= -n_2\, u' \end{aligned} \right. \quad \Longrightarrow \quad \begin{aligned} v_1' &= (n_1^2 - n_2^2)\; u' \\ v_2' &= -(2\, n_1\, n_2)\; u' \end{aligned}$$

with the solutions

$$\begin{aligned} v_1(t,s) &= (n_1^2 - n_2^2)\; u(t,s) + c_1 \\ v_2(t,s) &= -(2\, n_1\, n_2)\; u(t,s) + c_2 \end{aligned}$$

The nonlinear expression in (1.2) leads to

$$\begin{aligned} \operatorname{div}(u\, \vec{v}) &= (n_1^2 - n_2^2)\, (u^2)_x - 2\, n_1\, n_2^2\, (u^2)_y + c_1 u_x + c_2 u_y \\ &= \left(2\, (n_1^2 - n_2^2)\, n_1 - 4\, n_1\, n_2^2\right)\; u\, u' + (c_1\, n_1 + c_2\, n_2)\, u' \\ &= 2\, \kappa\, u\, u' + \beta\, u' \end{aligned}$$

where

$$\beta = c_1\, n_1 + c_2\, n_2 \quad \text{and} \quad \kappa = \kappa(\alpha) = (n_1^2 - n_2^2)\, n_1 - 2\, n_1\, n_2^2 = \cos(3\, \alpha)$$

The factor $\kappa = \cos(3\,\alpha)$ is further evidence for the threefold rotational invariance of the Novikov–Veselov equations. Now we examine the Novikov–Veselov equation

$$\begin{aligned} u_t &= -\frac{1}{4}u_{xxx} + \frac{3}{4}u_{xyy} + \frac{3}{4}\,\mathrm{div}(u\,\vec{v}) \\ &= -\frac{1}{4}\,n_1^3\,u''' + \frac{3}{4}\,n_1\,n_2^2\,u''' + \frac{6}{4}\,\kappa\,u\,u' + \frac{3}{4}\,\beta\,u' \\ &= -\frac{1}{4}\,\kappa\,u''' + \frac{6}{4}\,\kappa\,u\,u' + \frac{3}{4}\,\beta\,u' \end{aligned}$$

Thus, a planar solution of the Novikov–Veselov has to be a solution of the KdV-like equation

$$\frac{4}{\kappa}\,u_t = -u''' + 6\,u\,u' + \frac{3\,\beta}{\kappa}\,u' \tag{2.1}$$

The only essential modification is the contribution proportional to u' and thus it should not come as a surprise that the solutions are related. If $q(t,s)$ is a solution of the standard KdV equation

$$\dot{q}(t,x) = -q'''(t,x) + 6\,q(t,x)\,q'(t,x) \tag{2.2}$$

then

$$u(t,s) = q(\frac{\kappa}{4}\,t, s + \frac{3\,\beta}{4}\,t)$$

is a solution of the Novikov–Veselov equation (1.2), which can be verified as follows:

$$\begin{aligned} u_t &= \frac{\kappa}{4}q_t + \frac{3\,\beta}{4}q' \\ \frac{4}{\kappa}u_t + u''' - 6\,u\,u' - \frac{3\,\beta}{\kappa}\,u' &= q_t + \frac{3\,\beta}{\kappa}\,q' + q''' - 6\,q\,q' - \frac{3\,\beta}{\kappa}\,q' \\ &= q_t + q''' - 6\,q\,q' \;=0 \end{aligned}$$

For a solution $q(t,s)$ of KdV choose constants k_1 and k_2 with

$$\frac{3\,\beta}{4} = k_1 + k_2\,\frac{3\,\kappa}{2}$$

and verify that

$$u(t,s) = q(\frac{\kappa}{4}\,t\,,\;s + k_1\,t) - k_2$$

is a solution of the Novikov–Veselov equation:

$$\begin{aligned} u_t &= \frac{\kappa}{4}q_t + k_1\,q' \\ \frac{4}{\kappa}\,u_t + u''' - 6\,u\,u' - \frac{3\,\beta}{\kappa}\,u' &= q_t + \frac{k_1\,4}{\kappa}\,q' + q''' - 6\,(q - k_2)\,q' - \frac{3\,\beta}{\kappa}\,q' \\ &= q_t + q''' - 6\,q\,q' + \left(\frac{k_1\,4}{\kappa} + 6\,k_2 - \frac{3\,\beta}{\kappa}\right)\,q' \\ &= q_t + q''' - 6\,q\,q' + \frac{4}{\kappa}\left(k_1 + k_2\,\frac{3\,\kappa}{2} - \frac{3\,\beta}{4}\right)\,q' \;=0 \end{aligned}$$

Remarks:

- Consider the choice $\beta = k_1 = k_2 = 0$. Then $q(t,s) = u(\frac{\kappa}{4}\,t, s)$ is the KdV solution where the time scale is multiplied with $\kappa/4$. Since $\kappa = \kappa(\alpha) = \cos(3\,\alpha)$ the speed of the solution changes according to an angle dependent profile, as shown in Figure 1.
- The additive constant k_2 moves the KdV solution up or down in the graph.

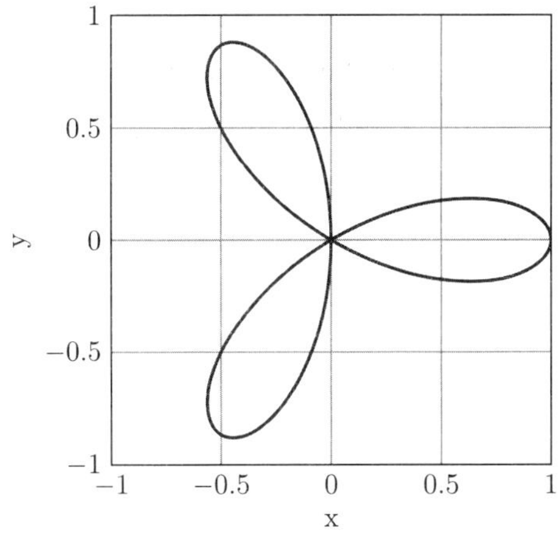

FIGURE 1. Speed profile for planar solutions of the Novikov–Veselov equation

- Replacing s by $s + k_1 t$ (with $k_1 = \frac{3\beta}{4} - k_2 \frac{3\kappa}{2}$) corresponds to observing the KdV solution in a moving frame, where the frame moves with velocity $-k_1$.

3. A Pseudo-Spectral method for the solution of (2+1) nonlinear wave equations

To numerically solve (1.2)–(1.4) we use a semi–implicit leap–frog spectral method based on the method in [**13**]. We restrict ourselves to a finite spatial domain $\Omega = [0, W_x] \times [0, W_y]$ with periodic boundary conditions. Thus we work on a torus topology and for the numerical results we have to observe the traveling waves across the boundary. We must also choose the domain large enough for the problem at hand.

This method uses the Fast Fourier Transform (FFT) to compute the spatial evolution and a leapfrog scheme to compute the time-stepping. The numerical scheme can be summarized as follows:

The Fourier transform of (1.2)–(1.4) is

$$4\,\hat{u}_t = i\,(\xi^3 - 3\,\xi\,\eta^2)\,\hat{u} + 3\,i\,\xi\,\mathcal{F}[(uv)] + 3\,i\,\eta\,\mathcal{F}[(uw)], \tag{3.1}$$

$$\xi\,\hat{u} = \xi\,\hat{v} - \eta\,\hat{w}, \tag{3.2}$$

$$\eta\,\hat{u} = -\eta\,\hat{v} - \nu\,\hat{w}. \tag{3.3}$$

As usual, $\hat{u}$ refers to the Fourier transform F of u and is given by

$$\hat{u}_{p,q} = \mathcal{F}[u_{l,m}] = \sum_{l=0}^{L-1}\sum_{m=0}^{M-1} u_{l,m}\, e^{-i(\xi_p x_l + \eta_q y_m)},$$

$$u_{l,m} = \mathcal{F}^{-1}[\hat{u}_{p,q}] = \frac{1}{L\,M}\sum_{p=-L/2}^{L/2-1}\ \sum_{q=-M/2}^{M/2-1} \hat{u}_{p,q}\, e^{i(\xi_p x_l + \eta_q y_m)}.$$

The parameters L and M are the number of grid points in $[0, W_x] \times [0, W_y]$, respectively. They need to be powers of two in order to use the standard FFT. The spatial grid is defined by $(x_l, y_m) = (l\Delta x, m\Delta y)$ where $l = 0, \ldots, L-1$, and

$m = 0, \ldots, M-1$. The spectral variables are $(\xi_p, \eta_q) = (2\pi p/W_x, 2\pi q/W_y)$, with $p = -L/2, \ldots, -1, 0, 1, \ldots, L/2$, and $q = -M/2, \ldots, -1, 0, 1, \ldots, M/2$.

Equations (3.2) and (3.3) can be solved in terms of $\hat{u}$

$$\hat{v} = \frac{\xi^2 - \eta^2}{\xi^2 + \eta^2}\,\hat{u}, \qquad \hat{w} = \frac{-2\eta\,\xi}{\eta^2 + \xi^2}\,\hat{u} \tag{3.4}$$

Special attention has to be paid to the case $\xi^2 + \eta^2 = 0$. The corresponding Fourier coefficients represent the average values of the functions v and w.

Let $\vec{c}(t)$ be the vector with the LM Fourier coefficients of the solution $u(x, y, t)$. Then solving (3.2), (3.3) and computing $-3\,i\,(\xi\,\mathcal{F}[(uv)] + \eta\,\mathcal{F}[(uw)])$ may be written as one nonlinear function $\vec{F}(\vec{c})$. With the appropriate diagonal matrix $\mathbf{D}$ the NV equation (3.1) reads as

$$4\,\frac{d}{dt}\,\vec{c}(t) = \mathbf{D}\,\vec{c}(t) + \vec{F}(\vec{c}(t)) \tag{3.5}$$

For the time integration we use a symmetric three-level difference method for the linear terms, and a leapfrog method for the nonlinear terms. For the parameter $0 \leq \theta \leq 1$ we use a superscript to refer to the time iteration and the above leads to

$$\begin{aligned} \frac{4}{2\,\Delta t}\,(\vec{c}^{\,n+1} - \vec{c}^{\,n-1}) &= \theta\,\mathbf{D}\,(\vec{c}^{\,n+1} + \vec{c}^{\,n-1}) + (1 - 2\,\theta)\,\mathbf{D}\,\vec{c}^{\,n} + \vec{F}(\vec{c}^{\,n}) \\ (2 - \theta\,\Delta t\,\mathbf{D})\,\vec{c}^{\,n+1} &= (1 - 2\,\theta)\,\Delta t\,\mathbf{D}\,\vec{c}^{\,n} + (2 + \theta\,\Delta t\,\mathbf{D})\,\vec{c}^{\,n-1} \\ &\quad + \Delta t\,\vec{F}(\vec{c}^{\,n}) \end{aligned} \tag{3.6}$$

Thus we have an implicit scheme for the linear contribution and an explicit scheme for the nonlinear contribution. Since the matrix $\mathbf{D}$ is diagonal we do not have to solve a system of linear equations at each time step. For the special case $\theta = 1/2$ we obtain a Crank Nicolson scheme

$$(4 - \Delta t\,\mathbf{D})\,\vec{c}^{\,n+1} = +(4 + \Delta t\,\mathbf{D})\,\vec{c}^{\,n-1} + 2\,\Delta t\,\vec{F}(\vec{c}^{\,n})$$

For the three level method we need a separate method for the first time step. For the sake of simplicity we may choose $\vec{c}^{\,-1} = \vec{c}^{\,0}$. For computations with known solutions, e.g. (1.6), we use the known values of the solution at time $-\Delta t$.

To examine the efficiency of the spectral method, we compare it to the finite difference (FD) approach presented in [**26**]:

- The FD approach can be used for periodic boundary conditions and for Dirichlet type conditions. The spectral approach is perfectly suited for periodic boundary conditions.
- For both methods, a linear system of equations has to be examined at each time step. The corresponding matrix is of size $N^2 \times N^2$.
  - For the spectral method the matrix $\mathbf{D}$ is diagonal, and so we do not have to solve a system of equations. The computational effort is N^2.
  - For the FD approach the matrix has a semi-bandwidth $2\,N$. Thus, a sparse LU factorization is required to store approximately $4\,N^3$ numbers. For large N this is prohibitive. Thus, an iterative solver has to be used, such as `bicgstab`, with or without a preconditioner. The computational effort for each iteration is proportional to N^2, with a sizable constant of proportionality.
- For both algorithms the nonlinear contribution requires the use of a $2D$ FFT. For the FD approach this is accomplished with Dirchlet BC's on an

enlarged domain. This is a sizable contribution to the total computation time.
- Both algorithms conserve the L_2 norm of the linear part and the stability behavior is similar.
- For both approaches a stability proof for the full, nonlinear equation is lacking. This should not come as a surprise, as the NV equations does have exponentially unstable solutions, examined in section 4.

Based on the above we conclude that the spectral method is a better approach for a problem with periodic boundary conditions. This is confirmed by a few test cases. For exact classical KdV soliton initial conditions, the spectral method reproduces the known exact solution with high accuracy on sizable time intervals.

3.1. Linear Numerical Stability Analysis. To gain insight into the stability of the spectral method, we include a linear stability analysis. The parameters determined here were used in the numerical experiments that follow. We examine the problem on a domain $\Omega = [0,W] \times [0,W] \subset \mathbb{R}^2$ and use periodic boundary conditions. Thus we examine the initial boundary value problem

$$\begin{aligned} 4\,u_t &= -u_{xxx} + 3\,u_{xyy} + 3\,\alpha\,(v_x + w_y), \\ u_x &= v_x - w_y, \qquad u_y = -w_x - v_y \end{aligned} \tag{3.7}$$

The case $\alpha = 0$ is the linearization of (1.2)–(1.4) about the zero solution. The utility of a linear stability analysis for a nonlinear system was addressed in [**13**] in the context of the KP and ZK equations, where they found that their results for the linear stability analysis were validated by numerical results and argued that while the analysis does not prove stability and convergence of the nonlinear scheme, the obtained stability conditions often suffice in practice. When implemented on a domain with periodic boundary condition (3.7) leads to

$$\begin{array}{ll} 4\,u_t = -u_{xxx} + 3\,u_{xyy} + 3\,\alpha\,(v_x + w_y), & (x,y) \times t \in \Omega \times \mathbb{R}, \\ u_x = v_x - w_y, \qquad u_y = -w_x - v_y & \\ u(x,y,0) = u_0(x,y), & (x,y) \in \Omega, \\ u(x,y,t) = u(x+W_x,y,t), & (x,y) \times t \in \mathbb{R}^3, \\ u(x,y,t) = u(x,y+W_y,t), & (x,y) \times t \in \mathbb{R}^3, \end{array} \tag{3.8}$$

Using Fourier series and (3.4) this leads to

$$4\,\hat{u}_t \;=\; i\,(\xi^3 - 3\,\xi\,\eta^2)\,\hat{u} + 3\,i\,\xi\,\alpha\,\frac{\xi^2 - \eta^2}{\eta^2 + \xi^2}\,\hat{u} + 3\,i\,\eta\,\alpha\,\frac{-2\eta\,\xi}{\eta^2 + \xi^2}\,\hat{u} = i\,\omega\,\hat{u} + i\,\frac{3\,\omega\,\alpha}{\eta^2 + \xi^2}\,\hat{u},$$

where $\omega = \xi^3 - \xi\,\eta^2$. This is an ODE of the form

$$4\,\frac{d}{dt}\,u(t) = i\,\lambda\,u(t) + i\,\gamma\,u(t), \tag{3.9}$$

where $\lambda = \omega = \xi^3 - 3\,\xi\,\eta^2$ and $\gamma = \omega\,\frac{3\,\alpha}{\eta^2+\xi^2}$. Using the numerical scheme (3.6), this leads to

$$(2 - i\,\theta\,\lambda\,\Delta t)\,u^{n+1} = i\,(1 - 2\,\theta)\,\lambda\,\Delta t\,u^n + (2 + i\,\theta\,\lambda\,\Delta t)\,u^{n-1} + i\,\gamma\,\Delta t\,u^n,$$

or with

$$b_1 = 2 + i\,\theta\,\lambda\,\Delta t \in \mathbb{C}, \qquad b_2 = i\,((1 - 2\,\theta)\,\lambda + \gamma)\,\Delta t \in \mathbb{C}, \tag{3.10}$$

to the iteration matrix

$$\left(\begin{array}{c} u^{n+1} \\ u^n \end{array}\right) = \mathbf{M} \left(\begin{array}{c} u^n \\ u^{n-1} \end{array}\right) = \left[\begin{array}{cc} \frac{b_2}{b_1} & \frac{\overline{b_1}}{b_1} \\ 1 & 0 \end{array}\right] \left(\begin{array}{c} u^n \\ u^{n-1} \end{array}\right).$$

For the system to be stable we have to verify that the norm of the eigenvalues $z_{1,2}$ of $\mathbf{M}$ are less than or equal to 1. The characteristic equation is given by

$$\Phi_{NV}(z) = z^2 - \frac{b_2}{\overline{b_1}}\, z - \frac{b_1}{\overline{b_1}} = 0$$

Thus, we conclude $|z_i \cdot z_2| = |\frac{b_1}{\overline{b_1}}| = 1$ and stability of the solution is equivalent to $|z_{1,2}| = 1$, i.e., we have $|u^{n+1}| = |u^{n-1}|$. Using the explicit solution for quadratic equations and

$$b_2 \in i\,\mathbb{R}, \qquad |b_2| \leq 2\,|b_1|, \tag{3.11}$$

we find $2\,\overline{b_1}\, z_{1,2} = b_2 \pm \sqrt{b_2^2 + 4\,|b_1|^2}$ and $|2\,\overline{b_1}|^2\,|z_{1,2}|^2 = |2\,b_1|^2$. Consequently, $|z_{1,2}| = 1$. With (3.10), the stability condition (3.11) for the ODE (3.9) becomes

$$((1-4\,\theta)\,\lambda^2 + 2\,(1-2\,\theta)\,\lambda\,\gamma + \gamma^2)\,(\Delta t)^2 \quad \leq \quad 16.$$

For $\gamma = \alpha = 0$ and $\frac{1}{4} \leq \theta \leq 1$ the stability condition is satisfied, independent of the step size $\Delta t > 0$, and we have a stability result for the initial boundary value problem (3.8). This proves the following theorem.

THEOREM 1. *The numerical scheme* (3.6) *applied to the linearization of* (1.2)–(1.4) *about the zero solution is stable for* $\frac{1}{4} \leq \theta \leq 1$.

Proving stability for (3.8) with $\alpha \neq 0$ requires a few more computations. Using the expressions for λ and γ in (3.9) results in the stability condition

$$(1-4\,\theta) + 2\,(1-2\,\theta)\,\frac{3\,\alpha}{\eta^2+\xi^2} + \left(\frac{3\,\alpha}{\eta^2+\xi^2}\right)^2 \quad \leq \quad \left(\frac{4}{\omega\,\Delta t}\right)^2.$$

For $\theta = \frac{1}{2}$ we find the sufficient conditions

$$\begin{aligned} -2 + \left(\frac{3\,\alpha}{\eta^2+\xi^2}\right)^2 \leq \left(\frac{3\,\alpha}{\eta^2+\xi^2}\right)^2 \quad &\leq \quad \left(\frac{4}{(\xi^3 - 3\,\xi\,\eta^2)\,\Delta t}\right)^2 \\ (\Delta t)^2 \quad &\leq \quad \left(\frac{4\,(\xi^2+\eta^2)}{3\,\alpha\,(\xi^3 - 3\,\xi\,\eta^2)}\right)^2. \end{aligned}$$

Since $|\xi^3 - 3\,\xi\,\eta^2| \leq (\xi^2+\eta^2)\cdot\max\{|\xi|,\,|\eta|\}$ we have the sufficient condition for Δt

$$\Delta t \leq \frac{4}{3\,|\alpha|\,\max\{|\xi|,\,|\eta|\}}.$$

For the DFT with $\Delta x = \Delta y = \frac{W_x}{L}$ we find $\max\{|\xi|,\,|\eta|\} = \frac{\pi}{\Delta x}$, and thus we have the stability condition

$$\Delta t \leq \frac{4}{3\,|\alpha|\,\pi}\,\Delta x. \tag{3.12}$$

As a consequence we have a stability result for the initial boundary value problem (3.8) with $\alpha \neq 0$, Theorem 2.

THEOREM 2. *The numerical scheme* (3.6) *applied to* (3.8) *is stable for* $\theta = \frac{1}{2}$ *if the time step* Δt *satisfies condition* (3.12).

REMARK 1. *The above results verify that the numerical scheme for the linear contributions is stable for small time steps for the Novikov–Veselov equation at nonzero energy $E \neq 0$, that is when equation (1.2) is replaced by*

$$4\,u_t = -u_{xxx} + 3\,u_{xyy} + 3\,(uv)_x + 3\,(uw)_y - E\,(v_x + w_y)$$

The divergence term $E\,(v_x + w_y)$ can either be integrated into the linear part of the scheme (i.e. in the matrix $\mathbf{D}$) or can be handled as part of the nonlinear contribution (i.e. in $\vec{F}(\vec{c})$).

4. Instability of traveling-wave solutions of the NV-equation to transverse perturbations

The **K**-expansion method presented by Allen and Rowlands considers long wavelength perturbations in the transversal direction. It was originally used to investigate the stability of solutions to the Zakharov–Kuznetsov equation, a non-integrable generalization of the KdV equation [**15**–**17**]. Since then the method has been applied to the KP equation and the modified ZK (mZK) equation [**27**]. This is the second application of the method to a system of PDE's, the first being by Bradley [**6**]. The **K**-expansion method is related to the method of Evans function [**1**, **9**–**12**] for determining soliton instability. The Evans function is an analytic function, the zeros of which correspond to the eigenvalues of the linearized operator. See [**21**] for explicit calculations of the Evans function in the stability analysis of pulses arising in nonlinear optics. Using the **K**-expansion method, we construct an eigenvalue with positive real part. This corresponds to a zero of the Evans function, also indicating instability.

4.1. The modified equations. To begin, we transform the system to move along with the soliton by using new independent and dependent variables. Using

$$(t, x, y) \mapsto \sqrt{c}\,(c\,t, x - c\,t, y) = (\tilde{t}, \tilde{x}, \tilde{y})$$

and

$$(u, v_1, v_2) \mapsto c\,(u, v_1, v_2)$$

or

$$\begin{aligned} u(t,x,y) &= c\,\tilde{u}(t\,c^{3/2}, (x - c\,t)\,c^{1/2}, y\,c^{1/2}) \\ v_1(t,x,y) &= c\,\tilde{v}_1(t\,c^{3/2}, (x - c\,t)\,c^{1/2}, y\,c^{1/2}) \\ v_2(t,x,y) &= c\,\tilde{v}_2(t\,c^{3/2}, (x - c\,t)\,c^{1/2}, y\,c^{1/2}), \end{aligned} \tag{4.1}$$

the NV equations (1.2)–(1.4) are transformed to

$$0 = 4u_t - 4u_x + u_{xxx} - 3u_{xyy} - 3(uv)_x - 3(uw)_y, \tag{4.2}$$

$$u_x = +v_x - w_y, \tag{4.3}$$

$$u_y = -w_x - v_y \tag{4.4}$$

For sake of a more readable notation we dropped the tildes on the new dependent and independent variables. The known solution

$$f_c(x - c\,t) = -2\,c\,\operatorname{sech}^2(\sqrt{c}\,(x - c\,t))$$

of the original system turns into a stationary solution of the modified system.

$$u_0(x) = v_0(x) = -2\,\operatorname{sech}^2(x), \qquad w_0(x) = 0$$

Now we examine perturbed functions of the form

$$u(x,y,t) = u_0(x) + \epsilon\, f(x)\, e^{iky+\gamma t}, \tag{4.5}$$

$$v(x,y,t) = v_0(x) + \epsilon\, g(x)\, e^{iky+\gamma t}, \tag{4.6}$$

$$w(x,y,t) = w_0(x) + \epsilon\, h(x)\, e^{iky+\gamma t}. \tag{4.7}$$

Thus we examine periodic, transversal perturbations with exponential growth. For numerical purposes we may also work with a purely real formulation.

$$u(x,y,t) = u_0(x) + \epsilon\, f(x)\, \cos(k\,y)\, e^{\gamma t}, \tag{4.8}$$

$$v(x,y,t) = v_0(x) + \epsilon\, g(x)\, \cos(k\,y)\, e^{\gamma t}, \tag{4.9}$$

$$w(x,y,t) = w_0(x) + \epsilon\, h(x)\, \sin(k\,y)\, e^{\gamma t}. \tag{4.10}$$

Substituting these expressions into (4.2)–(4.4) and dropping terms proportional to ϵ^2 leads to (use $'$ for $\frac{d}{dx}$)

$$\begin{aligned} 0 &= 4\,\gamma\, f - 4\, f' + f''' + 3\, k^2\, f' \\ &\quad -3\,(u_0\, g)' - 3\,(v_0\, f)' - 3\, u_0\, i\, k\, h \end{aligned} \tag{4.11}$$

$$f' = +g' - i\, k\, h \tag{4.12}$$

$$i\, k\, f = -h' - i\, k\, g \tag{4.13}$$

This is a system of linear ordinary differential equations with non-constant coefficients. Use $h(x) = \frac{1}{i\,k}\,(g'(x) - f'(x))$ and the above can be written in the form

$$\begin{aligned} f'''(x) &= (-4\,\gamma + 3\, u_0'(x))\, f(x) + (4 - 3\, k^2)\, f'(x) + \\ &\quad +3\, u_0'(x)\, g(x) + 6\, u_0(x)\, g'(x) \end{aligned} \tag{4.14}$$

$$g''(x) = k^2\, g(x) + f''(x) + k^2\, f(x) \tag{4.15}$$

4.2. Known solutions. Since the $f_c(x - c\,t + s)$ is a solution of the original NV equation (1.2)–(1.4), we conclude that $(u_0(x-s),\, v_0(x-s),\, w_0(x-s))$ solve the system (4.2)–(4.4) and thus their derivatives must solve the linearized problem (4.11)–(4.13). Since $\frac{d}{dx}\,\mathrm{sech}^2(x) = -2\,\mathrm{sech}^2(x)\,\tanh(x)$ we find solutions

$$f(x) = g(x) = \mathrm{sech}^2(x)\,\tanh(x)\,, \qquad h(x) = 0 \tag{4.16}$$

for $k = \gamma = 0$ to the system (4.14)–(4.15). For $k = 1$ and $\gamma = 0$ we found

$$\begin{aligned} f(x) &= +\mathrm{sech}^3(x) \\ g(x) &= -\,\mathrm{sech}(x)\,\tanh^2(x) \\ h(x) &= -i\,\mathrm{sech}(x)\,\tanh(x) \end{aligned} \tag{4.17}$$

as a second, explicit solution of (4.11)–(4.13).

With the help of these solutions we will construct unstable solution to the Novikov-Veselov equations, as function of the parameters k and γ. Thus we need nonzero solutions of (4.14)–(4.15).

4.3. Locating unstable solutions. Using the function $u_0(x) = -2\,\mathrm{sech}^2(x)$, a matrix notation translates the high order system (4.14)–(4.15) into a system of first order ordinary differential equations.

$$\frac{d}{dx}\begin{pmatrix} f \\ f' \\ f'' \\ g \\ g' \end{pmatrix} = \begin{bmatrix} 0 & 1 & 0 & 0 & 0 \\ 0 & 0 & 1 & 0 & 0 \\ -4\,\gamma + 3\,u_0' & (4-3\,k^2) & 0 & 3\,u_0'(x) & 6\,u_0 \\ 0 & 0 & 0 & 0 & 1 \\ k^2 & 0 & 1 & k^2 & 0 \end{bmatrix}\begin{pmatrix} f \\ f' \\ f'' \\ g \\ g' \end{pmatrix}$$

We seek nontrivial solutions satisfying decay conditions as $x \to \pm\infty$. We use a combination of asymptotic analysis and numerical methods. It is necessary to consider large and small $|x|$ separately. We take the following approach:

- Choose a large value $M > 0$. For $|x| > M$ we use $u_0(x) \approx 0$ and an elementary asymptotic analysis to construct solutions satisfying the decay conditions.
- For $|x| \leq M$ an asymptotic analysis is not feasible, since the contributions $u_0(x)$ and $u_0'(x)$ in (4.14)–(4.15) have to be taken into account. Thus we use a numerical solver on the domain $-M \leq x \leq +M$ to construct a fundamental matrix $\mathbf{T}$ containing the information for the solutions of the linear system for $-M \leq x \leq M$.
- Then the above two results are combined to find criterion (4.23) for the existence of nontrivial solutions.

4.3.1. *Behavior of solution for large $|x|$.* For large values of $|x|$ we use $u_0(x) \approx 0$ and arrive at a decoupled system of linear differential equations with constant coefficients.

$$\frac{d}{dx}\begin{pmatrix} f \\ f' \\ f'' \end{pmatrix} = \begin{bmatrix} 0 & 1 & 0 \\ 0 & 0 & 1 \\ -4\gamma & (4-3\,k^2) & 0 \end{bmatrix}\begin{pmatrix} f \\ f' \\ f'' \end{pmatrix} \tag{4.18}$$

$$\frac{d}{dx}\begin{pmatrix} g \\ g' \end{pmatrix} = \begin{bmatrix} 0 & 1 \\ k^2 & 0 \end{bmatrix}\begin{pmatrix} g \\ g' \end{pmatrix} + \begin{pmatrix} 0 \\ k^2\,f' + f'' \end{pmatrix} \tag{4.19}$$

The characteristic equation of the matrix in (4.18) is given by

$$\lambda^3 + (3\,k^2 - 4)\,\lambda + 4\,\gamma = 0, \tag{4.20}$$

and we will denote the eigenvalues λ_1, λ_2, and λ_3 of this system by p_1, p_2, and p_3, respectively. The inhomogeneous system in (4.19) leads to eigenvalues $\lambda_{4,5} = \pm k$. Using Cardano's formulas for zeros of polynomials of degree 3, we verify that (4.20) has three distinct real values if

$$4\,\gamma^2 + \frac{(3\,k^2-4)^3}{27} < 0.$$

For our domain of (k,γ) to be examined we may use

$$p_3 \leq 0 \leq p_1,\ p_2$$

and for small values of $|\gamma|$ we find

$$\lambda(\gamma) = \begin{cases} p_1 \approx 0 + \frac{4}{4-3\,k^2}\,\gamma \\ p_2 \approx +\sqrt{4-3\,k^2} - \frac{2}{4-3\,k^2}\,\gamma \\ p_3 \approx -\sqrt{4-3\,k^2} - \frac{2}{4-3\,k^2}\,\gamma \end{cases} \tag{4.21}$$

with the solution of the form $f(x) \approx c_1 e^{p_1 x} + c_2 e^{p_2 x} + c_3 e^{p_3 x}$. We examine $\gamma \geq 0$, and for $x \to +\infty$ we require the solution $f(x)$ of (4.18) to be bounded, and thus

$$f(x) \approx f_+(x) = c_3\, e^{p_3\, x}.$$

Next, examine (4.19) in the form

$$g_+''(x) - k^2\, g_+(x) = k^2\, f_+(x) + f_+''(x) = c_3\,(k^2 + p_3^2)\, e^{p_3\, x}$$

with the solution

$$g_+(x) = \beta_1\, e^{+k\, x} + \beta_2\, e^{-k\, x} + c_3\, \frac{k^2 + p_3^2}{p_3^2 - k^2}\, e^{p_3\, x}.$$

Since this solution has to remain bounded as $x \to +\infty$, we find

$$g_+(x) = \beta_2\, e^{-k\, x} + c_3\, \frac{k^2 + p_3^2}{p_3^2 - k^2}\, e^{p_3\, x}.$$

The case $\gamma \geq 0$ and $x \to -\infty$, leads in an analogous way to

$$f(x) \approx f_-(x) = c_1\, e^{p_1\, x} + c_2\, e^{p_2\, x}$$

and

$$g(x) \approx g_-(x) = \beta_1\, e^{+k\, x} + c_1\, \frac{k^2 + p_1^2}{p_1^2 - k^2}\, e^{p_1\, x} + c_2\, \frac{k^2 + p_2^2}{p_2^2 - k^2}\, e^{p_2\, x}.$$

For $0 < k^2 \ll 1$ we find

$$p_1^2 \approx \frac{4^2}{(4 - 3\,k^2)^2}\, \gamma^2 \approx \gamma^2 \tag{4.22}$$

and thus the above formula for $g(x)$ may not be valid for $\gamma^2 \approx k^2 \ll 1$. For $k = 0$ the equation for g simplifies to $g''(x) = f''(x)$, and the conditions at $x = \pm\infty$ imply $g(x) = f(x)$.

4.3.2. *Solutions for intermediate* $|x|$. Choosing a large value for $M > 0$, we construct nontrivial solutions to (4.14)–(4.15) for $-\infty < x \leq -M$, then for $-M \leq x \leq +M$, and then for $+M \leq x < +\infty$. We seek nonzero values of the parameters $(c_1,\, c_2,\, c_3,\, \beta_1,\, \beta_2)$, such that we find a nonzero solution.

Use the solutions f_- and g_- to define a matrix

$$\mathbf{T}_- = \begin{bmatrix} e^{-p_1\, M} & e^{-p_2\, M} & 0 & 0 & 0 \\ p_1\, e^{-p_1\, M} & p_2\, e^{-p_2\, M} & 0 & 0 & 0 \\ p_1^2\, e^{-p_1\, M} & p_2^2\, e^{-p_2\, M} & 0 & 0 & 0 \\ \frac{p_1^2+k^2}{p_1^2-k^2}\, e^{-p_1\, M} & \frac{p_2^2+k^2}{p_2^2-k^2}\, e^{-p_2\, M} & 0 & e^{-k\, M} & 0 \\ \frac{(p_1^2+k^2)\, p_1}{p_1^2-k^2}\, e^{-p_1\, M} & \frac{(p_2^2+k^2)\, p_2}{p_2^2-k^2}\, e^{-p_2\, M} & 0 & k\, e^{-k\, M} & 0 \end{bmatrix}$$

with

$$\mathbf{T}_- \begin{pmatrix} c_1 \\ c_2 \\ c_3 \\ \beta_1 \\ \beta_2 \end{pmatrix} = \begin{pmatrix} f(-M) \\ f'(-M) \\ f''(-M) \\ g(-M) \\ g'(-M) \end{pmatrix}$$

Use the solutions f_+ and g_+ to define a matrix

$$\mathbf{T}_+ = \begin{bmatrix} 0 & 0 & e^{p_3\,M} & 0 & 0 \\ 0 & 0 & p_3\,e^{p_3\,M} & 0 & 0 \\ 0 & 0 & p_3^2\,e^{p_3\,M} & 0 & 0 \\ 0 & 0 & \frac{p_3^2+k^2}{p_3^2-k^2}\,e^{p_3\,M} & 0 & e^{-k\,M} \\ 0 & 0 & \frac{(p_3^2+k^2)\,p_3}{p_3^2-k^2}\,e^{p_3\,M} & 0 & -k\,e^{-k\,M} \end{bmatrix}$$

with

$$\mathbf{T}_+ \begin{pmatrix} c_1 \\ c_2 \\ c_3 \\ \beta_1 \\ \beta_2 \end{pmatrix} = \begin{pmatrix} f(+M) \\ f'(+M) \\ f''(+M) \\ g(+M) \\ g'(+M) \end{pmatrix}$$

Then use an ODE solver to examine the system (4.14)–(4.15) on the interval $[-M, M]$ as an initial value problem. This leads to a matrix $\mathbf{T}$ such that

$$\mathbf{T} \begin{pmatrix} f(-M) \\ f'(-M) \\ f''(-M) \\ g(-M) \\ g'(-M) \end{pmatrix} = \begin{pmatrix} f(+M) \\ f'(+M) \\ f''(+M) \\ g(+M) \\ g'(+M) \end{pmatrix}$$

Now we have two methods to compute the values of the solution at $x = +M$, and the system (4.14)–(4.15) has a nonzero solution if and only if

$$D(k,\gamma) \;=\; \det(\mathbf{M}(k,\gamma)) = \det(\mathbf{T}\cdot\mathbf{T}_- - \mathbf{T}_+) = 0 \tag{4.23}$$

Thus we examine solutions of this equation as function of the parameters k and γ.

4.4. The special case $k = 0$. For $k = 0$, equation (4.15) reads as $g''(x) = f''(x)$ and the only solution satisfying $g(\pm\infty) = f(\pm\infty)$ is $g(x) = f(x)$. Then (4.14) leads to

$$\frac{d}{dx}\begin{pmatrix} f \\ f' \\ f'' \end{pmatrix} = \begin{bmatrix} 0 & 1 & 0 \\ 0 & 0 & 1 \\ -4\,\gamma + 6\,u_0' & 4 + 6\,u_0 & 0 \end{bmatrix} \begin{pmatrix} f \\ f' \\ f'' \end{pmatrix} \tag{4.24}$$

Using the same approach as in the previous section we define

$$\mathbf{T}_- \begin{pmatrix} c_1 \\ c_2 \\ c_3 \end{pmatrix} = \begin{pmatrix} f(-M) \\ f'(-M) \\ f''(-M) \end{pmatrix} = \begin{bmatrix} e^{-p_1\,M} & e^{-p_2\,M} & 0 \\ p_1\,e^{-p_1\,M} & p_2\,e^{-p_2\,M} & 0 \\ p_1^2\,e^{-p_1\,M} & p_2^2\,e^{-p_2\,M} & 0 \end{bmatrix} \begin{pmatrix} c_1 \\ c_2 \\ c_3 \end{pmatrix}$$

$$\mathbf{T}_+ \begin{pmatrix} c_1 \\ c_2 \\ c_3 \end{pmatrix} = \begin{pmatrix} f(+M) \\ f'(+M) \\ f''(+M) \end{pmatrix} = \begin{bmatrix} 0 & 0 & e^{p_3\,M} \\ 0 & 0 & p_3\,e^{p_3\,M} \\ 0 & 0 & p_3^2\,e^{p_3\,M} \end{bmatrix} \begin{pmatrix} c_1 \\ c_2 \\ c_3 \end{pmatrix}$$

$$\mathbf{T}_0 \begin{pmatrix} f(-M) \\ f'(-M) \\ f''(-M) \end{pmatrix} = \begin{pmatrix} f(+M) \\ f'(+M) \\ f''(+M) \end{pmatrix}$$

where $\mathbf{T}_0$ is constructed using an ODE solver for (4.24). Then the system (4.14)–(4.15) has a nonzero solution for $k = 0$ if and only if

$$D_0(\gamma) \;=\; \det(\mathbf{M}_0(\gamma)) = \det(\mathbf{T}_0\cdot\mathbf{T}_- - \mathbf{T}_+) = 0 \tag{4.25}$$

5. Numerical Results on the Instabilities of Plane-Wave Soliton Solutions

5.1. Locating unstable solutions. For our numerical test we used $M = 5$ and for $|x| > 5$ we know $|u_0(x)| = 2\cosh^{-2}(x) \leq 3.6 \cdot 10^{-4} \ll 2$ and $|u_0'(x)| \leq 7.3 \cdot 10^{-4} \ll 2$. Tests with larger values for M did not change the result significantly. We used the numerical ODE solver `lsode` provided by *Octave*. Based on expression (4.23) we generate plots of the function $D(k,\gamma) = \det(\mathbf{M}(k,\gamma))$ on a domain $0 \leq k \leq 1$ and $0 \leq \gamma \leq 0.5$, leading to Figure 2. In the corner $k \approx 1$ and $\gamma \approx 0.5$ the real part of the function vanishes, but a second plot verifies that the imaginary part is different from zero. Thus Figure 2 indicates that we have a clearly defined solution curve of $\det(\mathbf{M}(k,\gamma)) = 0$, away from the origin.

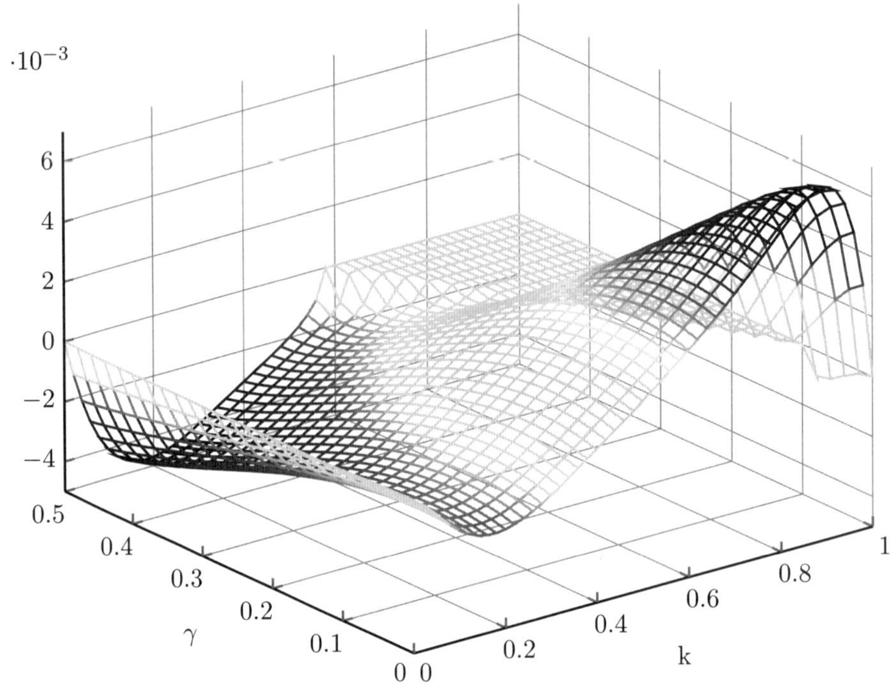

FIGURE 2. The real part of $\det(\mathbf{M})$ as a function of k and γ

Since the behavior close to $(k,\gamma) \approx (0,\,0)$ is critical, we examine this section with a finer resolution, leading to Figure 3(a). The obvious spikes are caused by the zeros in the denominator in condition (4.22). Figure 3(a) suggests the existence of a solution along the axis $k = 0$. Using (4.25) we generate Figure 3(b). As a consequence the only solution along the axis $k = 0$ is at $\gamma = 0$. Thus the known solution (4.16) is an isolated solution in the parameter space (k,γ) at $(0,\,0)$.

With the above preparation we can now construct values of (k,γ) leading to nonzero solutions of (4.14)–(4.15) and thus for $\gamma > 0$ to unstable soliton solutions of the NV equations (1.2)–(1.4).

We trace a solution curve of (4.23) by generating an arc length parametrization of the curve. Using an arbitrary initial point on the curve (use a contour plot of

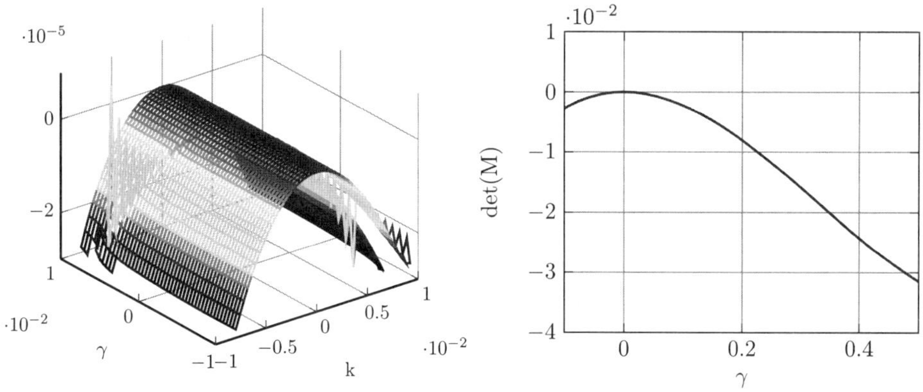

(a) det $\mathbf{M}(k, \gamma)$ close to origin

(b) det $\mathbf{M}_0(\gamma)$

FIGURE 3. Behavior at the origin and along the axis $k = 0$

Figure 2) and a starting direction, we minimize $|\det(\mathbf{M}(k, \gamma))|$ along a straight line segment orthogonal to the stepping direction. Using this local minimum, we adjust the stepping direction and then make a small step to follow the solution curve. While stepping along the curve we verify that we actually have a solution of (4.23), and not only a minimum. This proved to be a stable algorithm and generated Figure 4.

Observe that Figure 4(a) also displays negative values of γ, which do not lead to unstable solutions of (1.2)–(1.4). We display these values to confirm that we have a closed curve without branching points.

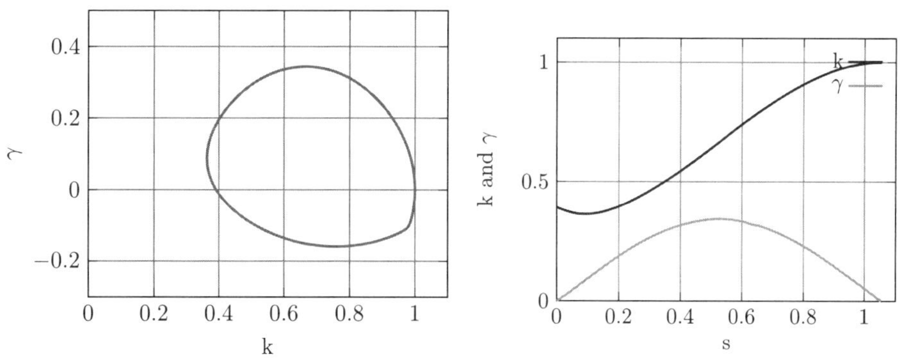

(a) solution curve of $|\det(\mathbf{M})| = 0$

(b) parameters leading to unstable solutions

FIGURE 4. Solution curve of $|\det(\mathbf{M})| = 0$ and parameters k and γ leading to unstable solutions

For the parameter values of (k, γ) in Figure 4(b) there are nonzero functions $f(x)$, $g(x)$ and $h(x)$, such that for small ϵ we have solutions of the equations (4.2)–(4.4) of the form (4.5)–(4.7). Since the y dependence of these functions is of the form $e^{i\,k\,y}$, the functions are periodic in y with a period of $L_y = \frac{2\,\pi}{k}$. The corresponding exponent is shown in Figure (4(b)). Thus the Novikov–Veselov equations (1.2)–(1.4) with initial condition $u(x,y,0) = -2\,c\,\mathrm{sech}^2(\sqrt{c}\,x)$ will lead to an unstable soliton solution.

THEOREM 3. *The soliton solutions* $u(x,y,t) = v(x,y,t) = -2\,c\,\mathrm{sech}^2(\sqrt{c}\,(x - c\,t))$ *and* $w(x,y,t) = 0$ *of the NV equations* (1.2)–(1.4) *are not stable. For values of* $0.363 < k \leq 1$ *there are* y*–periodic, unstable contributions with a period of* $L_y = \frac{2\,\pi}{k\,\sqrt{c}}$.

5.2. Constructing unstable solutions numerically. As an example, in this section we construct one of the above unstable solutions numerically, using the algorithm from Section 3 for periodic solutions in x and y. Since the soliton solution $\mathrm{sech}^2(x)$ decays rapidly, the above results still apply when working on a sufficiently large domain. The steps of the algorithm are as follows:

(1) Choose values of (k, γ) in Figure 4(b) and determine the eigenvector $(c_1, c_2, c_3, \beta_1, \beta_2)$ for the zero eigenvalue.
(2) Use the algorithm leading to the matrices $\mathbf{T}$, $\mathbf{T}_-$ and $\mathbf{T}_+$ to construct the nonzero functions $f(x), g(x), h(x)$.
(3) Pick a size domain such that $\frac{2\pi}{k}$ periodic functions in y are admissable.
(4) Construct initial values, using (4.8)–(4.10), use a small value of ϵ. Without taking the transformations (4.1) into account we obtain a speed of $c = 1$ of the unperturbed soliton.
(5) Solve the NV equations (1.2)–(1.4), using the algorithm presented in Section 3.
(6) The deviation from the single soliton solution should not change its shape, but the size is expected to be proportional to $e^{\gamma\,t}$.
(7) Solitions for speeds $c \neq 1$ can be constructed similarly, using the transformations (4.1).

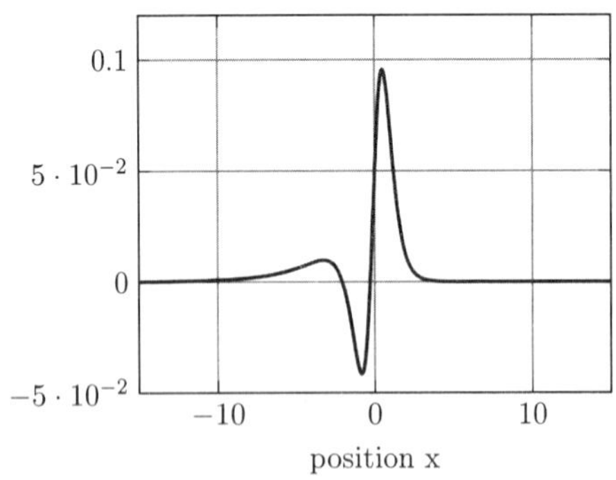

(a) Cross-section of the initial perturbation with $k \approx 0.5$ and $\gamma \approx 0.3$

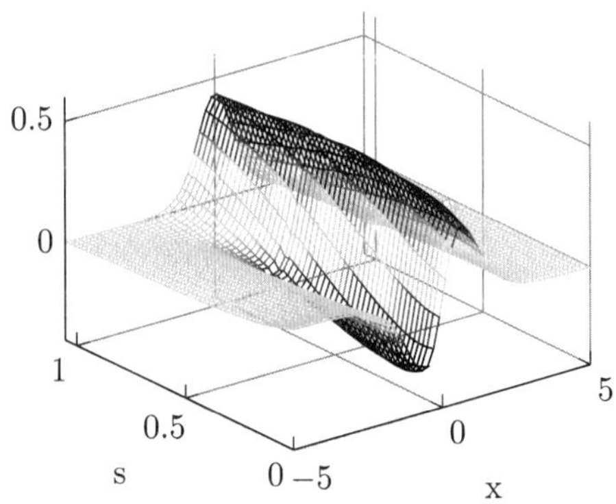

(b) Perturbation as a function of position x and arc length s

FIGURE 5. The shape of the perturbation function $f(x)$ from equation (4.5)

The evolution of a perturbed soliton from Theorem 3 was computed using the semi-implicit pseudo-spectral method. As initial data, we chose a perturbed KdV

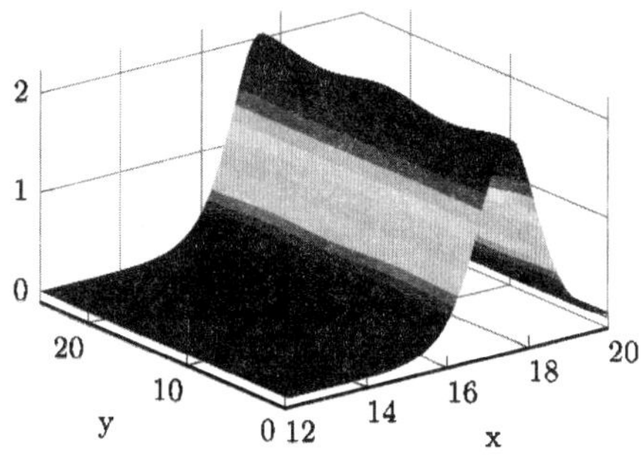

(a)

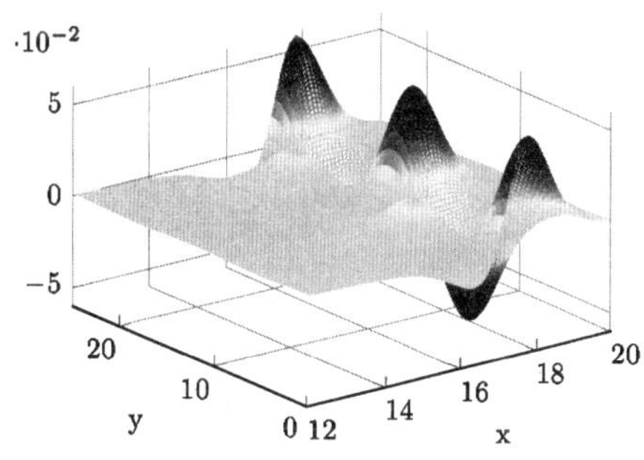

(b)

FIGURE 6. a) the perturbed solution $-u(x, y, t)$ at $t = 5$. b) difference from the unperturbed soliton at time $t = 5$.

soliton with speed $c = 1$, starting at $x = 10$. The perturbation had an initial amplitude of 0.0095. For the domain size we chose $W_x = W_y = \frac{4\pi}{k\,c} \approx 25$. The shape of the perturbation $f(x)$ is found in Figure 5(a), and the values $k = 0.504$ and $\gamma = 0.296$ were chosen in Figure 4(b). The values of the exact KdV soliton at the boundaries $x = 0$ and $x = W_x$ are smaller than $2 \cdot 10^{-8}$, and we apply periodic boundary conditions. A time-snapshot of the solution at time $t = 5$ and the difference to the unperturbed KdV soliton at time $t = 5$ are found in Figure 6. Note that in Figure 6 $-u$ is plotted for easier viewing. In Figure 6(a) the deviation from the KdV soliton $-2\cosh^{-2}(x - t - 10)$ is barely visible, but Figure 6(b) shows the difference. The L_2 norm of the difference of the computed solution and the known exact KdV soliton follows the expected exponential curve $e^{\gamma t}$ very closely. The front edge in Figure 6(b) matches the shape of the function $f(x)$ in Figure 5(a). Animations of the time evolution of the perturbed solution are available on the web site [**30**]. Similar computations with initially smaller perturbations confirm the above and thus the instability result in Theorem 3.

One can construct the shapes of the functions $f(x)$ for all positive values of γ along the arc in Figure 4 to obtain Figure 5(b). The solutions constructed for $(k, \gamma) = (1, 0)$ must match the known exact solutions (4.17), which is confirmed.

6. Conclusions

In this work a semi-implicit pseudo-spectral method was introduced for the numerical computation of evolutions of solutions to the NV equation and used to compute the evolution of an unstable soliton solution. A linear stability analysis yields a stability condition for the Crank-Nicolson scheme in the pseudo-spectral method on the linearized IBVP, which is also valid for the NV equation with nonzero energy. The instability of traveling wave solutions to transverse perturbations was established by the **K**-expansion method, and unstable soliton solutions were constructed.

References

[1] J. Alexander, R. Gardner, and C. Jones, *A topological invariant arising in the stability analysis of travelling waves*, J. Reine Angew. Math. **410** (1990), 167–212. MR1068805 (92d:58028)

[2] M. A. Allen and Phibanchon, *Time evolution of perturbed solitons of modified Kadomtsev-Petviashvili equations*, Computational Science and its Applications, 2007. ICCSA 2007. International Conference on (2007), 20–23.

[3] M. A. Allen and G. Rowlands, *Determination of the growth rate for the linearized Zakharov-Kuznetsov equation*, Journal of Plasma Physics **50** (1993), no. 03, 413–424.

[4] M. A. Allen and G. Rowlands, *On the transverse instabilities of solitary waves*, Phys. Lett. A **235** (1997), no. 2, 145–146, DOI 10.1016/S0375-9601(97)00618-X. MR1479603

[5] L. V. Bogdanov, *The Veselov-Novikov equation as a natural generalization of the Korteweg-de Vries equation* (Russian, with English summary), Teoret. Mat. Fiz. **70** (1987), no. 2, 309–314. MR894472 (88k:35170)

[6] R. M. Bradley, *Electromigration-induced soliton propagation on metal surfaces*, Phys. Rev. E **60** (1999), no. 4, 3736–3740.

[7] T. J. Bridges, *Transverse instability of solitary-wave states of the water-wave problem*, J. Fluid Mech. **439** (2001), 255–278, DOI 10.1017/S0022112001004530. MR1849635 (2003e:35257)

[8] B. Dubrovin, T. Grava, and C. Klein, *On universality of critical behavior in the focusing nonlinear Schrödinger equation, elliptic umbilic catastrophe and the tritronquée solution to the Painlevé-I equation*, J. Nonlinear Sci. **19** (2009), no. 1, 57–94, DOI 10.1007/s00332-008-9025-y. MR2480731 (2010b:35428)

[9] John W. Evans, *Nerve axon equations. I. Linear approximations*, Indiana Univ. Math. J. **21** (1971/72), 877–885. MR0292531 (45 #1616)

[10] John W. Evans, *Nerve axon equations. II. Stability at rest*, Indiana Univ. Math. J. **22** (1972/73), 75–90. MR0323372 (48 #1729)

[11] John W. Evans, *Nerve axon equations. III. Stability of the nerve impulse*, Indiana Univ. Math. J. **22** (1972/73), 577–593. MR0393890 (52 #14697)

[12] John W. Evans, *Nerve axon equations, IV: The stable and unstable impulse*, Indiana U. Math J. **24** (1975), 1169–1190.

[13] B.-F. Feng, T. Kawahara, and T. Mitsui, *A conservative spectral method for several two-dimensional nonlinear wave equations*, J. Comput. Phys. **153** (1999), no. 2, 467–487, DOI 10.1006/jcph.1999.6286. MR1705941 (2000d:65184)

[14] P. Frycz and E. Infeld, *Self-focusing of nonlinear ion-acoustic waves and solitons in magnetized plasmas. part 3. Arbitrary-angle perturbations, period doubling of waves*, J. Plasma Phys. **41** (1989), 441–446.

[15] P. Frycz and E. Infeld, *Self-focusing of nonlinear ion-acoustic waves and solitons in magnetized plasmas. part 3. Arbitrary-angle perturbations, period doubling of waves*, J. Plasma Phys. **41** (1989), no. 03, 441–446.

[16] E. Infeld, *Self-focusing of nonlinear ion-acoustic waves and solitons in magnetized plasmas*, J. Plasma Phys. **33** (1985), no. 02, 171–182.

[17] E. Infeld and P. Frycz, *Self-focusing of nonlinear ion-acoustic waves and solitons in magnetized plasmas. part 2. Numerical simulations, two-soliton collisions*, J. Plasma Phys. **37** (1987), no. 01, 97–106.

[18] Eryk Infeld and George Rowlands, *Nonlinear waves, solitons and chaos*, 2nd ed., Cambridge University Press, Cambridge, 2000. MR1780300 (2001f:76019)

[19] E. Infeld, G. Rowlands, and A. Senatorski, *Instabilities and oscillations of one- and two-dimensional Kadomtsev-Petviashvili waves and solitons*, R. Soc. Lond. Proc. Ser. A Math. Phys. Eng. Sci. **455** (1999), no. 1992, 4363–4381, DOI 10.1098/rspa.1999.0505. MR1809365 (2001j:35239)

[20] B.B. Kadomtsev and V.I. Petviashvili, *On the stability of solitary waves in weakly dispersing media*, Soviet Physics Doklady **15** (1970), 539–+.

[21] T. Kapitula, *Stability analysis of pulses via the Evans function: dissipative systems*, Dissipative solitons, Lecture Notes in Phys., vol. 661, Springer, Berlin, 2005, pp. 407–428, DOI 10.1007/10928028_16. MR2174303 (2007a:37096)

[22] C. Klein and K. Roidot, *Fourth order time-stepping for Kadomtsev-Petviashvili and Davey-Stewartson equations*, SIAM J. Sci. Comput. **33** (2011), no. 6, 3333–3356, DOI 10.1137/100816663. MR2862016

[23] C. Klein and C. Sparber, *Transverse stability of periodic travelling waves in KP equations: A numerical study*, Preprint (arXiv:1108.3363v1 16 Aug 2011) (2011).

[24] B. Konopelchenko and A. Moro, *Integrable equations in nonlinear geometrical optics*, Stud. Appl. Math. **113** (2004), no. 4, 325–352, DOI 10.1111/j.0022-2526.2004.01536.x. MR2094235 (2005e:78004)

[25] Boris G. Konopelchenko and Antonio Moro, *Geometrical optics in nonlinear media and integrable equations*, J. Phys. A **37** (2004), no. 10, L105–L111, DOI 10.1088/0305-4470/37/10/L01. MR2039848 (2004m:78001)

[26] M. Lassas, J. L. Mueller, S. Siltanen, and A. Stahel, *The Novikov-Veselov equation and the inverse scattering method: II. Computation*, Nonlinearity **25** (2012), no. 6, 1799–1818, DOI 10.1088/0951-7715/25/6/1799. MR2929603

[27] S. Munro and E. J. Parkes, *The derivation of a modified Zakharov-Kuznetsov equation and the stability of its solutions*, J. Plasma Phys. **62** (1999), no. 03, 305–317.

[28] A. P. Veselov and S. P. Novikov, *Finite-gap two-dimensional potential Schrödinger operators. Explicit formulas and evolution equations* (Russian), Dokl. Akad. Nauk SSSR **279** (1984), no. 1, 20–24. MR769198 (86d:58053)

[29] Peter Perry, *Miura maps and inverse scattering for the Novikov–Veselov equation*, Anal. PDE **7** (2014), no. 2, 311–343, DOI 10.2140/apde.2014.7.311. MR3218811

[30] A. Stahel, `staff.ti.bfh.ch/sha1/NovikovVeselov/NovikovVeselov.html`.

[31] Li-Yeng Sung, *An inverse scattering transform for the Davey-Stewartson II equations. I*, J. Math. Anal. Appl. **183** (1994), no. 1, 121–154, DOI 10.1006/jmaa.1994.1136. MR1273437 (95c:35237)

Department of Mathematics, Colorado State University, Fort Collins, 80523
Current address: Department of Applied Mathematics, University of Colorado, Boulder, Colorado 80309
E-mail address: `rpcroke@gmail.com`

Department of Mathematics and School of Biomedical Engineering, Colorado State University, Fort Collins, Colorado 80523
E-mail address: `mueller@math.colostate.edu`

Bern University of Applied Sciences, Engineering and Information Technology, Biel, Switzerland
E-mail address: `Andreas.Stahel@bfh.ch`

Contemporary Mathematics
Volume **635**, 2015
http://dx.doi.org/10.1090/conm/635/12717

Semiclassical soliton ensembles for the focusing nonlinear Schrödinger equation: Recent developments

Gregory D. Lyng

Abstract. We give an overview of the analysis of the semiclassical (zero-dispersion) limit of the focusing nonlinear Schrödinger equation via semiclassical soliton ensembles, and we describe some recent developments in this direction.

1. Introduction

1.1. The semiclassical limit problem. A semiclassical soliton ensemble (SSE) is a particular family of exact, multisoliton solutions of the focusing nonlinear Schrödinger (NLS) equation

$$\mathrm{i}\epsilon\frac{\partial u}{\partial t} + \frac{\epsilon^2}{2}\frac{\partial^2 u}{\partial x^2} + |u|^2 u = 0\,, \quad 0 < \epsilon \ll 1\,. \tag{1.1}$$

Given fixed data

$$u(x,0) = A_0(x)\,, \quad x \in \mathbb{R}\,; \tag{1.2}$$

where A_0 is an analytic, positive, bell-shaped function which decays rapidly to zero as $|x| \to \infty$, the corresponding SSE is generated by adding a distinguished, asymptotically negligible (as $\epsilon \downarrow 0$) sequence of perturbations to the data A_0 and solving the resulting sequence of initial-value problems. These perturbations are induced by a formal WKB analysis of the associated spectral problem, and we shall denote elements of this family of special solutions, indexed by (a sequence of decreasing values of) the real parameter $\epsilon > 0$, by

$$u^{(\epsilon)}(x,t)\,.$$

2010 *Mathematics Subject Classification.* Primary: 35Q55.

Figure 3 in this paper is reprinted from *Phys. D*, **241**, no. 21, Long Lee, Gregory Lyng, and Irena Vankova, "The Gaussian semiclassical soliton ensemble and numerical methods for the focusing nonlinear Schrödinger equation", 1767–1781, Copyright 2012, with permission from Elsevier.

Figure 4 in this paper is reprinted from *Phys. Lett A*, **377**, no. 16–17, Long Lee and Gregory D. Lyng, "A second look at the Gaussian semiclassical soliton ensemble for the focusing nonlinear Schrödinger equation", 1179–1188, Copyright 2013, with permission from Elsevier.

Figure 5 in this paper is reprinted with permission from *Journal of Mathematical Physics*, **55**, no. 8, Yeongjoh Kim, Long Lee, and Gregory Lyng, "The Wentzel–Kramers–Brillouin approximation of semiclassical eigenvalues of the Zakharov–Shabat problem", Copyright 2014–2015, AIP Publishing LLC..

A detailed study of the behavior of $u^{(\epsilon)}(x,t)$ in the limit $\epsilon \downarrow 0$ is a key element of one of the principal approaches—pioneered by Kamvissis, McLaughlin, & Miller in their groundbreaking monograph [**16**]—to the analysis of the semiclassical (zero-dispersion, $\epsilon \downarrow 0$) limit for equation (1.1). In its most basic form, the semiclassical limit problem is to provide a complete asymptotic description of the true solutions—indexed by $\epsilon > 0$ and here denoted by $u(x,t;\epsilon)$ to distinguish them from elements $u^{(\epsilon)}(x,t)$ of the SSE—of (1.1), (1.2). However, as we shall describe below, equation (1.1) has modulational instabilities whose exponential growth rates increase without bound in the semiclassical limit. Thus, the impact of even small perturbations to the initial data on the limiting behavior is far from clear. Indeed, one of the remaining open problems in the theory is to determine, rigorously, whether or not the semiclassical asymptotics of $u^{(\epsilon)}(\cdot,\cdot)$ coincide with those of the true solutions $u(\cdot,\cdot;\epsilon)$, say, in a neighborhood of fixed values of x and t. Our aim here is to give a panoramic view of the use of SSEs to study the semiclassical limit problem and, in particular, to describe several numerical studies of recent vintage which address the connection between the asymptotics of these two families of solutions. *These numerical experiments, taken together, give very strong support to the claim that the limiting behavior of these two families of solutions is the same.*

1. Remark (Data). Here, we shall only concern ourselves with bell-shaped real data; we sometimes say that this data is of *Klaus–Shaw* type, after the result of Klaus & Shaw [**18**] that restricts the point spectrum of the related nonselfadjoint Zakharov–Shabat eigenvalue problem (equation (2.1) below) with such a bell-shaped potential to the imaginary axis. However, it is also desirable to consider complex data in amplitude-phase form

$$u(x,0) = A_0(x)e^{iS_0(x)/\epsilon}\,. \tag{1.3}$$

Moreover, it is natural (from the point of view of applications) to consider data which is not analytic. In general, however, the semiclassical limit problem has largely resisted a completely rigorous analysis outside of two families of specific analytic data (for which the forward scattering problem (2.1) can be solved exactly [**27, 30**]); thus, an extension to an entire class of bell-shaped data may be regarded as an important first step toward a comprehensive theory for the semiclassical limit problem.

1.2. About the problem: modulational instability. Equation (1.1) is a universal model equation that arises naturally in multiscale analyses of various physical phenomena as an envelope equation governing the evolution of nearly monochromatic waves in a weakly dispersive nonlinear medium. In this setting, the real parameter $0 < \epsilon \ll 1$ is a measure of the competition between dispersion and nonlinearity, and it is precisely this competition—in which nonlinear effects amplify waves and dispersive effects pull them apart—that gives rise to the behaviors of interest. Indeed, the *focusing* nonlinearity ("+" sign in front of the nonlinear term) in (1.1) arises when the combination of nonlinear and dispersive effects destabilizes periodic wavetrains. This instability is known to occur, for example, in the theory of finite-amplitude plane waves on deep water where it is known as the *Benjamin–Feir* instability. See, e.g., [**28**] for a more detailed discussion of these points.

The task of obtaining detailed semiclassical asymptotics for solutions of (1.1), (1.2) have long been identified as one of the central challenges in the theory of singular asymptotics for integrable nonlinear wave equations (see, e.g., Bronski &

McLaughlin [7], Forest and Lee [14] for early reports on this problem), and this challenge has been the focus of a substantial body of work in recent years [2–4, 8, 16, 20–22, 24, 32, 33]. The mathematical interest in this problem has been due, in large part, to the aforementioned phenomenon of *modulational instability.* To describe this phenomenon in more detail, we simply note that it is a straightforward exercise to verify that a plane wave of the form $u(x,t;\epsilon) = U_0 \exp(\mathrm{i}(kx - \omega t))$ is a solution of (1.1) provided that $\omega = \omega(k;\epsilon)$ satisfies the dispersion relation

$$\omega = \frac{\epsilon}{2}k^2 - \frac{|U_0|^2}{\epsilon}\,. \tag{1.4}$$

But, a linear stability analysis, perturbing the amplitude and phase slightly, shows that *all* wave numbers are destabilized in the limit of vanishing dispersion [28]. This feature of equation (1.1) has a profound effect on the analysis of the semiclassical limit. Indeed, modulational instability manifests itself as ellipticity of the formal model (known as the Whitham or modulation equations) expected to describe the large-scale limiting dynamics. To obtain the Whitham system, we make the standard definitions $\rho = |u|^2$, $\mu = -\frac{\mathrm{i}\epsilon}{2}\left(u^*\frac{\partial u}{\partial x} - u\frac{\partial u^*}{\partial x}\right)$, and we obtain exactly—that is, with no approximation at all—from (1.1) the system

$$\frac{\partial \rho}{\partial t} + \frac{\partial \mu}{\partial x} = 0\,, \tag{1.5a}$$

$$\frac{\partial \mu}{\partial t} + \frac{\partial}{\partial x}\left(\frac{\mu^2}{\rho} - \frac{\rho}{2}\right) = \frac{\epsilon^2}{4}\frac{\partial}{\partial x}\left[\left(\frac{\partial^2}{\partial x^2}\log(\rho)\right)\rho\right]. \tag{1.5b}$$

The Whitham equations for (1.1) are then obtained from (1.5) simply by neglecting the formally small $O(\epsilon^2)$ term on the right-hand side of (1.5b). However, a standard calculation starting with the resulting first-order system reveals at once that the Whitham equations are of elliptic type, and are therefore expected to be generally ill-posed as an initial-value problem. This ill-posedness is troubling for several reasons. First and foremost, it casts doubt on the utility of the Whitham system in the first place; one hopes that the Whitham system should provide a relatively simple description of the oscillatory behavior of the semiclassical dynamics. Second, since one only expects to be able to solve the elliptic initial-value problem for short-time using the Cauchy–Kovalevskaya theorem for analytic data, the elliptic nature of the equations suggests the possibility that the limiting dynamics are extremely sensitive to the presence (or absence!) of analyticity in the data. As noted in Remark 1 above this is particularly troublesome given the importance of (1.1) in applications where nonanalytic data is ubiquitous. Finally, as indicated above, the formation of a SSE is based on an uncontrolled modification of the initial data, and the lack of well-posedness (associated with ellipticity) obscures the effect of introducing even small perturbations to the initial data.

1.3. Plan. In §2, to set the scene, we outline the rudiments of the inverse scattering transform (IST) for (1.1), (1.2). In particular, we recall the form of the Zakharov–Shabat problem [34] and the WKB analysis of [12] that is the basis for the construction of SSEs associated to bell-shaped initial data. For completeness, in §3 we briefly describe the asymptotic analysis of the Riemman–Hilbert problem for inverse scattering associated with a SSE. This analysis is fundamentally based on the steepest descent techniques of Deift & Zhou [9]. Notably, this analysis gives a clear mathematical interpretation of the *nonlinear caustics* in the semiclassical limit; these caustics are one of the dramatic features of the semiclassical behavior (see Figure 2 below). Indeed, in §3.3 we show new, previously unpublished, computations by the numerical inverse scattering technique originally proposed and implemented by Miller & Kamvissis [**26**] and later refined by Lyng & Miller [**22**]. These computations show, in addition to the previously studied primary and secondary caustics, the presence of a *tertiary caustic*. Lastly, in §4 we describe a coordinated collection of numerical experiments (taken from [**17**, **20**, **21**]) aimed at exploring the connections between the limiting behavior of the SSE and the corresponding true solutions, and, in §5, we discuss possible future developments related to SSEs and their role in the theory of the semiclassical limit problem.

2. Remark (Other aspects of the semiclassical focusing NLS equation). We shall confine our discussion here to the issues related to SSEs, and we shall not touch on other aspects of the fashionable semiclassical limit problem. However, we do note that substantial progress has been made in recent years on other fronts including, for example, the following.

(a) A detailed analysis of the nonlinear caustics and their "universal" properties has been undertaken by Bertola & Tovbis [**3**, **4**]; see also [**11**].
(b) Tovbis, Venakides, & Zhou [**31**–**33**] have written extensively about this problem. Much of their work is based on a particular one-parameter family of complex data (with nontrivial phase) identified by Tovbis & Venakides [**30**]. For certain parameter values, their data generates no solitons (a case in some ways the opposite of the case considered here), and in this setting they are able to establish the long-time ($t \to \infty$) behavior of the semiclassical limit.
(c) Tovbis & Venakides [**29**]—sidestepping WKB issues by using a kind of complexified Abel transform—have recently proposed an alternative approach to the forward scattering map in the semiclassical regime.

2. Background & framework

2.1. Inverse scattering transform & the nonselfadjoint Zakharov–Shabat problem. The construction of the SSE is intimately connected to the integrability of the nonlinear equation (1.1). We recall that Zakharov & Shabat [**34**] discovered a Lax pair for the NLS equation. That is, they showed that nonlinear equation (1.1) can be represented as the compatibility condition for two auxiliary linear problems, and this structure allows one to construct solutions of the initial-value problem by the IST. For (1.1), (1.2), the first of these auxiliary problems is the eigenvalue problem

$$\epsilon \frac{\mathrm{d}}{\mathrm{d}x}\mathbf{w} = \begin{bmatrix} -\mathrm{i}\lambda & A_0 \\ -A_0 & \mathrm{i}\lambda \end{bmatrix} \mathbf{w}\,. \tag{2.1}$$

In (2.1), we have written $\mathbf{w}(x;\lambda,\epsilon) = \big(w_1(x;\lambda,\epsilon), w_2(x;\lambda,\epsilon)\big)^{\top}$ for the solution and $\lambda \in \mathbb{C}$ is the eigenvalue parameter. As above, the real parameter ϵ is assumed to be positive but small. Our principal interest here are those values of $\lambda \in \mathbb{C}$ for which (2.1) has a solution in $L^2(\mathbb{R})^2$; these values comprise the discrete spectrum—the eigenvalues. In addition, to fully implement the IST, we recall that the essential properties of the data for the initial-value problem are encoded in the scattering data. In addition to the eigenvalues, this includes the norming constants, and the reflection coefficient. In the construction of the SSE, the reflection is assumed to be identically zero due to the smoothness of A_0, and for simplicity we shall largely ignore the role of the norming constants in our exposition here. As noted above, Klaus & Shaw [**18**] have shown that the eigenvalues (for bell-shaped potential A_0 in (2.1)) are purely imaginary. The evolution in time of the scattering data in the spectral domain is governed by the other half of the Lax pair, a linear problem, and is simple and explicit. Finally, the solution at times $t > 0$ is recovered by an inverse scattering map; that is, the solution $u(x, t; \epsilon)$ is recovered from the time-evolved scattering data (eigenvalues, norming constants, reflection). A complete and detailed description of this methodology for (1.1) can be found, for example, in the monographs [**1, 13**].

2.2. WKB analysis. In the case of a Klaus–Shaw potential, as in (2.1), one may obtain further, *formal* information about the precise location of the eigenvalues by performing a WKB analysis. We only summarize the results here; details may be found in the article of Ercolani et al. [**12**]. We begin with the density function

$$\rho^0(\eta) := \frac{\eta}{\pi} \int_{x_-(\eta)}^{x_+(\eta)} \frac{\mathrm{d}x}{\sqrt{A_0(x)^2 + \eta^2}} = \frac{1}{\pi} \frac{\mathrm{d}}{\mathrm{d}\eta} \int_{x_-(\eta)}^{x_+(\eta)} \sqrt{A_0(x)^2 + \eta^2}\, \mathrm{d}x\,, \tag{2.2}$$

defined for $\eta \in (0, \mathrm{i}A)$ where $x_\pm(\eta)$ are the two real turning points guaranteed to exist by our assumption that A_0 is bell-shaped and $A \stackrel{\text{def}}{:=} \max_{x\in\mathbb{R}} A_0(x)$; see Figure 1. From ρ^0 we obtain the function

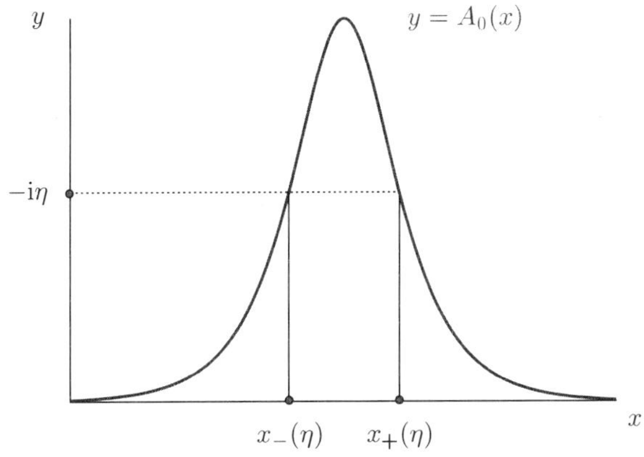

FIGURE 1. The turning points $x_\pm(\eta)$.

$$\theta^0(\lambda) := -\pi \int_{\lambda}^{\mathrm{i}A} \rho^0(\eta)\, \mathrm{d}\eta\,. \tag{2.3}$$

This function measures the number of WKB eigenvalues in the segment $(\lambda, \mathrm{i}A) \subset \mathrm{i}\mathbb{R}$. We then define for $N = 1, 2, 3, \ldots$

$$\epsilon_N := -\frac{1}{N} \int_0^{\mathrm{i}A} \rho^0(\eta)\, \mathrm{d}\eta = \frac{1}{\pi N} \int_{-\infty}^{\infty} A_0(x)\, \mathrm{d}x\,.$$

Finally, the WKB eigenvalues $\lambda_{N,k}^{\mathrm{wkb}}$ are defined (there are N of them for ϵ_N) by the formula

$$(2.4)\qquad -\int_{\lambda_{N,k}^{\mathrm{wkb}}}^{\mathrm{i}A} \rho^0(\eta)\,\mathrm{d}\eta = \epsilon_N\left(k+\frac{1}{2}\right) = \frac{\theta^0(\lambda_{N,k}^{\mathrm{wkb}})}{\pi}\,, \qquad k=0,\ldots,N-1\,.$$

Evidently, we may rewrite the left-hand side as

$$-\int_{\lambda_{N,k}^{\mathrm{wkb}}}^{\mathrm{i}A} \rho^0(\eta)\,\mathrm{d}\eta = \frac{2}{\pi}\int_0^{x_+(\lambda_{N,k}^{\mathrm{wkb}})} \sqrt{A_0(x)^2+(\lambda_{N,k}^{\mathrm{wkb}})^2}\,\mathrm{d}x\,.$$

Therefore, writing $\lambda_{N,k}^{\mathrm{wkb}} = \mathrm{i}\tau_{N,k}^{\mathrm{wkb}}$ for $\tau_{N,k}^{\mathrm{wkb}} \in (0,A)\subset\mathbb{R}$, the eigenvalue locations (important components of the scattering data for A_0) are approximately given by solutions to the equations

$$(2.5)\quad \int_0^{x_+(\mathrm{i}\tau_{N,k}^{\mathrm{wkb}})} \sqrt{A_0(x)^2-(\tau_{N,k}^{\mathrm{wkb}})^2}\,\mathrm{d}x = \frac{\pi\epsilon_N}{2}\left(k+\frac{1}{2}\right), \quad k=0,1,2,\ldots,N-1\,.$$

3. Semiclassical soliton ensembles

3.1. Origins. We are now in a position to describe precisely the origin of the SSE. As noted above, the WKB method applied to (2.1) suggests that the reflection coefficient is negligible and the purely imaginary eigenvalues are given by a quantization condition of Bohr–Sommerfeld type (2.5). Remarkably, this information is completely parallel to the information used by Lax & Levermore [**19**] in their analysis of the zero-dispersion limit of the Korteweg–de Vries (KdV) equation. Lax & Levermore neglected reflection and adopted the WKB eigenvalues[1] in place of the true eigenvalues. That is, they approximated the initial data by a family of reflectionless potentials which, in the limit, approach the true initial data. Since exact information about the true scattering data (eigenvalues of (2.1) and the reflection coefficient) is not known, the temptation to adopt Lax & Levermore's strategy is overwhelming, and we use the (formal) WKB scattering data in its place. This process, neglecting reflection and using the WKB eigenvalues, amounts to replacing the true initial data A_0 with some other initial condition, denoted by $A_0^{(\epsilon)}$, which depends on the small parameter ϵ and for which the WKB data is the *true* data. Thus, this process introduces perturbations to the initial data, and we write

$$(3.1)\qquad A_0^{(\epsilon)}(x) = A_0(x) + q^{(\epsilon)}(x)\,,$$

so that $q^{(\epsilon)}$ is the induced perturbation; see Figure 4 (b) for an example of such a perturbed initial condition. Because reflection is neglected, each solution of (1.1) with initial data $A_0^{(\epsilon)}$ is an N-soliton with $N\sim\epsilon^{-1}$. The collection of all these exact N-soliton solutions of (1.1) (with $N\to\infty$ and $\epsilon\downarrow 0$) is the *semiclassical soliton ensemble* associated with the bell-shaped data A_0.

It is important to note, however, that in the case of the KdV equation, the Whitham equation is *hyperbolic* and is therefore locally well posed. Lax & Levermore appealed to this local well-posedness to absorb the asymptotically vanishing family of perturbations induced by their reflectionless approximation. As described above, we are unable to appeal to the well-posedness of the Whitham system in the

[1] In their case, these eigenvalues are associated with the WKB analysis of the spectral problem for the stationary Schrödinger operator associated with the solution of KdV by IST not with (2.1).

setting of the focusing NLS equation, and we must instead investigate the effect of the perturbations $q^{(\epsilon)}$ directly; see §4.

3. REMARK (Notation). For initial data A_0, we denote by $A_0^{(\epsilon)}$ the data induced by the WKB eigenvalues as above. Similarly, we shall denote the value of the solution generated by data $A_0^{(\epsilon)}$ at the point (x,t) by $u^{(\epsilon)}(x,t)$. The corresponding solution born of the original data A_0 will be denoted by $u(x,t;\epsilon)$. Thus, we always use the superscript "(ϵ)" to signal the presence of a member of the SSE. Later, we shall use a superscript "ϵ" (without parentheses) to denote an ϵ-dependent family of perturbations and solutions *not* associated with the special WKB-induced modifications to data/solution. Similarly, we shall follow the same convention for the density and denote by ρ (or $\rho^{(\epsilon)}$ or ρ^ϵ as the case may be) the square modulus of the solution u (or $u^{(\epsilon)}$ or u^ϵ).

3.2. Riemann–Hilbert formulation and asymptotic analysis.

3.2.1. *RHP.* Given N imaginary eigenvalues

$$\{\lambda_{N,0}, \lambda_{N,1}, \ldots, \lambda_{N,N-1}\}$$

and the corresponding norming constants $\{\gamma_{N,k}\}_{k=0}^{N-1}$, we now utilize the fact that the inverse scattering step for the corresponding N-soliton can be formulated as meromorphic Riemann-Hilbert problem (RHP) for a 2×2 matrix-valued unknown with simple poles at the eigenvalues. The residues are completely specified. The variables x, t, and N enter the RHP in an explicit fashion as parameters. The relevant RHP is given below, and it may be helpful to visualize the WKB eigenvalues from (2.5) as the input into the RHP. Note that in the SSE framework $N\to\infty$ corresponds to the $\epsilon\downarrow 0$.

4. RIEMANN–HILBERT PROBLEM. Find a 2×2 matrix-valued function $\mathsf{m}(\lambda;x,t)$ with the following properties.

(1) $\mathsf{m}(\lambda;x,t)\to\mathsf{I}$ as $\lambda\to\infty$.
(2) $\mathsf{m}(\lambda;x,t)$ is a rational function of λ with poles confined to the values $\lambda_{N,k}$ and $\lambda_{N,k}^*$. At the singularities,

$$\operatorname{Res}_{\lambda=\lambda_{N,k}}\mathsf{m}(\lambda) = \lim_{\lambda\to\lambda_{N,k}}\mathsf{m}(\lambda)\sigma_1^{\frac{1-J}{2}}\begin{bmatrix}0 & 0\\ c_{N,k}(x,t) & 0\end{bmatrix}\sigma_1^{\frac{1-J}{2}}, \tag{3.2}$$

$$\operatorname{Res}_{\lambda=\lambda_{N,k}^*}\mathsf{m}(\lambda) = \lim_{\lambda\to\lambda_{N,k}^*}\mathsf{m}(\lambda)\sigma_1^{\frac{1-J}{2}}\begin{bmatrix}0 & -c_{N,k}(x,t)^*\\ 0 & 0\end{bmatrix}\sigma_1^{\frac{1-J}{2}}. \tag{3.3}$$

Here,

$$c_{N,k}(x,t) := \left(\frac{1}{\gamma_k}\right)^J \frac{\prod_{n=0}^{N-1}(\lambda_{N,k}-\lambda_{N,n}^*)}{\prod_{\substack{n=0\\ n\neq k}}^{N-1}(\lambda_{N,k}-\lambda_{N,n})}\exp\left(\frac{2iJ(\lambda_{N,k}x+\lambda_{N,k}t^2)}{\epsilon}\right), \tag{3.4}$$

and σ_1 is the Pauli matrix

$$\sigma_1 = \begin{bmatrix}0 & 1\\ 1 & 0\end{bmatrix}.$$

Finally, once the solution of RHP 4 is found, one recovers an N-soliton solution via the formula

$$u(x,t) = 2\mathrm{i} \lim_{\lambda\to\infty} \lambda m_{12}(\lambda; x, t)\,,$$

where m_{12} is the $(1,2)$ entry of m.

3.2.2. *Steepest Descent.* We condense the asymptotic analysis to its most abbreviated form[2]. Roughly, after a change of coordinates, the meromorphic RHP can be exchanged for a sectionally holomorphic RHP. This allows the study the $N \to \infty$ (equivalently, $\epsilon \downarrow 0$) problem using the "g-function mechanism" [**10**], a modification of the steepest descent method [**9**]. This method of asymptotic analysis hinges on being able to find an appropriate complex phase, or "g-function", which allows the RHP to go over to a tractable form in the limit. In the present setting, such a g selects certain contours in the complex plane on which the RHP should be posed; the contours consist of *bands* which are fixed and *gaps* which may lie anywhere a certain inequality is satisfied. If g and the appropriate contours (bands & gaps) can be found, the solution of the RHP (for small ϵ) can be approximately built from Riemann theta functions of the hyperelliptic surface with branch points at the band end points. (Note: In [**16**, **22**] g and the band/gap contours are found by an ansatz about the number of bands and numerical calculation. In contrast to direct simulation of (1.1) for small ϵ, these calculations are extremely well conditioned; "bad" behavior has been factored out.) The upshot is that the semiclassical asymptotics for focusing NLS are described by modulated multiphase waves written in terms of theta functions, where the number of phases is related to the genus G of the surface. The nonlinear caustics (phase transitions) correspond to a change in the genus of the surface. An essential feature of the analysis, one that bears on the problems below, is that this global parametrix for the RHP (uniformly close to the true solution on all of $\mathbb{C}$) is built by matching an "inner" part (near branch points) to an "outer" part (elsewhere).

The above description outlines the strategy employed by Kamvissis et al. [**16**]. This was the first analytic breakthrough in the analysis of the semiclassical NLS equation, and variations on this basic scheme has been used in much of the subsequent work on this problem; see, e.g., [**2**, **22**, **31**–**33**].

3.3. Nonlinear Caustics. As described in the previous subsection, nonlinear caustics are curves in space-time which separate qualitatively different modulated oscillations due to differences in the number of phases. The sharpening boundaries are clearly visible in well resolved simulations of the initial-value problem; see Figure 2. Certainly, any reasonable theory of the semiclassical limit for (1.1) should be able to predict the number and location of such important features of the dynamics. However, much remains to be done before these phenomena are completely understood (but see the interesting recent work of Bertola & Tovbis [**3**, **4**]).

Applying the methodology of Kamvissis et al. [**16**] described above, the search for caustics thus amounts to a search, as parameters x and t vary, for topological changes in the region of the complex plane where a certain inequality must be satisfied in order for a gap contour to be located there. Using these ideas, Kamvissis et al. showed that a $G = 0$ ansatz correctly describes the semiclassical behavior for $t \neq 0$ and for t sufficiently small (but independent of ϵ). They also showed that the

[2]A less abbreviated survey (from a slightly different point of view) can be found in the survey article of Miller [**25**].

failure of the $G = 0$ ansatz is due to the pinching off of the gap inequality region; this is the primary caustic. They were also able to show that the semiclassical dynamics after the primary caustic are described by repairing the ansatz to be $G = 2$, as expected.

The secondary caustic was studied by Lyng & Miller [**22**], and the analysis is distinct from that used to study the primary caustic. In particular, the dual-interpolant method of [**24**] was used in an essential way, and Lyng & Miller showed that the mathematical mechanism for the secondary caustic was also due to a pinch-off event. However, in this case, the pinch-off was related to the failure of a *different* variational inequality (for a "different" gap) that arises due to the use of the dual-interpolant method. In related work, Belov [**2**] identified a "catastrophic break" after the primary caustic. In his analysis, a relevant level curve collides (as t increases, say) with a logarithmic branch point, and this is an obstacle to continuing the asymptotic analysis. The relationship between the catastrophic break and the failure of the variational inequality of Lyng & Miller is not entirely clear. Indeed, we know of no other results for second and higher caustics. However, the numerical simulations shown in Figure 2 clearly suggest a *tertiary nonlinear caustic.* The total number of caustics is unknown. Are there infinitely many? Can their locations in space-time be predicted? What are the salient features of the initial data that determine the location of the caustics? These basic questions about the semiclassical limit for the focusing NLS equation remain open.

4. Connection to the true initial-value problem

4.1. Justification of the WKB approximation. The outline in §3 above illustrates the advantages of substituting $A_0^{(\epsilon)}$ for A_0; powerful mathematical tools can then be brought to bear on the asymptotic analysis of the semiclassical limit problem. However, as noted above, the effect of the induced perturbations is *a priori* far from clear. In this section we aim to understand the effect of the effect of the perturbation $q^{(\epsilon)}$ on the semiclassical dynamics. On the one hand, Miller [**24**] has shown that the WKB approximation and the initial data are asymptotically pointwise close.

5. THEOREM (Miller [**24**]). *In the situation described above, there is a sequence* $(\epsilon_N)_{N=1}^{\infty}$ *such that*

$$\lim_{N\to\infty} \epsilon_N = 0, \tag{4.1}$$

and such that for each $x \neq 0$ *there exists a constant* K *(depending on* x*) such that*

$$|A_0^{(\epsilon_N)}(x) - A_0(x)| \leq K\epsilon_N^{1/7-\nu}, \quad N = 1, 2, 3, \dots \tag{4.2}$$

for all $\nu > 0$.

However, on the other hand, as noted above and as Miller himself points out, the focusing NLS equation has modulational instabilities whose exponential growth rates become arbitrarily large as $\epsilon \downarrow 0$. Thus, Theorem 5 does not allow us to conclude anything about the size of

$$|u^{(\epsilon)}(x,t) - u(x,t;\epsilon)|$$

for any $t > 0$. To investigate this question, we do a series of numerical experiments. The results of this section are reported in greater detail in [**17, 20, 21**].

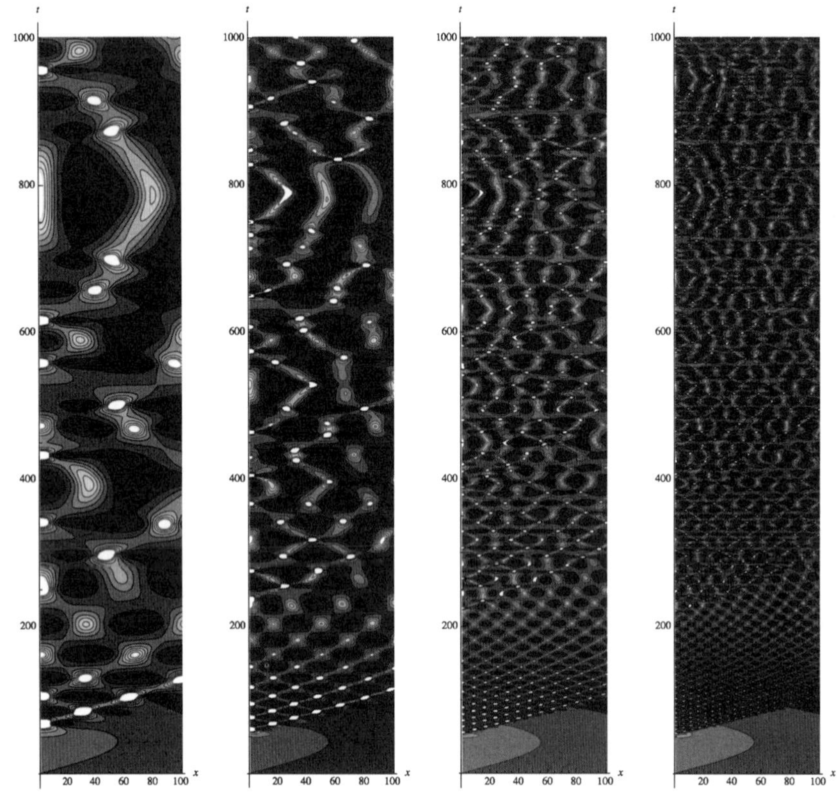

FIGURE 2. Plots of the square modulus $|u(x,t;\epsilon_N)|^2$ for $(x,t) \in [0,1] \times [0,5]$ corresponding to data $A_0(x) = 2\,\mathrm{sech}\, x$ for $\epsilon_N = 2/N$ with $N = 10, 20, 40, 60$. The primary and secondary caustics are clearly visible. The tertiary caustic can also be seen.

4.2. The Gaussian SSE. For our numerical experiment, we restricted ourselves to the Gaussian SSE. That is, we considered the problem (1.1) with fixed initial data given by

$$u_0(x) = A_0(x) = \mathrm{e}^{-x^2}. \tag{4.3}$$

Then, from (2.4)

$$\epsilon_N = \frac{1}{\pi N} \int_{-\infty}^{\infty} \mathrm{e}^{-x^2}\, \mathrm{d}x = \frac{1}{\sqrt{\pi} N}.\,, \tag{4.4}$$

and formula (2.5) may be reduced to

$$\int_0^{\sqrt{-\ln \tau_{N,k}^{\mathrm{wkb}}}} \sqrt{\mathrm{e}^{-2x^2} - (\tau_{N,k}^{\mathrm{wkb}})^2}\, \mathrm{d}x = \frac{\sqrt{\pi}}{2N}\left(k + \frac{1}{2}\right), \quad k = 0, 1, 2, \ldots N-1\,. \tag{4.5}$$

It is the solutions $\tau_{N,k}^{\mathrm{wkb}}$, $k = 0, \ldots, N-1$ of (4.5) which will generate the exact N-soliton solution of (1.1). The collection of these solutions for $N \in \mathbb{N}$ is the *Gaussian SSE.*

We first solved (4.5) to very high precision[3]. With the numerically computed WKB spectral data in hand, we then used the numerical linear algebra routines of [**26**] and [**22**] to reconstruct various members of the SSE at $t = 0$ (and later times too). We then compared the numerical reconstructions of members of the Gaussian SSE at $t = 0$ to the true initial data $A_0 = \mathrm{e}^{-x^2}$ and with approximations to the evolution for $t > 0$. We systematically computed the 2-norm difference between the square modulus of the SSE solutions (computed by numerical inverse scattering) and those of a finite difference method (representing the true evolution of the initial data $A_0(x) = \exp(-x^2)$) for the ranges of $N = 5$ to $N = 20$ and $t = 0.0$ to $t = 0.5$. Figure 3 shows the data; the markers show the 2-norm differences versus N for times $t = 0.0$, 0.1, 0.2, 0.3, 0.4, and 0.5, respectively. The figure shows a remarkable consistency in the decay of the error as N increases (ϵ decreases). *In fact, a least squares fit of the data suggests that the SSE solutions converge to the true solutions at a $O(\epsilon)$ rate! Indeed, this rate of convergence persists for positive times, even including times after the first caustic.* Figure 4 (panels (a) and (b)) show a representative illustration of this phenomenology. Despite the presence of acute modulational instabilities, the small oscillations induced by the perturbation $q^{(\epsilon)}$ almost immediately disperse, and the large-scale structure of the solutions are virtually indistinguishable from each other.

6. Remark (Numerical methods). We used two distinct numerical methods for approximating solutions of (1.1). First, to compute multisoliton solutions (elements of the SSE), we adopted the numerical inverse scattering approach used previously [**22, 26**]; this method was also used to generate the solutions shown in Figure 2. On the other hand, the construction of accurate numerical methods for approximating the solution of (1.1) more generally when ϵ is small is a notoriously difficult problem. We refer the interested reader to the recent survey of Jin et al. [**15**] for a comprehensive discussion of the various challenges that occur even in the linear case; see also [**21**] and the references therein for additional details about numerical methods for (1.1) when ϵ is small. The results presented here were obtained by the use of a second-order implicit finite-difference algorithm developed in [**21**] and shown there to be a suitable numerical method for solving the semiclassical focusing nonlinear Schrödinger equation when ϵ is small.

4.3. "Nearby" Initial data & stability of the semiclassical limit. As a kind of control experiment, we next considered the family of functions[4]

$$q^{\epsilon}(x) = 0.3\epsilon\cos\left(\frac{x}{0.54\epsilon}\right)\exp(-x^2) \tag{4.6}$$

indexed by $\epsilon > 0$. We proposed q^{ϵ} as an ϵ-dependent perturbation of the initial data A_0, and we call the members of this family of functions "cosine perturbations." Figure 4 (panel (c)) shows the modification to the initial data using the cosine perturbation q^{ϵ} in equation (4.6) for $\epsilon = 1/(15\sqrt{\pi})$. From a rough, qualitative point of view, these data look quite similar to the modified data for the SSE in Figure 4 (panel (b)). Indeed, without the Gaussian template, the differences are practically indiscernible.

[3]High precision knowledge of the eigenvalue locations is necessary due to fact that the solution is obtained by solving a poorly conditioned linear system [**26**].

[4]The values 0.3 and 0.54 that appear in (4.6) are chosen so that the amplitude and frequency of the cosine perturbations q^{ϵ} track their counterparts $q^{(\epsilon)}$; compare Figure 4 (panels (b) and (c)).

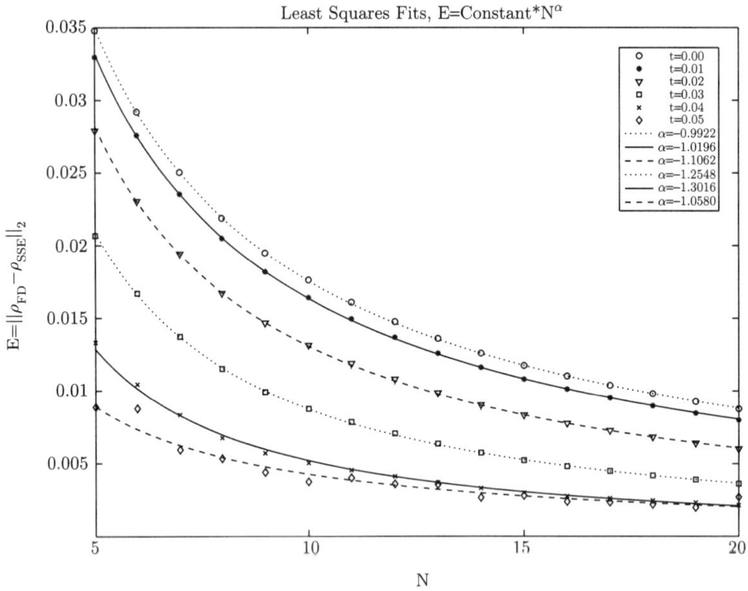

FIGURE 3. The 2-norm differences versus N for $t = 0.0$, 0.1, 0.2, 0.3, 0.4, and 0.5. The data points are the computed values of the 2-norm error $\mathrm{E} = \|\rho_{\mathrm{FD}} - \rho_{\mathrm{SSE}}\|_2$ for $-1 \le x \le 1$. The plotted curves are of the form $\mathrm{E} = C \cdot N^{\alpha}$ where the constants C, α are determined by a least squares fit to the data points. The values for α in the legend show a $O(1/N) = O(\epsilon)$ rate of convergence even for the times $t = 0.4$ and $t = 0.5$ which are after the breaking time. Figure taken from [**21**].

It is a simple matter to show that if $0 < \epsilon < 1$, then

$$\int_{-\infty}^{\infty} |q^{\epsilon}(x)|^2 \, \mathrm{d}x \le (0.09)\epsilon^2 \sqrt{\frac{\pi}{2}}, \tag{4.7}$$

and thus $\|q^{\epsilon}\|_{L^2(\mathbb{R})} = O(\epsilon)$ as $\epsilon \downarrow 0$. We note that the cosine perturbation q^{ϵ} is also even (like $q^{(\epsilon)}$) and oscillatory with frequency depending on ϵ (like $q^{(\epsilon)}$). Moreover, for each fixed $\epsilon > 0$, q^{ϵ} is the composition of real analytic functions hence is real analytic. Now, the basic question is whether or not the evolution from the family of initial data

$$A_0^{\epsilon} = A_0 + q^{\epsilon}$$

has structure as ϵ tends to zero that matches the remarkable structure, shown in Figure 3, of the solutions obtained from the data $A_0^{(\epsilon)} = A_0 + q^{(\epsilon)}$. The answer is a definitive no; see Figure 4 (panel (c)) for an illustrative visualization that demonstrates how the small oscillations in q^{ϵ} are immediately detected, and the modulational instabilities are triggered. We took this result as evidence, in support of the claim from [**21**], that the observed $O(\epsilon)$ decay in the case of the SSE is definitively a consequence of the special, integrable structure embedded in $A_0^{(\epsilon)}$ and is not due to its superficial features (namely, those present in A_0^{ϵ} as well). Said differently, these results lead us to suspect that the WKB approximations to the

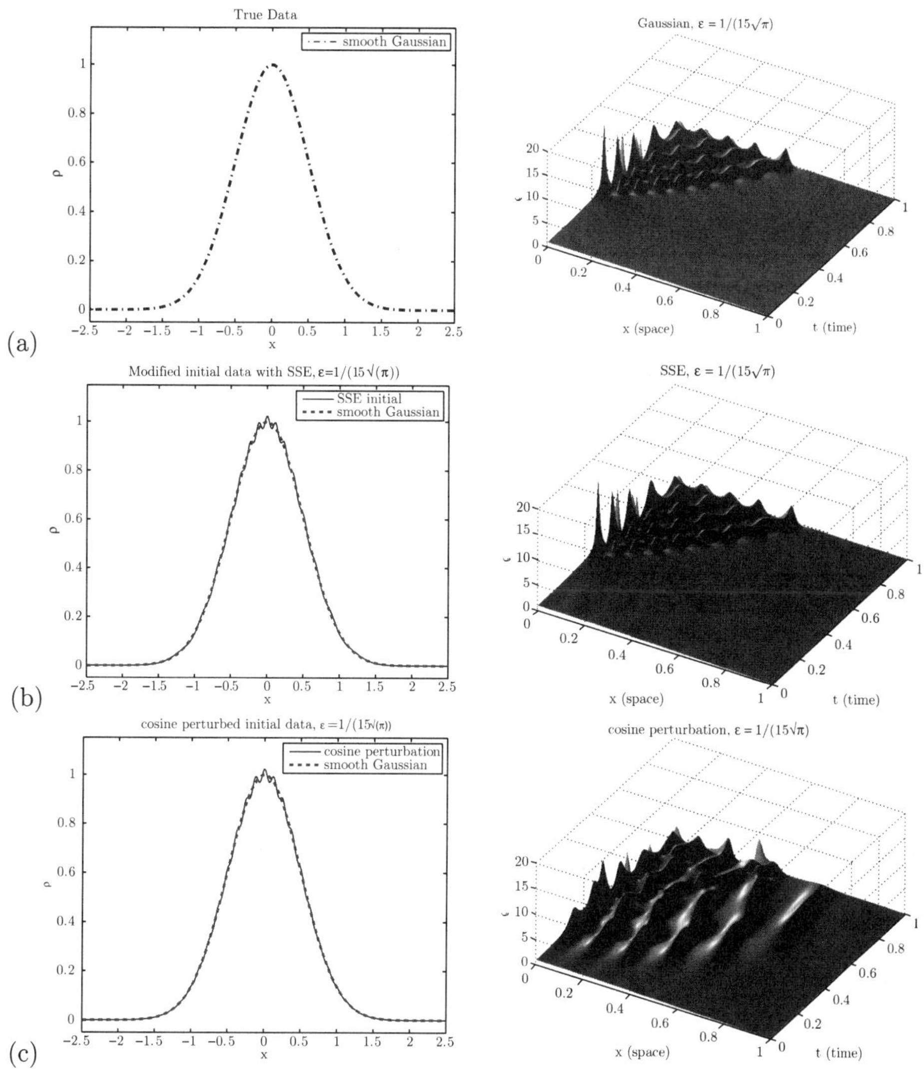

FIGURE 4. LEFT: Initial data (a) Pure Gaussian, $A_0(x) = \exp(-x^2)$; (b) SSE at $t = 0$, $A_0^{(\epsilon)}(x) = A_0(x) + q^{(\epsilon)}(x)$; (c) Cosine perturbation, $A_0^\epsilon(x) = A_0(x)+q^\epsilon(x)$. RIGHT: The time-space plot for (a) Gaussian initial data, $|u(x,t;\epsilon)|^2$; (b) the SSE modified initial data, $|u^{(\epsilon)}(x,t)|^2$; and (c) the square modulus of the solution evolved from the cosine perturbed initial data, $|u^\epsilon(x,t)|^2$. In this figure $\epsilon = 1/(15\sqrt{\pi}) \approx 0.01175$ for all three cases. Picture taken from [**20**].

eigenvalues of (2.1) used to construct $A_0^{(\epsilon)}$ are indeed remarkably good approximations to the true (unknown) eigenvalues associated with the Gaussian. In fact, we can measure the quality of the approximation directly, and we briefly summarize on these results in the next subsection.

4.4. SSEs in the spectral domain. We used the numerical protocol of Bronski [**5**, **6**] to compute the eigenvalues of (2.1) with a Gaussian potential,

$$A_0(x) = \mathrm{e}^{-x^2},$$

for a variety of values of ϵ. Our aim was to compute the rate of convergence of the approximate (WKB) eigenvalues used in the SSE formulation to the true ones. We did this to both to verify directly our expectation of the high-quality of the WKB approximation (we merely inferred this from our previously described simulations) and to support an eventual proof of the rate which we believe will be a necessary ingredient in any rigorous extension of the analysis of Kamvissis et al. [**16**] which accounts for the use of the WKB eigenvalues in the asymptotic analysis for (1.1). In summary, we numerically measured the difference directly in the spectral plane, and our experiments suggest convergence at a rate of $O(\epsilon^2)$ as $\epsilon \downarrow 0$. Details of this computation can be found in [**17**].

Briefly, we performed a least squares fit of the data in the form of

$$D^N = C \cdot N^\alpha, \tag{4.8}$$

for some constants C and α to be determined. Here D denotes the data, a collection of computed differences between the WKB eigenvalues used to construct the SSE and the (numerically computed) true eigenvalues. Now for each $N = 10, \cdots, 20$, we have data for $k = 0, \cdots, N-1$, and for $N = 21$ and 22, we have data for the largest ($k = 0$) eigenvalue. Therefore we have total 167 data points. Figure 5(a) shows the least square fit for all 167 data points. The triangles are the computed differences. For example, there are 10 eigenvalues for $N = 10$, and hence there are 10 computed differences. The solid line is the computed least square curve, which indicates the overall trend of decay of D^N versus N. The rate of decay is $\alpha = -2.00848$. Another way to monitor the rate of decay is to track the differences between the largest eigenvalues for each N. That is, we computed

$$D_{N,0} = |\lambda^{\mathrm{app}}_{N,0} - \lambda^{\mathrm{wkb}}_{N,0}|, \quad N = 10, \cdots, 22, \tag{4.9}$$

and again perform a least squares fit. Figure 5(b) shows the data and the least squares curve for this collection of 13 data points. In this case the rate of decay is given by $\alpha = -2.0135$. These experiments support a rate of decay that is of the order of the square of the small parameter.

4.5. Summary: A_0 versus $A_0^{(\epsilon)}$. To summarize, the numerical experiments described above strongly support the notion that the family of perturbations, $q^{(\epsilon)}$, used to generate the SSE associated to a particular bell-shaped initial condition for (1.1) are particularly well suited to the analysis of the semiclassical limit. Indeed, as shown in Figure 3, the extreme modulational instabilities of the focusing NLS equation (which get worse in the semiclassical limit) do not seem to be triggered at all by these perturbations, and we are led to conjecture that the semiclassical asymptotics for the solutions emanating from A_0 agree with those emanating from the limit of $A_0^{(\epsilon)}$. Presumably, this is due in large part to the fact the these perturbations are strongly interlaced with the integrable structure of the equation. Indeed, an additional experiment shows that qualitatively similar perturbations which have no special link to the integrable structure of the equation are detected immediately (Figure 4). Finally, a numerical computation confirms that the approximate (WKB) eigenvalues used in the construction of the modified data approach

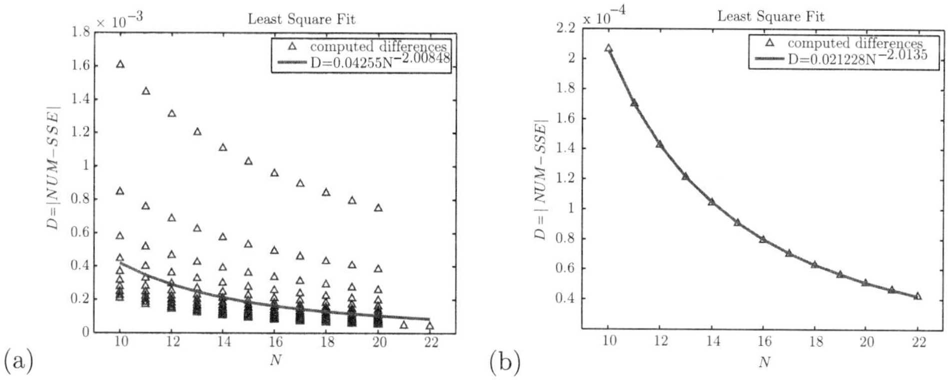

FIGURE 5. (a) Least square fit for 167 data points, for which $N = 10, \cdots, 20$ with $k = 0, \ldots, N-1$, and $N = 21, 22$ with $k = 0$. (b) Least square fit for $k = 0$, $N = 10, \ldots, 22$. Figure taken from [**17**].

the true eigenvalues at a rate which is quadratic in ϵ; this is another indication of the viability of the WKB approximation in the context of this problem.

5. Looking ahead

In this final section, we briefly indicate a handful of possible directions for future research building on this work with SSEs.

5.1. Estimates. A natural place to start would be to attempt to establish rigorous estimates for the deviation of the true eigenvalues associated with Klaus–Shaw data A_0 from their WKB approximations given by (2.5). Assuming one could accomplish this, the next step would be to incorporate the resulting control over the eigenvalues into the overall asymptotic analysis. One immediate question that arises is due to the fact that the initial step in the analytic framework of [**16**] is a coordinate change which exchanges a meromorphic RHP for a sectionally holomorphic one. This coordinate change is based on exact knowledge of the locations of the poles (eigenvalue locations). But, if these are known only approximately, the entire process is stymied at the first step. However, assuming these challenges can be overcome, the result would be the development of a rigorous theory for semiclassical dynamics for an entire family of suitable Klaus–Shaw data. This would be a great extension of the current state of the art. Indeed, the corresponding problem for general data with nontrivial phase, as in (1.3), remains almost entirely open.

5.2. Higher caustics. As discussed in §3.3, relatively little is known about the higher caustics (those nonlinear caustic curves that occur—or do not occur as the case may be—after the primary caustic). The results of Lyng & Miller suggests that the mathematical mechanisms for these features of the semiclassical dynamics may differ from caustic to caustic due to the introduction of new variational inequalities. In fact, we do not know the answers to seemingly basic questions, like the following.

- How many caustics are there? Infinitely many?
- Can their locations be predicted?
- Under what circumstances can we compute the long-time limit?

5.3. Stability. Finally, we mention one other possible avenue for further investigation stimulated by the experiments in [**20**]. Figure 4 suggests that the sensitivity properties of the semiclassical limit for (1.1) (even for real data in the analytic class) may be quite delicate. One interesting follow-up problem would be to begin investigate the sensitivity by studying, for example, the continuity properties of one or more of the important features of the semiclassical limit (e.g., the t coordinate of the primary caustic at $x = 0$) with respect to perturbations in the space F_K—the space of band-limited functions[5]—proposed by Clarke & Miller [**8**].

Acknowledgement

The author thanks Christopher Curtis, Anton Dzhamay, Willy Hereman, & Barbara Prinari for organizing a stimulating special session at the AMS Sectional Meeting in Boulder, CO in May 2013; this special session provided the impetus for writing this summary of recent results for SSEs. In addition, the author gratefully acknowledges the support of the National Science Foundation under grant number DMS-0845127.

References

[1] M. J. Ablowitz, B. Prinari, and A. D. Trubatch, *Discrete and continuous nonlinear Schrödinger systems*, London Mathematical Society Lecture Note Series, vol. 302, Cambridge University Press, Cambridge, 2004. MR2040621 (2005c:37117)

[2] Sergey M. Belov, *Breaking in the semiclassical solution of the focusing nonlinear Schrodinger equation*, ProQuest LLC, Ann Arbor, MI, 2008. Thesis (Ph.D.)–Duke University. MR2713942

[3] Marco Bertola and Alexander Tovbis, *Universality in the profile of the semiclassical limit solutions to the focusing nonlinear Schrödinger equation at the first breaking curve*, Int. Math. Res. Not. IMRN **11** (2010), 2119–2167, DOI 10.1093/imrn/rnp196. MR2646356 (2011e:35348)

[4] Marco Bertola and Alexander Tovbis, *Universality for the focusing nonlinear Schrödinger equation at the gradient catastrophe point: rational breathers and poles of the tritronquée solution to Painlevé I*, Comm. Pure Appl. Math. **66** (2013), no. 5, 678–752, DOI 10.1002/cpa.21445. MR3028484

[5] Jared C. Bronski, *Semiclassical eigenvalue distribution of the Zakharov-Shabat eigenvalue problem*, Phys. D **97** (1996), no. 4, 376–397, DOI 10.1016/0167-2789(95)00311-8. MR1412550 (98c:81062)

[6] J. C. Bronski, *Spectral instability of the semiclassical Zakharov-Shabat eigenvalue problem*, Phys. D **152/153** (2001), 163–170, DOI 10.1016/S0167-2789(01)00167-1. Advances in nonlinear mathematics and science. MR1837906

[7] Jared C. Bronski and David W. McLaughlin, *Semiclassical behavior in the NLS equation: optical shocks—focusing instabilities*, Singular limits of dispersive waves (Lyon, 1991), NATO Adv. Sci. Inst. Ser. B Phys., vol. 320, Plenum, New York, 1994, pp. 21–38. MR1321192

[8] S. R. Clarke and P. D. Miller, *On the semi-classical limit for the focusing nonlinear Schrödinger equation: sensitivity to analytic properties of the initial data*, R. Soc. Lond. Proc. Ser. A Math. Phys. Eng. Sci. **458** (2002), no. 2017, 135–156, DOI 10.1098/rspa.2001.0862. MR1879462 (2002k:35288)

[9] P. Deift and X. Zhou, *A steepest descent method for oscillatory Riemann-Hilbert problems. Asymptotics for the MKdV equation*, Ann. of Math. (2) **137** (1993), no. 2, 295–368, DOI 10.2307/2946540. MR1207209 (94d:35143)

[10] P. Deift, S. Venakides, and X. Zhou, *New results in small dispersion KdV by an extension of the steepest descent method for Riemann-Hilbert problems*, Internat. Math. Res. Notices **6** (1997), 286–299, DOI 10.1155/S1073792897000214. MR1440305 (98b:35155)

[5] F_K is the linear subspace of $L^2(\mathbb{R})$ consisting of functions ϕ whose Fourier transforms satisfy $\hat{\phi}(k) = 0$ for $|k| > K$.

[11] B. Dubrovin, T. Grava, and C. Klein, *On universality of critical behavior in the focusing nonlinear Schrödinger equation, elliptic umbilic catastrophe and the tritronquée solution to the Painlevé-I equation*, J. Nonlinear Sci. **19** (2009), no. 1, 57–94, DOI 10.1007/s00332-008-9025-y. MR2480731 (2010b:35428)
[12] Nicholas M. Ercolani, Shan Jin, C. David Levermore, and Warren D. MacEvoy Jr., *The zero-dispersion limit for the odd flows in the focusing Zakharov-Shabat hierarchy*, Int. Math. Res. Not. **47** (2003), 2529–2564, DOI 10.1155/S1073792803130851. MR2007539 (2005h:37158)
[13] Ludwig D. Faddeev and Leon A. Takhtajan, *Hamiltonian methods in the theory of solitons*, Reprint of the 1987 English edition, Classics in Mathematics, Springer, Berlin, 2007. Translated from the 1986 Russian original by Alexey G. Reyman. MR2348643 (2008f:37145)
[14] M. Gregory Forest and Jong Eao Lee, *Geometry and modulation theory for the periodic nonlinear Schrödinger equation*, Oscillation theory, computation, and methods of compensated compactness (Minneapolis, Minn., 1985), IMA Vol. Math. Appl., vol. 2, Springer, New York, 1986, pp. 35–69, DOI 10.1007/978-1-4613-8689-6_3. MR869821
[15] Shi Jin, Peter Markowich, and Christof Sparber, *Mathematical and computational methods for semiclassical Schrödinger equations*, Acta Numer. **20** (2011), 121–209, DOI 10.1017/S0962492911000031. MR2805153 (2012e:81070)
[16] Spyridon Kamvissis, Kenneth D. T.-R. McLaughlin, and Peter D. Miller, *Semiclassical soliton ensembles for the focusing nonlinear Schrödinger equation*, Annals of Mathematics Studies, vol. 154, Princeton University Press, Princeton, NJ, 2003. MR1999840 (2004h:37115)
[17] Y. Kim, L. Lee, and G. Lyng, *The Wentzel–Kramers–Brillouin approximation of semiclassical eigenvalues of the Zakharov–Shabat problem*, J. Math. Phys. **55** (2014), no. 083516, 18 pages.
[18] M. Klaus and J. K. Shaw, *Purely imaginary eigenvalues of Zakharov-Shabat systems*, Phys. Rev. E (3) **65** (2002), no. 3, 036607, 5, DOI 10.1103/PhysRevE.65.036607. MR1905252 (2003c:34144)
[19] Peter D. Lax and C. David Levermore, *The small dispersion limit of the Korteweg-de Vries equation. II*, Comm. Pure Appl. Math. **36** (1983), no. 5, 571–593, DOI 10.1002/cpa.3160360503. MR716197 (85g:35105b)
[20] Long Lee and Gregory D. Lyng, *A second look at the Gaussian semiclassical soliton ensemble for the focusing nonlinear Schrödinger equation*, Phys. Lett. A **377** (2013), no. 16-17, 1179–1188, DOI 10.1016/j.physleta.2013.03.009. MR3040497
[21] Long Lee, Gregory Lyng, and Irena Vankova, *The Gaussian semiclassical soliton ensemble and numerical methods for the focusing nonlinear Schrödinger equation*, Phys. D **241** (2012), no. 21, 1767–1781, DOI 10.1016/j.physd.2012.08.006. MR2982055
[22] Gregory D. Lyng and Peter D. Miller, *The N-soliton of the focusing nonlinear Schrödinger equation for N large*, Comm. Pure Appl. Math. **60** (2007), no. 7, 951–1026, DOI 10.1002/cpa.20162. MR2319053 (2008k:35454)
[23] Peter D. Miller, *Some remarks on a WKB method for the nonselfadjoint Zakharov-Shabat eigenvalue problem with analytic potentials and fast phase*, Phys. D **152/153** (2001), 145–162, DOI 10.1016/S0167-2789(01)00166-X. Advances in nonlinear mathematics and science. MR1837905 (2003d:34187)
[24] Peter D. Miller, *Asymptotics of semiclassical soliton ensembles: rigorous justification of the WKB approximation*, Int. Math. Res. Not. **8** (2002), 383–454, DOI 10.1155/S1073792802109020. MR1884077 (2003j:30060)
[25] Peter D. Miller, *Riemann-Hilbert problems with lots of discrete spectrum*, Integrable systems and random matrices, Contemp. Math., vol. 458, Amer. Math. Soc., Providence, RI, 2008, pp. 163–181, DOI 10.1090/conm/458/08935. MR2411906 (2010a:37147)
[26] Peter D. Miller and Spyridon Kamvissis, *On the semiclassical limit of the focusing nonlinear Schrödinger equation*, Phys. Lett. A **247** (1998), no. 1-2, 75–86, DOI 10.1016/S0375-9601(98)00565-9. MR1650432 (99e:81058)
[27] Junkichi Satsuma and Nobuo Yajima, *Initial value problems of one-dimensional self-modulation of nonlinear waves in dispersive media*, Progr. Theoret. Phys. Suppl. No. 55 (1974), 284–306. MR0463733 (57 #3675)
[28] Catherine Sulem and Pierre-Louis Sulem, *The nonlinear Schrödinger equation*, Applied Mathematical Sciences, vol. 139, Springer-Verlag, New York, 1999. Self-focusing and wave collapse. MR1696311 (2000f:35139)

[29] Alexander Tovbis and Stephanos Venakides, *Semiclassical limit of the scattering transform for the focusing nonlinear Schrödinger equation*, Int. Math. Res. Not. IMRN **10** (2012), 2212–2271, DOI 10.1093/imrn/rnr092. MR2923165

[30] Alexander Tovbis and Stephanos Venakides, *The eigenvalue problem for the focusing nonlinear Schrödinger equation: new solvable cases*, Phys. D **146** (2000), no. 1-4, 150–164, DOI 10.1016/S0167-2789(00)00126-3. MR1787410 (2001i:35266)

[31] Alexander Tovbis, Stephanos Venakides, and Xin Zhou, *Semiclassical focusing nonlinear Schrödinger equation I: inverse scattering map and its evolution for radiative initial data*, Int. Math. Res. Not. IMRN **22** (2007), Art. ID rnm094, 54, DOI 10.1093/imrn/rnm094. MR2376209 (2009a:37146)

[32] Alexander Tovbis, Stephanos Venakides, and Xin Zhou, *On the long-time limit of semiclassical (zero dispersion limit) solutions of the focusing nonlinear Schrödinger equation: pure radiation case*, Comm. Pure Appl. Math. **59** (2006), no. 10, 1379–1432, DOI 10.1002/cpa.20142. MR2248894 (2007e:37041)

[33] Alexander Tovbis, Stephanos Venakides, and Xin Zhou, *On semiclassical (zero dispersion limit) solutions of the focusing nonlinear Schrödinger equation*, Comm. Pure Appl. Math. **57** (2004), no. 7, 877–985, DOI 10.1002/cpa.20024. MR2044068 (2005c:35269)

[34] V. E. Zakharov and A. B. Shabat, *Exact theory of two-dimensional self-focusing and one-dimensional self-modulation of waves in nonlinear media* (Russian, with English summary), Ž. Èksper. Teoret. Fiz. **61** (1971), no. 1, 118–134; English transl., Soviet Physics JETP **34** (1972), no. 1, 62–69. MR0406174 (53 #9966)

Department of Mathematics, University of Wyoming, Laramie, Wyoming 82071-3036
E-mail address: glyng@uwyo.edu

Contemporary Mathematics
Volume **635**, 2015
http://dx.doi.org/10.1090/conm/635/12714

Relative-Periodic Elastic Collisions of Water Waves

Jon Wilkening

ABSTRACT. We compute time-periodic and relative-periodic solutions of the free-surface Euler equations that take the form of overtaking collisions of unidirectional solitary waves of different amplitude on a periodic domain. As a starting guess, we superpose two Stokes waves offset by half the spatial period. Using an overdetermined shooting method, the background radiation generated by collisions of the Stokes waves is tuned to be identical before and after each collision. In some cases, the radiation is effectively eliminated in this procedure, yielding smooth soliton-like solutions that interact elastically forever. We find examples in which the larger wave subsumes the smaller wave each time they collide, and others in which the trailing wave bumps into the leading wave, transferring energy without fully merging. Similarities notwithstanding, these solutions are found quantitatively to lie outside of the Korteweg-de Vries regime. We conclude that quasi-periodic elastic collisions are not unique to integrable model water wave equations when the domain is periodic.

1. Introduction

A striking feature of multiple-soliton solutions of integrable model equations such as the Korteweg-de Vries equation, the Benjamin-Ono equation, and the nonlinear Schrödinger equation is that they interact elastically, leading to time-periodic, relative-periodic, or quasi-periodic dynamics. By contrast, the interaction of solitary waves for the free-surface Euler equations is inelastic. However, it has been observed many times in the literature [**10**, **13**, **15**, **35**, **37**, **38**, **44**, **53**] that the residual radiation after a collision of such solitary waves can be remarkably small. In the present paper we explore the possibility of finding nearby time-periodic and relative-periodic solutions of the Euler equations using a collision of unidirectional Stokes waves as a starting guess. Such solutions demonstrate that recurrent elastic collisions of solitary waves in the spatially periodic case do not necessarily indicate that the underlying system is integrable.

A relative-periodic solution is one that returns to a spatial phase shift of its initial condition at a later time. This only makes sense on a periodic domain, where the waves collide repeatedly at regular intervals in both time and space, with the locations of the collisions drifting steadily in time. They are special cases (with $N = 2$) of quasi-periodic solutions, which have the form $u(x,t) = U(\vec{\kappa}x + \vec{\omega}t + \vec{\alpha})$

2010 *Mathematics Subject Classification.* Primary 76B15, 37G15, 70K43, 65K10, 37K10.

This research was supported in part by the Director, Office of Science, Computational and Technology Research, U.S. Department of Energy under Contract No. DE-AC02-05CH11231, and by the National Science Foundation through grant DMS-0955078.

with U an N-periodic continuous function, i.e. $U \in C(\mathbb{T}^N)$, and $\vec{\kappa}$, $\vec{\omega}$, $\vec{\alpha} \in \mathbb{R}^N$. Throughout the manuscript, we will use the phrase "solitary waves" in a broad sense to describe waves that, most of the time, remain well-separated from one another and propagate with nearly constant speed and shape. "Stokes waves" will refer to periodic progressive solutions of the free-surface Euler equations of permanent form, or waves that began at $t = 0$ as a linear superposition of such traveling waves. They comprise a special class of solitary waves. "Solitons" will refer specifically to superpositions of sech^2 solutions of the KdV equation on the whole line, while "cnoidal solutions" will refer to their spatially periodic, multi-phase counterparts; see §5 for elaboration.

It was found in [**49**] that decreasing the fluid depth causes standing waves to transition from large-scale symmetric sloshing behavior in deep water to pairs of counter-propagating solitary waves that collide repeatedly in shallow water. In the present work, we consider unidirectional waves of different amplitude that collide due to taller waves moving faster than shorter ones. We present two examples of solutions of this type: one where the resulting dynamics is fully time-periodic; and one where it is relative-periodic, returning to a spatial phase shift of the initial condition at a later time. Both examples exhibit behavior typical of collisions of KdV solitons. In the first, one wave is significantly larger than the other, and completely subsumes it during the interaction. In the second, the waves have similar amplitude, with the trailing wave bumping into the leading wave and transferring energy without fully merging.

Despite these similarities, the amplitude of the waves in our examples are too large for the assumptions in the derivation of the KdV equation to hold. In particular, the larger wave in the first example is more than half the fluid depth in height, and there is significant vertical motion of the fluid when the waves pass by. A detailed comparison of the Euler and KdV equations for waves with these properties is carried out in §5. A review of the literature on water wave collisions and the accuracy of the KdV model of water waves is also given in that section.

Rather than compute such solutions by increasing the amplitude from the linearized regime via numerical continuation, as was done for counter-propagating waves in [**49**], we use collisions of right-moving Stokes waves as starting guesses. The goal is to minimally "tune" the background radiation generated by the Stokes collisions so that the amount coming out of each collision is identical to what went into it. In the first example of §4, we find that the tuned background radiation takes the form of a train of traveling waves of smaller wavelength moving to the right more slowly than either solitary wave. By contrast, in the counter-propagating case studied in [**49**], it consists of an array of smaller-wavelength standing waves oscillating rapidly relative to the time between collisions of the primary waves. In the second example of §4, the background radiation is essentially absent, which is to say that the optimized solution is free from high-frequency, low-amplitude disturbances in the trough, and closely resembles a relative-periodic cnoidal solution of KdV. We call the collisions in this solution "elastic" as they repeat forever, unchanged up to spatial translation, and there are no features to distinguish radiation from the waves themselves. This process of tuning parameters to minimize or eliminate small-amplitude oscillations in the wave troughs is reminiscent of Vanden-Broeck's work [**45**] in which oscillations at infinity could be eliminated from solitary capillary-gravity waves by choosing the amplitude appropriately.

To search for relative periodic solutions, we use a variant of the overdetermined shooting method developed by the author and collaborators in previous work to study several related problems: time-periodic solutions of the Benjamin-Ono equation [**3**, **5**] and the vortex sheet with surface tension [**4**, **6**]; Hopf bifurcation and stability transitions in mode-locked lasers [**50**]; cyclic steady-states in rolling treaded tires [**21**]; self-similarity (or lack thereof) at the crests of large-amplitude standing water waves [**47**]; harmonic resonance and spontaneous nucleation of new branches of standing water waves at critical depths [**49**]; and three-dimensional standing water waves [**42**]. The three approaches developed in these papers are the adjoint continuation method [**5**, **50**], a Newton-Krylov shooting method [**21**], and a trust region shooting method [**49**] based on the Levenberg-Marquardt algorithm [**39**]. We adopt the latter method here to exploit an opportunity to consolidate the work in computing the Dirichlet-Neumann operator for many columns of the Jacobian simultaneously, in parallel.

One computational novelty of this work is that we search directly for large-amplitude solutions of a nonlinear two-point boundary value problem, without using numerical continuation to get there. This is generally difficult. However, in the present case, numerical continuation is also difficult due to non-smooth bifurcation "curves" riddled with Cantor-like gaps [**41**], and the long simulation times that occur between collisions in the unidirectional case. Our shooting method has proven robust enough to succeed in finding time-periodic solutions, when they exist, with a poor starting guess. False positives are avoided by resolving the solutions spectrally to machine accuracy and overconstraining the minimization problem. Much of the challenge is in determining the form of the initial condition and the objective function to avoid wandering off in the wrong direction and falling into a nonzero local minimum before locking onto a nearby relative-periodic solution.

2. Equations of motion

The equations of motion of a free surface $\eta(x,t)$ evolving over an ideal fluid with velocity potential $\phi(x,y,t)$ may be written [**17**, **18**, **27**, **46**]

$$\begin{aligned} \eta_t &= \phi_y - \eta_x \phi_x, \\ \varphi_t &= P\left[\phi_y \eta_t - \frac{1}{2}\phi_x^2 - \frac{1}{2}\phi_y^2 - g\eta\right], \end{aligned} \tag{2.1}$$

where subscripts denote partial derivatives, $\varphi(x,t) = \phi(x,\eta(x,t),t)$ is the restriction of ϕ to the free surface, $g = 1$ is the acceleration of gravity, $\rho = 1$ is the fluid density, and P is the projection

$$Pf = f - \frac{1}{2\pi}\int_0^{2\pi} f(x)\,dx, \tag{2.2}$$

where we assume a 2π-periodic domain. The velocity components $u = \phi_x$ and $v = \phi_y$ at the free surface can be computed from φ via

$$\begin{pmatrix} \phi_x \\ \phi_y \end{pmatrix} = \frac{1}{1+\eta'(x)^2}\begin{pmatrix} 1 & -\eta'(x) \\ \eta'(x) & 1 \end{pmatrix}\begin{pmatrix} \varphi'(x) \\ \mathcal{G}\varphi(x) \end{pmatrix}, \tag{2.3}$$

where a prime denotes a derivative and $\mathcal{G}$ is the Dirichlet-Neumann operator [**16**]

$$\mathcal{G}\varphi(x) = \sqrt{1+\eta'(x)^2}\,\frac{\partial\phi}{\partial n}(x+i\eta(x)) = \phi_y - \eta_x\phi_x \tag{2.4}$$

for the Laplace equation, with periodic boundary conditions in x, Dirichlet conditions ($\phi = \varphi$) on the upper boundary, and Neumann conditions ($\phi_y = 0$) on the lower boundary, assumed flat. We have suppressed t in the notation since time is frozen in the Laplace equation. We compute $\mathcal{G}\varphi$ using a boundary integral collocation method [**7**, **8**, **30**, **34**, **36**] and advance the solution in time using an 8th order Runge-Kutta scheme [**23**] with 36th order filtering [**25**]. See [**49**] for details.

3. Computation of relative-periodic solutions

Traveling waves have the symmetry that

$$\eta(x,0) \text{ is even}, \qquad \varphi(x,0) \text{ is odd}. \tag{3.1}$$

This remains true if x is replaced by $x - \pi$. As a starting guess for a new class of time-periodic and relative-periodic solutions, we have in mind superposing two traveling waves, one centered at $x = 0$ and the other at $x = \pi$. Doing so will preserve the property (3.1), but the waves will now interact rather than remain pure traveling waves. A solution will be called *relative periodic* if there exists a time T and phase shift θ such that

$$\eta(x, t+T) = \eta(x-\theta, t), \qquad \varphi(x, t+T) = \varphi(x-\theta, t) \tag{3.2}$$

for all t and x. Time-periodicity is obtained as a special case, with $\theta \in 2\pi\mathbb{Z}$. We can save a factor of 2 in computational work by imposing the alternative condition

$$\eta(x+\theta/2, T/2) \text{ is even}, \qquad \varphi(x+\theta/2, T/2) \text{ is odd}. \tag{3.3}$$

From this, it follows that

$$\begin{aligned}
\eta(x+\theta/2, T/2) &= \eta(-x+\theta/2, T/2) = \eta(x-\theta/2, -T/2), \\
\varphi(x+\theta/2, T/2) &= -\varphi(-x+\theta/2, T/2) = \varphi(x-\theta/2, -T/2).
\end{aligned}$$

But then both sides of each equation in (3.2) agree at time $t = -T/2$. Thus, (3.2) holds for all time.

In the context of traveling-standing waves in deep water [**48**], it is natural to define T as twice the value above, replacing all factors of $T/2$ by $T/4$. That way a pure standing wave returns to its original configuration in time T instead of shifting in space by π in time T. In the present work, we consider pairs of solitary waves moving to the right at different speeds, so it is more natural to define T as the first (rather than the second) time there exists a θ such that (3.2) holds.

3.1. Objective function.

We adapt the overdetermined shooting method of [**47**, **49**] to compute solutions of (3.1)–(3.3). This method employs the Levenberg-Marquardt method [**39**] with delayed Jacobian updates [**49**] to solve the nonlinear least squares problem described below.

For (3.1), we build the symmetry into the initial conditions over which the shooting method is allowed to search: we choose an integer n and consider initial conditions of the form

$$\hat{\eta}_k(0) = c_{2|k|-1}, \qquad \hat{\varphi}_k(0) = \pm i c_{2|k|}, \tag{3.4}$$

where $k \in \{\pm 1, \pm 2, \ldots, \pm\frac{n}{2}\}$ and $\hat{\eta}_k(t)$, $\hat{\varphi}_k(t)$ are the Fourier modes of $\eta(x,t)$, $\varphi(x,t)$. The numbers $c_1, \ldots, c_n$ are assumed real and all other Fourier modes (except $\hat{\eta}_0$) are zero. We set $\hat{\eta}_0$ to the fluid depth so that $y = 0$ is a symmetry line corresponding to the bottom wall. This is convenient for computing the Dirichlet-Neumann operator [**49**]. In the formula for $\hat{\varphi}_k$, the minus sign is taken if $k < 0$ so that $\hat{\varphi}_{-k} = \overline{\hat{\varphi}_k}$. We also solve for the period,

$$T = c_{n+1}. \tag{3.5}$$

The phase shift θ is taken as a prescribed parameter here. Alternatively, in a study of traveling-standing waves [**48**], the author defines a traveling parameter β and varies $\theta = c_{n+2}$ as part of the algorithm to obtain the desired value of β. This parameter β is less meaningful for solitary wave collisions in shallow water, so we use θ itself as the traveling parameter in the present study. We also need to specify the amplitude of the wave. This can be done in various ways, e.g. by specifying the value of the energy,

$$E = \frac{1}{2\pi}\int_0^{2\pi} \tfrac{1}{2}\varphi\mathcal{G}\varphi + \tfrac{1}{2}g\eta^2\,dx,$$

by constraining a Fourier mode such as $\hat{\eta}_1(0)$, or by specifying the initial height of the wave at $x = 0$:

$$\eta(0,0) = \hat{\eta}_0 + \sum_{k=1}^{n/2} 2c_{2k-1}.$$

Thus, to enforce (3.3), we can minimize the objective function

$$f(c) = \frac{1}{2}r(c)^T r(c), \tag{3.6}$$

where

$$\begin{aligned} &r_1 = \big(\text{ choose one:} \quad E - a \quad , \quad \hat{\eta}_1(0) - a \quad , \quad \eta(0,0) - a \;\big), \\ &r_{2j} = \operatorname{Im}\{e^{ij\theta/2}\hat{\eta}_j(T/2)\}, \qquad r_{2j+1} = \operatorname{Re}\{e^{ij\theta/2}\hat{\varphi}_j(T/2)\}, \qquad (1 \le j \le M/2). \end{aligned} \tag{3.7}$$

Here a is the desired value of the chosen amplitude parameter. Alternatively, we can impose (3.2) directly by minimizing

$$\tilde{f} = \frac{1}{2}r_1^2 + \frac{1}{4\pi}\int_0^{2\pi}\Big(\big[\eta(x,T) - \eta(x-\theta,0)\big]^2 + \big[\varphi(x,T) - \varphi(x-\theta,0)\big]^2\Big)\,dx, \tag{3.8}$$

which also takes the form $\frac{1}{2}r^T r$ if we define r_1 as above and

$$\begin{aligned} r_{4j-2} + ir_{4j-1} &= \sqrt{2}\left[\hat{\eta}_j(T) - e^{-ij\theta}\hat{\eta}_j(0)\right], \\ r_{4j} + ir_{4j+1} &= \sqrt{2}\left[\hat{\varphi}_j(T) - e^{-ij\theta}\hat{\varphi}_j(0)\right], \end{aligned} \qquad (1 \le j \le M/2). \tag{3.9}$$

Note that f measures deviation from evenness and oddness of $\eta(x + \theta/2, T/2)$ and $\varphi(x + \theta/2, T/2)$, respectively, while $\tilde{f}$ measures deviation of $\eta(x + \theta, T)$ and $\varphi(x + \theta, T)$ from their initial states. In the first example of §4, we minimize $\tilde{f}$ directly, while in the second we minimize f and check that $\tilde{f}$ is also small, as a means of validation. The number of equations, $m = M + 1$ for f and $m = 2M + 1$ for $\tilde{f}$, is generally larger than the number of unknowns, $n + 1$, due to zero-padding of the initial conditions. This adds robustness to the shooting method and causes all Fourier modes varied by the algorithm, namely those in (3.4), to be well-resolved on the mesh.

3.2. Computation of the Jacobian. To compute the kth column of the Jacobian $J = \nabla_c r$, which is needed by the Levenberg-Marquardt method, we solve the linearized equations along with the nonlinear ones:

$$\frac{\partial}{\partial t}\begin{pmatrix} q \\ \dot{q} \end{pmatrix} = \begin{pmatrix} F(q) \\ DF(q)\dot{q} \end{pmatrix}, \qquad \begin{array}{l} q(0) = q_0 = (\eta_0, \varphi_0), \\ \dot{q}(0) = \dot{q}_0 = \partial q_0/\partial c_k. \end{array} \tag{3.10}$$

Here $q = (\eta, \varphi)$, $\dot{q} = (\dot{\eta}, \dot{\varphi})$, $F(q)$ is given in (2.1), DF is its derivative (see [**49**] for explicit formulas), and a dot represents a variational derivative with respect to perturbation of the initial conditions, not a time derivative. To compute $\partial r_i/\partial c_k$ for $i \geq 2$ and $k \leq n$, one simply puts a dot over each Fourier mode on the right-hand side of (3.7) or (3.9), including $\dot{\hat{\eta}}_j(0)$ and $\dot{\hat{\varphi}}_j(0)$ in (3.9). If $k = n+1$, then $c_k = T$ and

$$\frac{\partial r_{2j}}{\partial T} = \operatorname{Im}\{e^{ij\theta/2}(1/2)\partial_t\hat{\eta}_j(T/2)\}, \qquad \frac{\partial(r_{4j} + ir_{4j+1})}{\partial T} = \sqrt{2}\big[\partial_t\hat{\varphi}_j(T)\big]$$

in (3.7) and (3.9), respectively, with similar formulas for $\partial(r_{4j-2} + ir_{4j-1})/\partial T$ and $\partial r_{2j+1}/\partial T$. The three possibilities for r_1 are handled as follows:

$$\text{case 1:} \quad \frac{\partial r_1}{\partial c_k} = \dot{E} = \frac{1}{2\pi}\int_0^{2\pi} [\dot{\varphi}\eta_t - \dot{\eta}\varphi_t]_{t=0}\, dx, \quad (k \leq n), \qquad \frac{\partial r_1}{\partial c_{n+1}} = 0,$$

$$\text{case 2:} \quad \frac{\partial r_1}{\partial c_k} = \dot{\eta}_1^{\wedge}(0) = \delta_{k,1}, \quad (k \leq n+1),$$

$$\text{case 3:} \quad \frac{\partial r_1}{\partial c_k} = \dot{\eta}(0,0) = 2\delta_{k,\text{odd}}, \quad (k \leq n), \qquad \frac{\partial r_1}{\partial c_{n+1}} = 0,$$

where $\delta_{k,j}$ and $\delta_{k,\text{odd}}$ equal 1 if $k = j$ or k is odd, respectively, and equal zero otherwise. The vectors $\dot{q}$ in (3.10) are computed in batches, each initialized with a different initial perturbation, to consolidate the work in computing the Dirichlet-Neumann operator during each timestep. See [**48, 49**] for details.

4. Numerical results

As mentioned in the introduction, our idea is to use collisions of unidirectional Stokes (i.e. traveling) waves as starting guesses to find time-periodic and relative periodic solutions of the Euler equations. We begin by computing traveling waves of varying wave height and record their periods. This is easily done in the framework of §3. We set $\theta = \pi/64$ (or any other small number) and minimize $\tilde{f}$ in (3.8). The resulting "period" T will give the wave speed via $c = \theta/T$. Below we report $T = 2\pi c$, i.e. T is rescaled as if θ were 2π. We control the amplitude by specifying $\hat{\eta}_1(0)$, which is the second option listed in §3 for defining the first component r_1 of the residual. A more conventional approach for computing traveling waves is to substitute $\eta(x - ct)$, $\varphi(x - ct)$ into (2.1) and solve the resulting stationary problem (or an equivalent integral equation) by Newton's method [**11, 12, 37**]. Note that the wave speed c here is unrelated to the vector c of unknowns in (3.4).

With traveling waves in hand, out next goal is to collide two of them and search for a nearby time-periodic solution, with $\theta = 0$. As shown in Figure 1, varying $\hat{\eta}_1(0)$ from 0 to 7.4×10^{-4} causes the period of a Stokes wave with wavelength $\lambda = 2\pi$ and mean fluid depth $h = 0.05$ to decrease from $T_O = 28.1110$ to $T_A = 22.9739$, and the wave height (vertical crest-to-trough distance) to increase from 0 to 0.02892. Solution C is the closest among the Stokes waves we computed to satisfying $5T_C = 6T_A$, where $p = 5$ is the smallest integer satisfying $\frac{p+1}{p}T_A < T_O$. We then combine

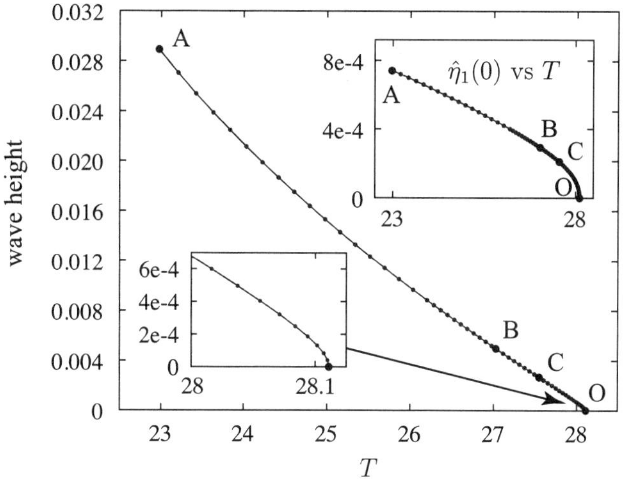

FIGURE 1. Plots of wave height and first Fourier mode versus period for Stokes waves with wavelength 2π and fluid depth $h = 0.05$. The temporal periods are $6T_A = 137.843 \approx 137.738 = 5T_C$.

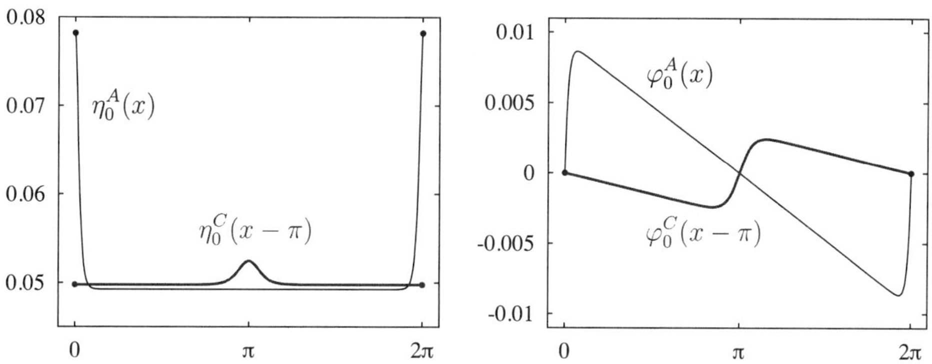

FIGURE 2. Wave profile and velocity potential of Stokes waves labeled A and C in Fig. 1, plotted over one spatial period at $t = 0$.

solution A with a spatial phase shift of solution C at $t = 0$. The resulting initial conditions are

$$\begin{aligned} \eta_0^{A+C}(x) &= h + \big[\eta_0^A(x) - h\big] + \big[\eta_0^C(x-\pi) - h\big], \\ \varphi_0^{A+C}(x) &= \varphi_0^A(x) + \varphi_0^C(x-\pi), \end{aligned} \tag{4.1}$$

where $h = 0.05$ is the mean fluid depth. Plots of $\eta_0^A(x)$, $\eta_0^C(x-\pi)$, $\varphi_0^A(x)$ and $\varphi_0^C(x-\pi)$ are shown in Figure 2. If the waves did not interact, the combined solution would be time-periodic (to the extent that $5T_C = 6T_A$, i.e. to about 0.076%). But the waves do interact. In addition to the complicated interaction that occurs when they collide, each slows the other down between collisions by introducing a negative gradient in the velocity potential between its own wave crests. Indeed, as shown in

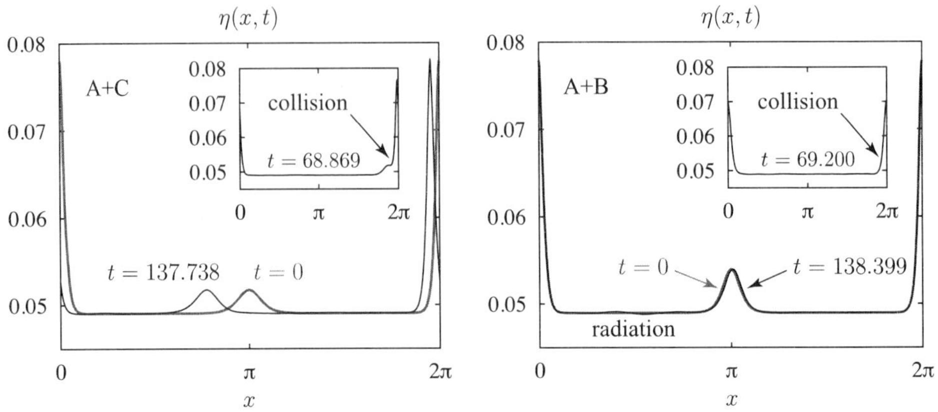

FIGURE 3. Collision of two right-moving Stokes waves that nearly return to their initial configuration after the interaction. (left) Solutions A and C were combined via (4.1) and evolved through one collision to $t = 137.738$. (right) Through trial and error, we adjusted the amplitude of the smaller Stokes wave and the simulation time to obtain a nearly time-periodic solution.

the right panel of Figure 2, the velocity potential increases rapidly across a right-moving solitary wave and decreases elsewhere to achieve spatial periodicity. The decreasing velocity potential induces a background flow opposite to the direction of travel of the other wave. In the left panel of Figure 3, we see that the net effect is that neither of the superposed waves has returned to its starting position at $t = 5T_C$, and the smaller wave has experienced a greater net decrease in speed. However, as shown in the right panel, by adjusting the amplitude of the smaller wave (replacing solution C by B) and increasing T slightly to 138.399, we are able to hand-tune the Stokes waves to achieve $\tilde{f} \approx 5.5 \times 10^{-8}$, where θ is set to zero in (3.8). Note that as t varies from 0 to $T/10$ in the left panel of Figure 4, the small wave advances by π units to the right while the large wave advances by 1.2π units. The waves collide at $t = T/2$. This generates a small amount of radiation, which can be seen at $t = T$ in the right panel of Figure 3. Some radiation behind the large wave is present for all $t > 0$, as shown in Figure 5.

Before minimizing $\tilde{f}$, we advance the two Stokes waves to the time of the first collision, $t = T/2$. At this time, the larger solitary wave has traversed the domain 3 times and the smaller one 2.5 times, so their peaks lie on top of each other at $x = 0$. The reason to do this is that when the waves merge, the combined wave is shorter, wider, and smoother than at any other time during the evolution. Quantitatively, the Fourier modes of $\hat{\eta}_k(t)$ and $\hat{\varphi}_k(t)$ decay below 10^{-15} for $k \geq 600$ at $t = 0$, and $k \geq 200$ when $t = T/2$. Thus, the number of columns needed in the Jacobian is reduced by a factor of 3, and the problem becomes more overdetermined, hence more robust. For the calculation of a time-periodic solution, we let $t = 0$ correspond to this merged state, which affects the time labels when comparing Figures 4 and 6. As a final initialization step, we project onto the space of initial conditions satisfying (3.4) by zeroing out the imaginary parts of $\hat{\eta}_k(0)$ and the real parts of $\hat{\varphi}_k(0)$, which

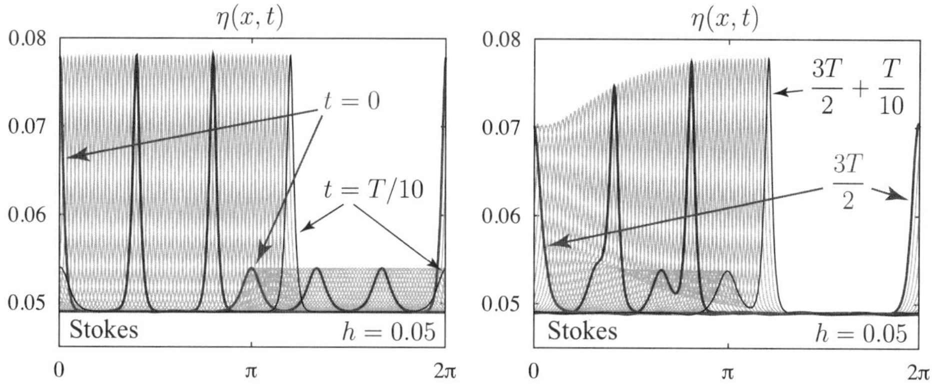

FIGURE 4. Evolution of two Stokes waves that collide repeatedly, at times $t \approx T/2 + kT$, $k \geq 0$. (left) Traveling solutions A and B in Figure 1 were initialized with wave crests at $x = 0$ and $x = \pi$, respectively. The combined solution is approximately time-periodic, with period $T = 138.399$. (right) The same solution, at later times, starting with the second collision ($t = 3T/2$).

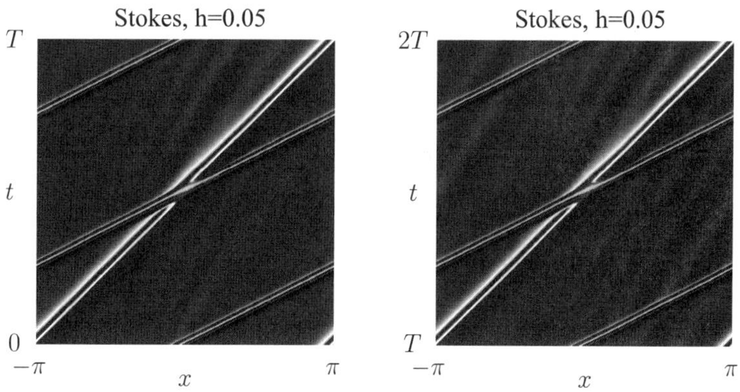

FIGURE 5. A different view of the solutions in Figure 4 shows the generation of background waves. Shown here are the functions $\eta(x + 8\pi t/T, t)$, which give the dynamics in a frame moving to the right fast enough to traverse the domain four times in time T. In a stationary frame, the smaller and larger solitary waves traverse the domain 5 and 6 times, respectively.

are already small. Surprisingly, this improves the time-periodicity of the initial guess in (3.8) to $\tilde{f} = 2.3 \times 10^{-8}$.

We emphasize that our goal is to find *any* nearby time-periodic solution by adjusting the initial conditions to drive $\tilde{f}$ to zero. Energy will be conserved as the solution evolves from a given initial condition, but is only imposed as a constraint (in the form of a penalty) on the search for initial conditions when the first component of the residual in (3.7) is set to $r_1 = E - a$. In the present calculation, we use $r_1 = \eta(0, 0) - a$ instead. In the second example, presented below, we will constrain

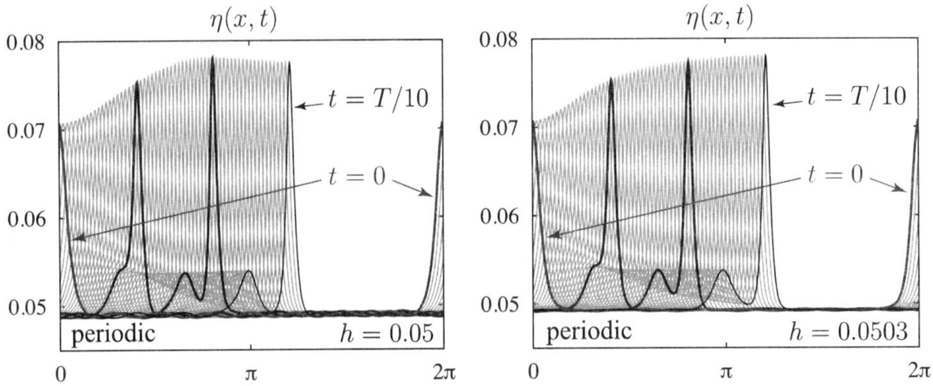

FIGURE 6. Time-periodic solutions near the Stokes waves of Figure 4. (left) $h = 0.05$, $\eta(0,0) = 0.0707148$, $T = 138.387$, $\tilde{f} = 4.26 \times 10^{-27}$. (right) $h = 0.0503$, $\eta(0,0) = 0.0707637$, $T = 138.396$, $\tilde{f} = 1.27 \times 10^{-26}$. The background radiation was minimized by hand in the right panel by varying h and $\eta(0,0)$.

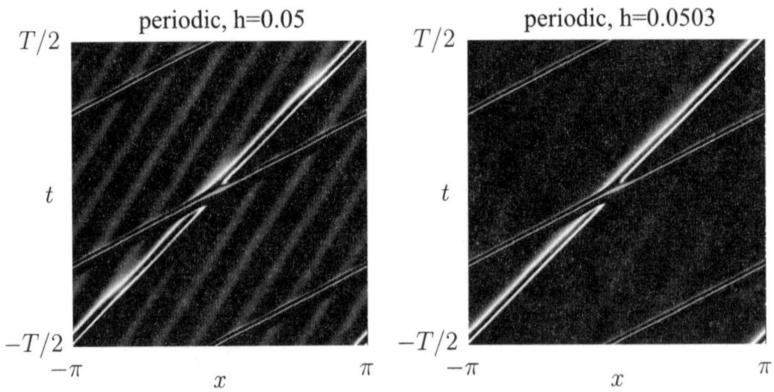

FIGURE 7. Same as Figure 5, but showing the time-periodic solutions of Figure 6 instead of the Stokes waves of Figure 4. The Stokes waves generate new background radiation with each collision while the time-periodic solutions are synchronized with the background waves to avoid generating additional disturbances.

energy. In either case, projecting onto the space of initial conditions satisfying (3.4) can cause r_1 to increase, but it will decrease to zero in the course of minimizing $\tilde{f}$. This projection is essential for the symmetry arguments of §3.1 to work.

We minimize $\tilde{f}$ subject to the constraint $\eta(0,0) = 0.0707148$, the third case described in §3 for specifying the amplitude. This causes $\tilde{f}$ to decrease from 2.3×10^{-8} to 4.26×10^{-27} using $M = 1200$ grid points and $N = 1200$ time-steps (to $t = T$). The results are shown in the left panel of Figures 6 and 7. The main difference between the Stokes collision and this nearby time-periodic solution is that the Stokes waves generate additional background ripples each time they collide while the time-periodic solution contains an equilibrium background wave configuration that does not grow in amplitude after the collision. While the background waves in the counter-propagating case (studied in [**49**]) look like small-amplitude standing waves, these background waves travel to the right, but slower than either solitary wave. After computing the $h = 0.05$ time-periodic solution, we computed 10 other solutions with nearby values of h and $\eta(0,0)$ to try to decrease the amplitude of the background radiation. The best solution we found (in the sense of small background radiation) is shown in the right panel of Figures 6 and 7, with $h = 0.0503$ and $\eta(0,0) = 0.0707637$. The amplitude of the background waves of this solution are comparable to that of the Stokes waves after two collisions.

Our second example is a relative periodic solution in which the initial Stokes waves (the starting guess) are B and C in Figure 1 instead of A and C. As before, solution C is shifted by π when the waves are combined initially, just as in (4.1). Because the amplitude of the larger wave has been reduced, the difference in wave speeds is smaller, and it takes much longer for the waves to collide. If the waves did not interact, we would have

$$c_{B,0} = 0.23246089, \quad c_{C,0} = 0.22808499, \quad T_0 = \frac{2\pi}{c_{B,0} - c_{A,0}} = 1435.86, \tag{4.2}$$

where wave B crosses the domain 53.1230 times in time T_0 while wave C crosses the domain 52.1230 times. The subscript 0 indicates that the waves are assumed not to interact. Since the waves do interact, we have to evolve the solution numerically to obtain useful estimates of T and θ. We arbitrarily rounded T_0 to 1436 and made plots of the solution at times $\Delta t = T_0/1200$. We found that η is nearly even (up to a spatial phase shift) for the first time at $463\Delta t = 554.057$. This was our initial guess for $T/2$. The phase shift required to make $\eta(x+\theta/2, T/2)$ approximately even and $\varphi(x+\theta/2, T/2)$ approximately odd was found by graphically solving $\varphi(x, T/2) = 0$. This gives the initial guess $\theta/2 = 2.54258$. This choice of T and θ (with η^{B+C} and φ^{B+C} as initial conditions) yields $f = 2.0 \times 10^{-11}$ and $\tilde{f} = 1.5 \times 10^{-10}$. We then minimize f holding E and θ constant, which gives $f = 2.1 \times 10^{-29}$ and $\tilde{f} = 3.0 \times 10^{-26}$. We note that $\tilde{f}$ is computed over $[0, T]$, twice the time over which the solution was optimized by minimizing f, and provides independent confirmation of the accuracy of the solution and the symmetry arguments of §3.1.

The results are plotted in Figure 8. We omit a plot of the initial guess (the collision of Stokes waves) as it is indistinguishable from the minimized solution. In fact, the relative change in the wave profile and velocity potential is about 0.35 percent,

$$\left(\frac{\|\eta_{\text{Stokes}} - \eta_{\text{periodic}}\|^2 + \|\varphi_{\text{Stokes}} - \varphi_{\text{periodic}}\|^2}{\|\eta_{\text{Stokes}} - h\|^2 + \|\varphi_{\text{Stokes}}\|^2} \right)^{1/2} \leq 0.0035, \tag{4.3}$$

and $T/2$ changes even less, from 554.057 (Stokes) to 554.053 (periodic). By construction, E and $\theta/2$ do not change at all. It was not necessary to evolve the Stokes waves to $T/2$, shift space by $\theta/2$, zero out Fourier modes that violate the symmetry condition (3.1), and reset $t = 0$ to correspond to this new initial state. Doing so increases the decay rate of the Fourier modes (slope of $\ln|\hat{\eta}_k|$ vs k) by a factor of 1.24 in this example, compared to 3.36 in the previous example, where it is definitely worthwhile.

The large change from $T_0/2 = 717.93$ to $T/2 = 554.053$ is due to nonlinear interaction of the waves. There are two main factors contributing to this change in

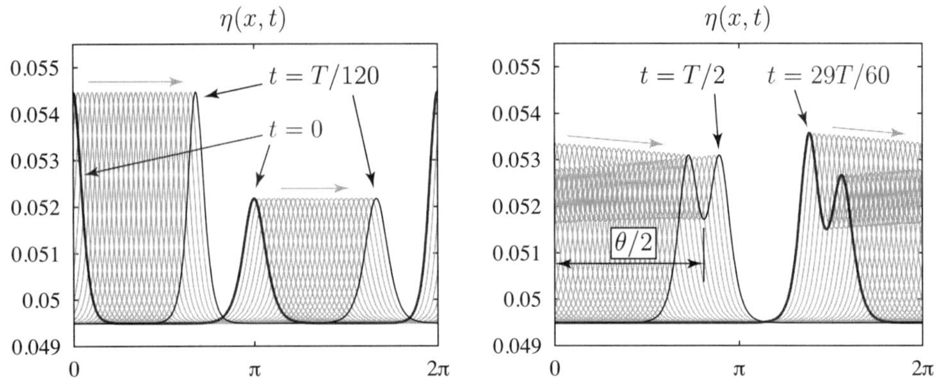

FIGURE 8. A relative-periodic solution found using a superposition of the Stokes waves labeled B and C in Figure 1 as a starting guess. Unlike the previous case, the waves do not fully merge at $t = T/2$.

period. The first is that the waves do not fully combine when they collide. Instead, the trailing wave runs into the leading wave, passing on much of its amplitude and speed. The peaks remain separated by a distance of $d = 0.52462$ at $t = T/2$, the transition point where the waves have the same amplitude. Thus, the peak separation changes by $\pi - d$ rather than π in half a period. The second effect is that the larger wave slows down the smaller wave more than the smaller slows the larger. Recall from Fig. 2 that each wave induces a negative potential gradient across the other wave that generates a background flow opposing its direction of travel. Quantitatively, when the waves are well separated, we find that the taller and smaller waves travel at speeds $c_B = 0.231077 = 0.994049c_{B,0}$ and $c_C = 0.226153 = 0.991531c_{C,0}$, respectively. The relative speed is then $(c_B - c_C) = 1.12526(c_{B,0} - c_{C,0})$. Thus,

$$\frac{\pi - d}{c_B - c_C} < \frac{T}{2} < \frac{\pi - d}{c_{B,0} - c_{C,0}} < \frac{T_0}{2} = \frac{\pi}{c_{B,0} - c_{C,0}}, \tag{4.4}$$

with numerical values $531.5 < 554.1 < 598.0 < 717.9$. This means that both effects together have overestimated the correction needed to obtain T from T_0. This is because the relative speed slows down as the waves approach each other, which is expected since the amplitude of the trailing wave decreases and the amplitude of the leading wave increases in this interaction regime. Indeed, the average speed of the waves is

$$\overline{c_B} = \frac{\theta/2 - d/2}{T/2} = 0.993388c_{B,0}, \qquad \overline{c_C} = \frac{\theta/2 + d/2 - \pi}{T/2} = 0.991737c_{C,0}, \tag{4.5}$$

which are slightly smaller and larger, respectively, than their speeds when well separated. Note that $T/2$ in (4.4) may be written $T/2 = (\pi - d)/(\overline{c_B} - \overline{c_C})$. We used $\theta/2 = 2.54258 + 40\pi$ in (4.5) to account for the 20 times the waves cross the domain $(0, 2\pi)$ in time $T/2$ in addition to the offset shown in Figure 8.

5. Comparison with KdV

In the previous section, we observed two types of overtaking collisions for the water wave: one in which the larger wave completely subsumes the smaller wave for a time, and one where the two waves remain distinct throughout the interaction. Similar behavior has of course been observed for the Korteweg-de Vries equation, which was part of our motivation for looking for such solutions. Lax [**32**] classified overtaking collisions of two KdV solitons as bimodal, mixed, or unimodal. Unimodal and bimodal waves are analogous to the ones we computed above, while mixed mode collisions have the larger wave mostly subsume the smaller wave at the beginning and end of the interaction, but with a two-peaked structure re-emerging midway through the interaction. Lax showed that if $1 < c_1/c_2 < A = (3+\sqrt{5})/2$, the collision is bimodal; if $c_1/c_2 > 3$, the collision is unimodal; and if $A < c_1/c_2 < 3$, the collision is mixed. Here c_1 and c_2 are the wave speeds of the trailing and leading waves, respectively, at $t = -\infty$. Leveque [**33**] has studied the asymptotic dynamics of the interaction of two solitons of nearly equal amplitude. Zou and Su [**53**] performed a computational study of overtaking water wave collisions, compared the results to KdV interactions, and found that the water wave collisions ceased to be elastic at third order. Craig *et. al.* [**15**] also found that solitary water waves collide inelastically. This does not conflict with our results since we optimize the initial conditions to make the collision elastic. Head on collisions have been studied numerically by Su and Mirie [**38, 44**], experimentally by Maxworthy [**35**], and by a mixture of analysis and computation by Craig *et. al.* [**15**].

Validation of KdV as a model of water waves has also been studied extensively. A formal derivation may be found in Ablowitz and Segur [**1**]. Rigorous justification has been given by Bona, Colin and Lannes [**9**], building on earlier work by Craig [**14**] as well as Schneider and Wayne [**43**]. According to [**9**], some gaps still exist in the theory in the spatially periodic case. Experimental studies of the validity of KdV for describing surface waves have been performed by Zabusky and Galvin [**52**] as well as Hammack and Segur [**24**]. Recently, Ostrovsky and Stepanyants [**40**] have compared internal solitary waves in laboratory experiments to the predictions of various model equations, including KdV, and give a comprehensive overview of the literature on this subject [**40**].

Our objective in this section is to determine quantitatively whether the solutions of the water wave equations that we computed in §4 are well-approximated by the KdV equation. Following Ablowitz and Segur [**1**], we introduce a small parameter ε and dimensionless variables

$$\hat{y} = \frac{y}{h}, \qquad \hat{x} = \sqrt{\varepsilon}\frac{x}{h}, \qquad \hat{t} = \sqrt{\frac{\varepsilon g}{h}}t, \qquad \hat{\eta} = \frac{\eta}{\varepsilon h}, \qquad \hat{\phi} = \frac{\phi}{\sqrt{\varepsilon g h^3}},$$

where h is the fluid depth. We assume the bottom boundary is at $y = -h$ rather than 0 in this derivation, so that $\hat{y}$ runs from -1 to $\varepsilon\hat{\eta}$. The Laplacian becomes $\Delta_\varepsilon = h^{-2}\big(\varepsilon\partial_{\hat{x}}^2 + \partial_{\hat{y}}^2\big)$, which allows for $\hat{\phi} = \hat{\phi}_0 + \varepsilon\hat{\phi}_1 + \varepsilon^2\hat{\phi}_2 + \cdots$ to be computed order by order, with leading terms satisfying

$$\hat{\phi}_{0,\hat{y}} = 0, \qquad \hat{\phi}_1 = -\frac{1}{2}(1+\hat{y})^2\hat{\phi}_{0,\hat{x}\hat{x}}, \qquad \hat{\phi}_2 = \frac{1}{24}(1+\hat{y})^4\hat{\phi}_{0,\hat{x}\hat{x}\hat{x}\hat{x}}.$$

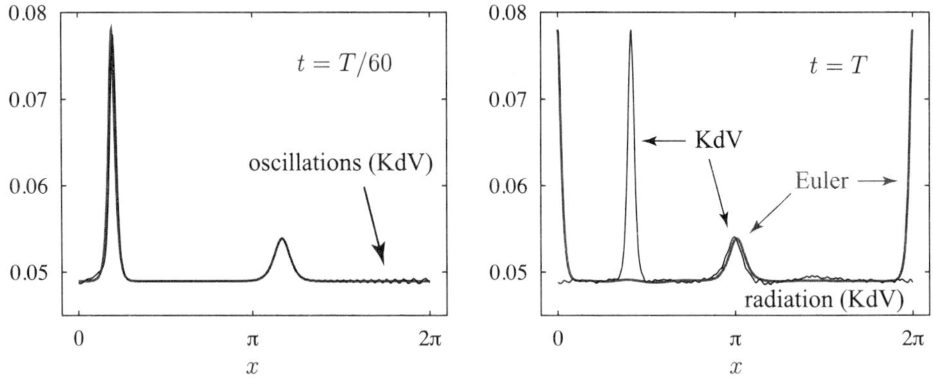

FIGURE 9. Comparison of the solutions of the KdV and Euler equations, initialized identically with the superposition of Stokes waves labeled A and B in Figure 1. The final time T is set to 138.399, as in Fig. 3, when the Euler solution nearly returns to its initial configuration after a single overtaking collision.

Here we used $\Delta\phi = 0$ and $\phi_y(x,-h) = 0$. Note that $\hat{\phi}_0$ is independent of $\hat{y}$, and agrees with the velocity potential ϕ on the bottom boundary, up to rescaling:

$$\hat{\phi}_0(\hat{x},\hat{t}) = (\varepsilon g h^3)^{-1/2}\phi(x,-h,t).$$

From the equations of motion, $\eta_t = \phi_y - \eta_x\phi_x$ and $\phi_t + \frac{1}{2}\phi_x^2 + \frac{1}{2}\phi_y^2 + g\eta = 0$, one finds that

$$\hat{\eta}_{\hat{t}} + \hat{u}_{\hat{x}} = \varepsilon\left\{\tfrac{1}{6}\hat{u}_{\hat{x}\hat{x}\hat{x}} - (\hat{\eta}\hat{u})_{\hat{x}}\right\} + O(\varepsilon^2),$$
$$\hat{u}_{\hat{t}} + \hat{\eta}_{\hat{x}} = \varepsilon\left\{\tfrac{1}{2}\hat{u}_{\hat{x}\hat{x}\hat{t}} - \tfrac{1}{2}\partial_{\hat{x}}(\hat{u})^2\right\} + O(\varepsilon^2),$$

where $\hat{u}(\hat{x},\hat{t}) = \partial_{\hat{x}}\hat{\phi}_0(\hat{x},\hat{t})$. Expanding $\hat{\eta} = \hat{\eta}_0 + \varepsilon\hat{\eta}_1 + \cdots$, $\hat{u} = \hat{u}_0 + \varepsilon\hat{u}_1 + \cdots$, we find that

$$\hat{\eta}_0 = f(\hat{x}-\hat{t};\tau) + g(\hat{x}+\hat{t};\tau), \qquad 2f_\tau + 3ff_r + (1/3)f_{rrr} = 0,$$
$$\hat{u}_0 = f(\hat{x}-\hat{t};\tau) - g(\hat{x}+\hat{t};\tau), \qquad -2g_\tau + 3gg_l + (1/3)g_{lll} = 0,$$

where we have introduced characteristic coordinates $r = \hat{x} - \hat{t}$, $l = \hat{x} + \hat{t}$ as well as a slow time scale $\tau = \varepsilon\hat{t}$ to eliminate secular growth in the solution with respect to r and l at first order in ε; see [**1**] for details. The notational conflict of $g(l,\tau)$ with the acceleration of gravity, g, is standard, and will not pose difficulty below.

In our case, the waves travel to the right, so we may set $g(l,\tau) = 0$ in the formulas above. Returning to dimensional variables, we then have

$$\eta(x,t) = h\varepsilon f\left(\sqrt{\varepsilon}\left(\frac{x}{h} - \sqrt{\frac{g}{h}}t\right), \sqrt{\frac{g}{h}}\varepsilon^{3/2}t\right),$$

which satisfies

$$\eta_t + \alpha\eta_x + \frac{3\sqrt{gh}}{2h}\eta\eta_x + \frac{1}{6}\sqrt{gh}\,h^2\eta_{xxx} = 0, \tag{5.1}$$

where $\alpha = \sqrt{gh}$. Note that ε drops out. For comparison with the results of §4, we will add h to η and set $\alpha = -\frac{1}{2}\sqrt{gh}$ instead. In Figure 9, we compare the solution of (5.1), with initial condition $\eta(x,0) = \eta_0^{A+B}(x)$, defined similarly to $\eta_0^{A+C}(x)$ in (4.1), to the solution of the free-surface Euler equations shown in Figs. 3

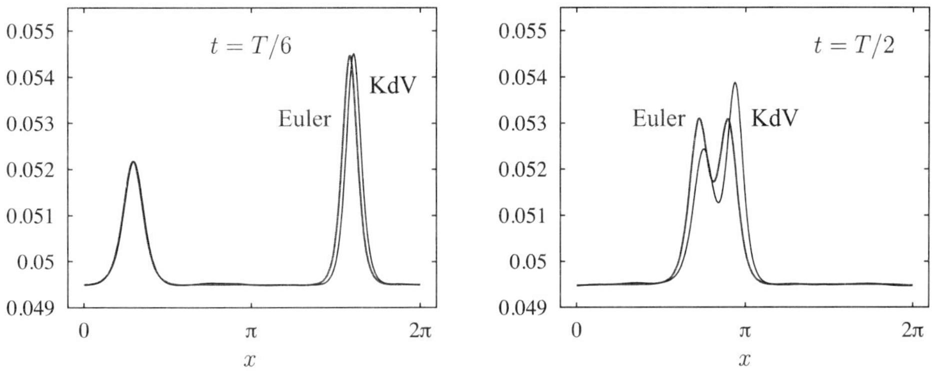

FIGURE 10. Comparison of the solutions of the KdV and Euler equations, both initialized with the superposition of Stokes waves labeled B and C in Figure 1. $T = 1108.11$ here.

and 4. Shortly after the waves are set in motion, the KdV solution develops high-frequency oscillations behind the larger peak that travel left and quickly fill up the computational domain with radiation. The solution of the Euler equations remains much smoother. The large peak of the KdV solution also travels 3.4% faster, on average, than the corresponding peak of the Euler solution, so that at $t = 138.399$, when the taller Euler wave has traversed the domain 6 times, the taller KdV wave has traversed it 6.206 times. For our purposes, these discrepancies are much too large for KdV to be a useful model, and we conclude that the first example in §4 is well outside of the KdV regime.

In this comparison, timestepping the KdV equation was done with the 8 stage, 5th order implicit/explicit Runge-Kutta method of Kennedy and Carpenter [**28**]. Spatial derivatives were computed spectrally using the 36th order filter of Hou and Li [**25**]. We found that 2048 spatial grid points and 96000 timesteps was sufficient to reduce the error at $t = 138.399$ below 5×10^{-6} near the larger peak and below 6×10^{-7} elsewhere, based on comparing the solution to one with 3072 grid points and 192000 timesteps. Our solutions of the Euler equations are much more accurate since there are no second or third spatial derivative terms present to make the equations stiff. Thus, we can use 8th or 15th order explicit timestepping rather than 5th order implicit/explicit timestepping. Monitoring energy conservation and performing mesh refinement studies suggests that we obtain 13–14 digits of accuracy in the solutions of the Euler equations, at which point roundoff error prevents further improvement in double-precision arithmetic.

In Figure 10, we repeat this computation using initial conditions corresponding to the superposition of Stokes waves $\eta_0^{B+C}(x)$, which was used as a starting guess for the second example of §4. This time the KdV solution does not develop visible high-frequency radiation in the wave troughs, and the solutions of KdV and Euler remain close to each other for much longer. However, the interaction time for a collision also increases, from $T = 138.399$ in the first example to $T = 1108.11$ here. In Fig. 10, by $t = T/6$, the taller KdV and Euler waves have visibly separated from each other, and by $t = T/2$, when the Euler waves have reached their minimum approach distance, the KdV solution is well ahead of the Euler solution. Thus, while there is good qualitative agreement between the KdV and Euler solutions, they do not agree quantitatively over the time interval of interest. From this point of view, the second example of §4 also lies outside of the KdV regime.

An alternative measure of the agreement between KdV and Euler is to compare the solutions from §4 with nearby relative-periodic solutions of KdV. In other words, we wish to quantify how much the initial conditions and period have to be perturbed to convert a relative-periodic solution of the Euler equations into one for the KdV equations. Since we used a superposition of Stokes waves for the initial guess to find time-periodic and relative-periodic solutions of the Euler equations, we will use a similar superposition (of cnoidal waves) for KdV. The vertical crest-to-trough heights of the three Stokes waves considered in §4 are

$$H_A = 0.028918699, \qquad H_B = 0.004973240, \qquad H_C = 0.002683648. \tag{5.2}$$

Well-known [**20**, **29**] periodic traveling wave solutions of (5.1) are given by

$$\eta(x,t) = h - H + \frac{H}{m}\left(1 - \frac{E(m)}{K(m)}\right) + H\,\mathrm{cn}^2\left(2K(m)\frac{x-ct}{\lambda}\middle|\, m\right),$$

$$\lambda = \sqrt{\frac{16mh^3}{3H}}\,K(m), \qquad c = \left[1 - \frac{H}{2h} + \frac{H}{mh}\left(1 - \frac{3E(m)}{2K(m)}\right)\right]\sqrt{gh},$$

where we added h to η to match the change in α from $\sqrt{gh}$ to $-\frac{1}{2}\sqrt{gh}$ in (5.1). Here $K(m)$ and $E(m)$ are the complete elliptic integrals of the first and second kind, respectively, and $\mathrm{cn}(z|m)$ is one of the Jacobi elliptic functions [**20**, **22**]. In our case $\lambda = 2\pi$, $g = 1$ and $h = 0.05$. For each H in (5.2), we solve the λ equation for m using Mathematica [**51**], and then evaluate $\eta(x,0)$ on a uniform grid that is fine enough that its Fourier coefficients decay below machine roundoff. The values of $m' = 1 - m$ are

$$m'_A = 1.81924\times 10^{-35}, \qquad m'_B = 1.98689\times 10^{-14}, \qquad m'_C = 1.79643\times 10^{-10}.$$

This approach requires extended precision arithmetic to compute m and evaluate η, but the running time takes only a few seconds on a typical laptop. A periodized version of the simpler sech^2 formula could be used for the first two waves, but decays too slowly for wave C to be spatially periodic to roundoff accuracy. Once these cnoidal waves have been computed, we superpose their initial conditions to form η_0^{A+B} and η_0^{B+C}, just as in §4. It is well-known that a superposition of N cnoidal waves retain this form when evolved via KdV, with N amplitude and N phase parameters governed by an ODE describing pole dynamics in the complex plane [**2**, **19**, **31**]. In the $N = 2$ case, the solutions are relative-periodic.

Figures 11 and 12 compare the time-periodic and relative-periodic solutions of the Euler equations, computed in §4, to these cnoidal solutions of KdV. Since the periods are different, only the initial conditions are compared. In the larger-amplitude example, shown in Fig. 11, the Euler solution is not as flat in the wave trough as the cnoidal solution due to an additional oscillatory component (the "tuned" radiation). From the difference plot in the right panel, we see that the crest-to-trough amplitude of these higher frequency oscillations is roughly 6×10^{-4}, or 2.1% of the wave height H_A. The Euler solution is time-periodic with period $T_{\text{Euler}} = 138.387$ while the cnoidal solution is relative-periodic, returning to a spatial phase shift of its initial condition at $T_{\text{KdV}} = 113.079$, which differs from T_{Euler} by 18%. In the smaller-amplitude example, shown in Fig. 12, both solutions have smooth, flat wave troughs, and it is difficult to distinguish one from the other in the left panel. The crest-to-trough amplitude of the difference in the right panel is roughly 5.5×10^{-5}, or 1.1% of H_B. The relative change in period is $(T_{\text{Euler}} - T_{\text{KdV}})/T_{\text{Euler}} = 3.6\%$. While the left panels of Figures 11 and 12 show close agreement between relative-periodic solutions of the Euler and KdV equations at $t = 0$, it should be noted that the wave amplitudes of the cnoidal solutions were chosen to minimize the discrepancy in these figures. The change in period by 18% and 3.6%, respectively, is perhaps a better measure of agreement.

A final comparison of the two equations is made in Fig. 13, where we evolve the Euler equations with the KdV initial conditions. This requires an initial condition for $\varphi(x) = \phi(x, \eta(x))$, where we have suppressed t in the notation for this discussion since it is held fixed at 0. Based on the derivation presented above, we first solve $\phi_x(x, 0) = \sqrt{g/h}[\eta(x) - h]$ for ϕ on the bottom boundary. We then use the approximation

$$\varphi(x) \approx \phi(x, 0) - \frac{\eta(x)^2}{2}\phi_{xx}(x, 0) + \frac{\eta(x)^4}{24}\phi_{xxx}(x, 0)$$

to evaluate ϕ on the free surface. In the left panel of Figure 13, the larger wave grows and overturns before $t = T/400$ when evolved under the Euler equations, instead of traveling to the right when evolved via KdV. To handle wave breaking, we switched to an angle-arclength formulation of the free-surface Euler equations [**4**, **26**]. In the small-amplitude example in the right panel, the Euler solution develops visible radiation and falls slightly behind the KdV solution, although the phases are closer at $T/2$ than the result of evolving the Stokes waves under KdV in Figure 10. We also tried evaluating

$$\phi(x, y) = \sqrt{g/h}\sum_{k=1}^{\infty} 2k^{-1}\hat{\eta}_k \sin(kx)\cosh(ky)$$

at $y = \eta(x)$ to obtain the initial condition for $\varphi(x)$, where $\hat{\eta}_k$ are the Fourier modes of $\eta(x)$ at $t = 0$, but the results were worse for the large-amplitude example — the wave breaks more rapidly — and were visually indistinguishable in the small-amplitude example from the results plotted in Fig. 13.

In summary, the large-amplitude time-periodic solution of the Euler equations found in §4 is well outside of the KdV regime by any measure, and the small-amplitude relative-periodic solution is closer, but not close enough to achieve quantitative agreement over the entire time interval of interest.

6. Conclusion

We have demonstrated that the small amount of background radiation produced when two Stokes waves interact in shallow water can often be tuned to obtain time-periodic and relative-periodic solutions of the free-surface Euler equations. Just as for the Korteweg-de Vries equation, the waves can fully merge when they collide or remain well-separated. However, the comparison is only qualitative as the waves are too large to be well-approximated by KdV theory.

In future work, we will study the stability of these solutions using Floquet theory. Preliminary results suggest that the first example considered above is unstable to harmonic perturbations while the second example is stable. In the stable case, an interesting open question is whether the Stokes waves used as a starting guess for the minimization algorithm, which have the same energy as the relative-periodic solution found, might remain close to it forever, executing almost-periodic oscillations around it. Presumably θ would need to be varied slightly for this to be

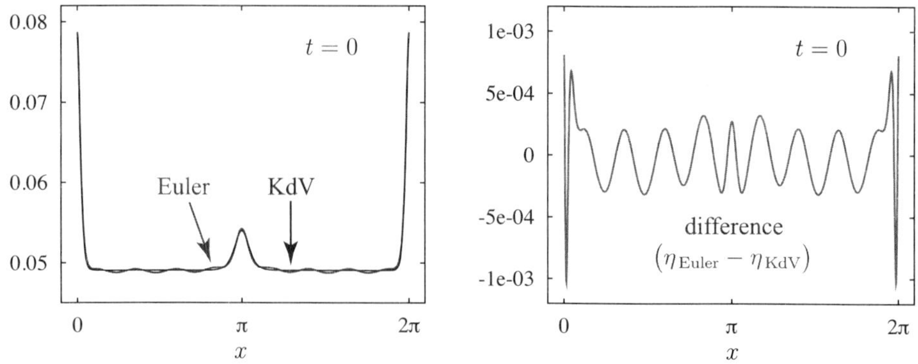

FIGURE 11. Comparison of time-periodic solution found in §4 to nearby relative-periodic two phase cnoidal solution of KdV. The periods are $T = 138.387$ and 113.079, respectively.

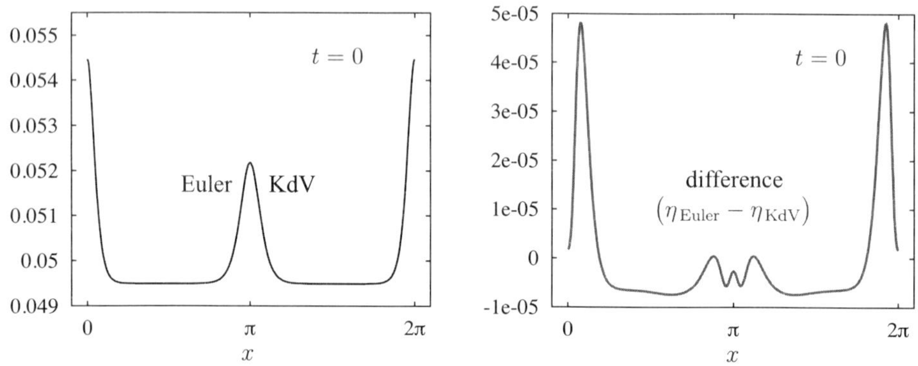

FIGURE 12. Comparison of relative-periodic solution found in §4 to nearby relative-periodic two phase cnoidal solution of KdV. The periods are $T = 1108.11$ and 1068.73, respectively.

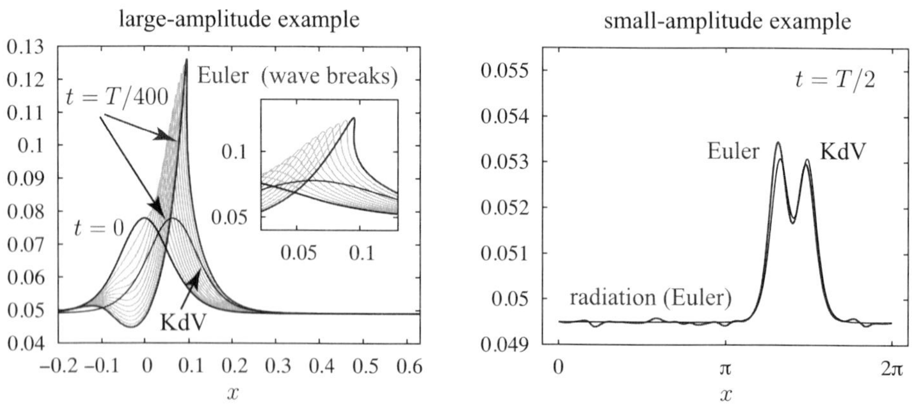

FIGURE 13. Comparison of KdV and Euler solutions, both initialized with a 2-phase cnoidal wave with peaks matching the heights of the Stokes waves labeled A and B (left) or B and C (right) in Fig. 1. Here $T = 138.387$ (left) and $T = 1068.73$ (right).

true, since θ is a free parameter that we selected by hand to obtain a small value of $\tilde{f}$ for the initial guess. Another open question is whether there are analogues for the Euler equations of N-phase quasi-periodic solutions of the KdV equation with $N \geq 3$. We are confident that the methods of this paper could be used to construct degenerate cases of $N \geq 3$ solitary water waves colliding elastically in a time-periodic or relative-periodic fashion, along the lines of what was done for the Benjamin-Ono equation in [**3**]. Computing more general quasi-periodic dynamics of the form $\eta(x,t) = H(\vec{\kappa}x + \vec{\omega}t + \vec{\alpha})$, $\varphi(x,t) = \Phi(\vec{\kappa}x + \vec{\omega}t + \vec{\alpha})$ with $H, \Phi \in C(\mathbb{T}^N)$ and $\vec{\kappa}$, $\vec{\omega}$, $\vec{\alpha} \in \mathbb{R}^N$ seems possible in principle using a more sophisticated shooting method to determine H, Φ and $\vec{\omega}$. Existence of such solutions for the Euler equations would show that non-integrable equations can also support recurrent elastic collisions even if they cannot be represented as N-phase superpositions of elliptic functions.

References

[1] Mark J. Ablowitz and Harvey Segur, *Solitons and the inverse scattering transform*, SIAM Studies in Applied Mathematics, vol. 4, Society for Industrial and Applied Mathematics (SIAM), Philadelphia, Pa., 1981. MR642018 (84a:35251)

[2] H. Airault, H. P. McKean, and J. Moser, *Rational and elliptic solutions of the Korteweg-de Vries equation and a related many-body problem*, Comm. Pure Appl. Math. **30** (1977), no. 1, 95–148. MR0649926 (58 #31214)

[3] David M. Ambrose and Jon Wilkening, *Global paths of time-periodic solutions of the Benjamin-Ono equation connecting pairs of traveling waves*, Commun. Appl. Math. Comput. Sci. **4** (2009), 177–215, DOI 10.2140/camcos.2009.4.177. MR2565874 (2010j:35437)

[4] David M. Ambrose and Jon Wilkening, *Computation of symmetric, time-periodic solutions of the vortex sheet with surface tension*, Proc. Nat. Acad. Sci. **107** (2010), no. 8, 3361–3366.

[5] David M. Ambrose and Jon Wilkening, *Computation of time-periodic solutions of the Benjamin-Ono equation*, J. Nonlinear Sci. **20** (2010), no. 3, 277–308, DOI 10.1007/s00332-009-9058-x. MR2639896 (2011d:65146)

[6] David M. Ambrose and Jon Wilkening, *Dependence of time-periodic votex sheets with surface tension on mean vortex sheet strength*, Procedia IUTAM **11** (2014), 15–22.

[7] Gregory R. Baker, Daniel I. Meiron, and Steven A. Orszag, *Generalized vortex methods for free-surface flow problems*, J. Fluid Mech. **123** (1982), 477–501, DOI 10.1017/S0022112082003164. MR687014 (84a:76002)

[8] Gregory R. Baker and Chao Xie, *Singularities in the complex physical plane for deep water waves*, J. Fluid Mech. **685** (2011), 83–116, DOI 10.1017/jfm.2011.283. MR2844303

[9] Jerry L. Bona, Thierry Colin, and David Lannes, *Long wave approximations for water waves*, Arch. Ration. Mech. Anal. **178** (2005), no. 3, 373–410, DOI 10.1007/s00205-005-0378-1. MR2196497 (2007a:76012)

[10] R. K.-C. Chan and R. Street, *A computer study of finite amplitude water waves*, J. Comput. Phys. **6** (1970), 68–94.

[11] G. A. Chandler and I. G. Graham, *The computation of water waves modelled by Nekrasov's equation*, SIAM J. Numer. Anal. **30** (1993), no. 4, 1041–1065, DOI 10.1137/0730054. MR1231326 (94e:65144)

[12] B. Chen and P. G. Saffman, *Numerical evidence for the existence of new types of gravity waves of permanent form on deep water*, Stud. Appl. Math. **62** (1980), no. 1, 1–21. MR557677 (80k:76014)

[13] M. J. Cooker, P. D. Weidman, and D. S. Bale, *Reflection of a high-amplitude solitary wave at a vertical wall*, J. Fluid Mech. **342** (1997), 141–158, DOI 10.1017/S002211209700551X. MR1460657 (98d:76026)

[14] Walter Craig, *An existence theory for water waves and the Boussinesq and Korteweg-de Vries scaling limits*, Comm. Partial Differential Equations **10** (1985), no. 8, 787–1003, DOI 10.1080/03605308508820396. MR795808 (87f:35210)

[15] W. Craig, P. Guyenne, J. Hammack, D. Henderson, and C. Sulem, *Solitary water wave interactions*, Phys. Fluids **18** (2006), no. 5, 057106, 25, DOI 10.1063/1.2205916. MR2259317 (2007i:76013)

[16] W. Craig and C. Sulem, *Numerical simulation of gravity waves*, J. Comput. Phys. **108** (1993), no. 1, 73–83, DOI 10.1006/jcph.1993.1164. MR1239970 (94h:76064)

[17] Alex D. D. Craik, *The origins of water wave theory*, Annual review of fluid mechanics. Vol. 36, Annu. Rev. Fluid Mech., vol. 36, Annual Reviews, Palo Alto, CA, 2004, pp. 1–28, DOI 10.1146/annurev.fluid.36.050802.122118. MR2062306 (2005a:01012)

[18] Alex D. D. Craik, *George Gabriel Stokes on water wave theory*, Annual review of fluid mechanics. Vol. 37, Annu. Rev. Fluid Mech., vol. 37, Annual Reviews, Palo Alto, CA, 2005, pp. 23–42, DOI 10.1146/annurev.fluid.37.061903.175836. MR2115337 (2005i:76001)

[19] Bernard Deconinck and Harvey Segur, *Pole dynamics for elliptic solutions of the Korteweg-de Vries equation*, Math. Phys. Anal. Geom. **3** (2000), no. 1, 49–74, DOI 10.1023/A:1009830803696. MR1781440 (2003i:37073)

[20] Maarten W. Dingemans, *Water wave propagation over uneven bottoms, part 2. non-linear wave propagation*, World Scientific, Singapore, 1997.

[21] S. Govindjee, T. Potter, and J. Wilkening, *Cyclic steady states of treaded rolling bodies*, Int. J. Numer. Meth. Engng. (2014), (accepted).

[22] I. S. Gradshteyn and I. M. Ryzhik, *Table of integrals, series, and products*, 7th ed., Elsevier/Academic Press, Amsterdam, 2007. Translated from the Russian; Translation edited and with a preface by Alan Jeffrey and Daniel Zwillinger; With one CD-ROM (Windows, Macintosh and UNIX). MR2360010 (2008g:00005)

[23] E. Hairer, S. P. Nørsett, and G. Wanner, *Solving ordinary differential equations. I*, Springer Series in Computational Mathematics, vol. 8, Springer-Verlag, Berlin, 1987. Nonstiff problems. MR868663 (87m:65005)

[24] Joseph L. Hammack and Harvey Segur, *The Korteweg-de Vries equation and water waves. II. Comparison with experiments*, J. Fluid Mech. **65** (1974), 289–313. MR0366198 (51 #2446)

[25] Thomas Y. Hou and Ruo Li, *Computing nearly singular solutions using pseudo-spectral methods*, J. Comput. Phys. **226** (2007), no. 1, 379–397, DOI 10.1016/j.jcp.2007.04.014. MR2356364 (2008f:76135)

[26] Thomas Y. Hou, John S. Lowengrub, and Michael J. Shelley, *Removing the stiffness from interfacial flows with surface tension*, J. Comput. Phys. **114** (1994), no. 2, 312–338, DOI 10.1006/jcph.1994.1170. MR1294935 (95e:76069)

[27] R. S. Johnson, *A modern introduction to the mathematical theory of water waves*, Cambridge Texts in Applied Mathematics, Cambridge University Press, Cambridge, 1997. MR1629555 (99m:76017)

[28] Christopher A. Kennedy and Mark H. Carpenter, *Additive Runge-Kutta schemes for convection-diffusion-reaction equations*, Appl. Numer. Math. **44** (2003), no. 1-2, 139–181, DOI 10.1016/S0168-9274(02)00138-1. MR1951292 (2003m:65111)

[29] D. J. Korteweg and G. de Vries, *On the change of form of long waves advancing in a rectangular canal, and on a new type of long stationary waves*, Philosophical Magazine **39** (1895), 422–443.

[30] Robert Krasny, *A study of singularity formation in a vortex sheet by the point-vortex approximation*, J. Fluid Mech. **167** (1986), 65–93, DOI 10.1017/S0022112086002732. MR851670 (87g:76028)

[31] Martin D. Kruskal, *The Korteweg-de Vries equation and related evolution equations*, Nonlinear wave motion (Proc. AMS-SIAM Summer Sem., Clarkson Coll. Tech., Potsdam, N.Y., 1972), Amer. Math. Soc., Providence, R.I., 1974, pp. 61–83. Lectures in Appl. Math., Vol. 15. MR0352741 (50 #5228)

[32] Peter D. Lax, *Integrals of nonlinear equations of evolution and solitary waves*, Comm. Pure Appl. Math. **21** (1968), 467–490. MR0235310 (38 #3620)

[33] Randall J. LeVeque, *On the interaction of nearly equal solitons in the KdV equation*, SIAM J. Appl. Math. **47** (1987), no. 2, 254–262, DOI 10.1137/0147017. MR881349 (89c:35138)

[34] M. S. Longuet-Higgins and E. D. Cokelet, *The deformation of steep surface waves on water. I. A numerical method of computation*, Proc. Roy. Soc. London Ser. A **350** (1976), no. 1660, 1–26. MR0411355 (53 #15091)

[35] T. Maxworthy, *Experiments on collisions between solitary waves*, J. Fluid Mech. **76** (1976), 177–185.

[36] G. N. Mercer and A. J. Roberts, *Standing waves in deep water: their stability and extreme form*, Phys. Fluids A **4** (1992), no. 2, 259–269, DOI 10.1063/1.858354. MR1146134 (92k:76014)
[37] Paul A. Milewski, J.-M. Vanden-Broeck, and Zhan Wang, *Dynamics of steep two-dimensional gravity–capillary solitary waves*, J. Fluid Mech. **664** (2010), 466–477, DOI 10.1017/S0022112010004714. MR2747492 (2011m:76026)
[38] R. M. Mirie and C. H. Su, *Collisions between two solitary waves, Part II*, J. Fluid Mech. **115** (1982), 475–492.
[39] Jorge Nocedal and Stephen J. Wright, *Numerical optimization*, Springer Series in Operations Research, Springer-Verlag, New York, 1999. MR1713114 (2001b:90002)
[40] L. A. Ostrovsky and Y. A. Stepanyants, *Internal solitons in laboratory experiments: comparison with theoretical models*, Chaos **15** (2005), no. 3, 037111, 28, DOI 10.1063/1.2107087. MR2184887 (2006h:76025)
[41] P. I. Plotnikov and J. F. Toland, *Nash-Moser theory for standing water waves*, Arch. Ration. Mech. Anal. **159** (2001), no. 1, 1–83, DOI 10.1007/PL00004246. MR1854060 (2002k:76019)
[42] Chris H. Rycroft and Jon Wilkening, *Computation of three-dimensional standing water waves*, J. Comput. Phys. **255** (2013), 612–638, DOI 10.1016/j.jcp.2013.08.026. MR3109807
[43] Guido Schneider and C. Eugene Wayne, *The long-wave limit for the water wave problem. I. The case of zero surface tension*, Comm. Pure Appl. Math. **53** (2000), no. 12, 1475–1535, DOI 10.1002/1097-0312(200012)53:12⟨1475::AID-CPA1⟩3.0.CO;2-V. MR1780702 (2002c:76025a)
[44] C. H. Su and Rida M. Mirie, *On head-on collisions between solitary waves*, J. Fluid Mech. **98** (1980), no. 3, 509–525, DOI 10.1017/S0022112080000262. MR583060 (81g:76023)
[45] J.-M. Vanden-Broeck, *Elevation solitary waves with surface tension*, Phys. Fluids A **3** (1991), 1989–1993.
[46] G. B. Whitham, *Linear and nonlinear waves*, Wiley-Interscience [John Wiley & Sons], New York-London-Sydney, 1974. Pure and Applied Mathematics. MR0483954 (58 #3905)
[47] J. Wilkening, *Breakdown of self-similarity at the crests of large amplitude standing water waves*, Phys. Rev. Lett **107** (2011), 184501.
[48] ———, *Traveling-standing water waves*, (2014), (in preparation).
[49] J. Wilkening and J. Yu, *Overdetermined shooting methods for computing standing water waves with spectral accuracy*, Comput. Sci. Disc. **5** (2012), 014017:1–38.
[50] M. O. Williams, J. Wilkening, E. Shlizerman, and J. N. Kutz, *Continuation of periodic solutions in the waveguide array mode-locked laser*, Physica D **240** (2011), no. 22, 1791–1804.
[51] Wolfram Research, Inc, *Mathematica, version 8.0*, Champaign, IL, 2010.
[52] N. J. Zabusky and C. J. Galvin, *Shallow-water waves, the Korteweg-de Vries equation and solitons*, J. Fluid Mech. **47** (1970), 811–824.
[53] Qisu Zou and Chau Hsing Su, *Overtaking collision between two solitary waves*, Phys. Fluids **29** (1986), no. 7, 2113–2123, DOI 10.1063/1.865599. MR848314 (87h:76033)

Department of Mathematics, University of California, Berkeley, California 94720-3840

E-mail address: `wilkening@berkeley.edu`

Contemporary Mathematics
Volume **635**, 2015
http://dx.doi.org/10.1090/conm/635/12716

The Instabilities of Periodic Traveling Water Waves with Respect to Transverse Perturbations

Katie Oliveras and Bernard Deconinck

ABSTRACT. We show that the exact reformulation of the classical surface water wave problem due to Ablowitz, Fokas and Musslimani, provides a convenient framework to investigate the instabilities of one-dimensional stationary periodic waves, with respect to transverse perturbations. Such perturbations have trigonometric dependence on the transverse variable, and are bounded (typically quasi periodic) in the longitudinal direction. Using the new formulation, we examine waves in both deep and shallow water, confirming previous results about their instabilities due to McLean and Francius & Kharif, among others.

1. Introduction

The objective of this paper is to examine the stability of one-dimensional periodic surface gravity waves with respect to two-dimensional perturbations using a recent nonlocal reformulation of the water wave problem due to Ablowitz, Fokas and Musslimani [**AFM06**]. Of course, there exists a great wealth of literature on the stability of traveling water waves, and by no means can this be completely covered, nor replicated, within the context of this paper. We discuss the major results that are fundamental to our investigation.

Perhaps the earliest results on the stability of surface water waves are those discussing the instability of small-amplitude waves in deep water with respect to long-wave perturbations. This modulational (or Benjamin-Feir) instability was discovered in 1967 by Benjamin [**Ben67**], Benjamin & Feir [**BF67**] and Whitham [**Whi67**]. Benjamin[**Ben67**] considered a Stokes wave solution perturbed by two sidebands. By examining the Fourier mode expansion of the perturbed solution, he determined that waves of period L in water of dimensionless depth $kh > 1.363$ ($k = 2\pi/L$), are unstable with respect to long-wave perturbations.

Since the 1960s, researchers have been using numerical techniques to determine the stability properties of traveling waves due to the complicated nature of the underlying equations of motion. Examples of in-depth numerical studies on one-dimensional solutions perturbed by one-dimensional disturbances include the early work of Longuet-Higgins[**LH78a, LH78b**], and the current work of Nicholls[**Nic09**]. Recently, we have examined the spectral stability of periodic

2010 *Mathematics Subject Classification.* Primary 54C40, 14E20; Secondary 46E25, 20C20.
Key words and phrases. Fluid flow, nonlinear waves, stability.
The first author was supported in part by NSF Grant NSF-DMS-VIGRE-0354131. NSF-DMS-0604546.

traveling waves in the one-dimensional setting[**DO**] . In addition to recovering expected results (i.e. waves in deep water are modulationally unstable with respect to long-wave perturbations), we found there are specific classes of quasi-periodic perturbations that yield instabilities in shallow water, even for small-amplitude waves.

As stated, the above results are limited to one-dimensional periodic traveling waves perturbed by one-dimensional periodic or quasi-periodic disturbances. As an intermediate step between investigating the one-dimensional and the two-dimensional problem, different researchers have considered the effects of two- dimensional perturbations on one-dimensional traveling wave solutions. In other words, they have investigated the transverse instabilities of the one-dimensional solutions. Early notable work on the transverse instabilities of solutions of the Euler equations was conducted by Bryant[**Bry78**]. Using a set of truncated nonlinear approximations to the Euler equations for water of finite depth, he explored the instabilities using a small number of wave interactions. Several years later, McLean presented the first results on the transverse instabilities of one-dimensional solutions to the Euler equations for water of finite depth. McLean et al. (Refs. [**McL82a, McL82b, MMM⁺81**]) considered perturbations of the form

$$\eta_1 = e^{-\sigma t} e^{ipx+iqy} \sum_{j=-\infty}^{\infty} a_j e^{ijx}, \text{ and}$$

$$\phi_1 = e^{-\sigma t} e^{ipx+iqy} \sum_{j=-\infty}^{\infty} b_j e^{ijx} \frac{\cosh(\kappa_j(z+h))}{\cosh(\kappa_j h)}.$$

Here η_1 and ϕ_1 are the perturbations of the surface elevation and the velocity potential in the interior of the fluid, respectively. Substituting these expansions into the original equations of motion, McLean calculated the growth rates $-\sigma$ associated with the linearized problem for waves in deep water. He found that in deep water, small-amplitude one-dimensional waves are most unstable to one-dimensional perturbations. However, as the amplitude of the wave is increased, the wave becomes more unstable with respect to two-dimensional perturbations. More recently, McLean's stability calculations have been extended to waves of even greater amplitude by Kharif & Ramamonjiarisoa[**KR90**] and to waves in water of shallower depths by Francius & Kharif[**FK06**].

The purpose of this paper is to demonstrate that we can replicate these transverse instability results using the nonlocal formulation of Ablowitz, Fokas, and Musslimani [**AFM06**] with less computational effort. In order to numerically calculate the spectral instabilities, we must determine (i) one-dimensional traveling wave solutions, and (ii) the growth rates associated with two-dimensional perturbations of these one-dimensional traveling waves. We have recently shown that one-dimensional periodic traveling waves solutions can be determined by numerically solving the single equation

$$\int_0^L e^{-ikx} \sqrt{(1+\eta_x^2)(c^2-2g\eta)} \sinh\left(k(\eta+h)\right)\, dx = 0,$$

$\forall k \in \frac{2\pi}{L}\mathbb{Z}_0$, for the surface elevation $\eta(x)$ moving with wave-speed c, see Ref. [**DO**]. Using the numerical scheme outlined in Deconinck & Oliveras[**DO**], we determine one-dimensional traveling wave solutions for various wave amplitudes and water depths. These one-dimensional solutions are perturbed with two-dimensional (quasi-) periodic disturbances. Substituting the two-dimensional perturbations into the equations of motion, we calculate the linear growth rates using a Fourier-Floquet decomposition, see for example Ref. [**DK06**]. We find that our stability results are in general agreement with those of Francius & Kharif [**FK06**] and the other sources mentioned above. Although a direct comparison is not possible (see Section 5), it appears that our method produces more accurate results at a significantly reduced computational cost.

2. Equations of Motion

We consider Euler's equations describing the dynamics of the surface of an ideal fluid in three dimensions. For the periodic problem, this requires us to solve Laplace's equation inside the fluid domain D with periodic boundary conditions in both horizontal directions x and y. The fluid domain D is shown in Figure 1. The equations of motion are

$$\phi_{xx} + \phi_{yy} + \phi_{zz} = 0, \qquad (x, y, z) \in D, \tag{2.1}$$

$$\phi_z = 0, \qquad z = -h, \tag{2.2}$$

$$\eta_t + \phi_x \eta_x + \phi_y \eta_y = \phi_z, \qquad z = \eta(x, y, t), \tag{2.3}$$

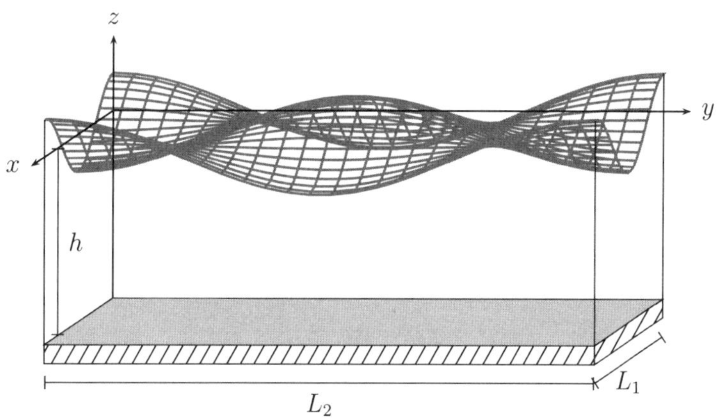

FIGURE 1. The fluid domain for a water wave surface with periodic boundary conditions.

$$\phi_t + \frac{1}{2}\left(\phi_x^2 + \phi_y^2\right) + g\eta = 0, \qquad z = \eta(x,y,t), \tag{2.4}$$

where $\phi = \phi(x,y,z,t)$ is the velocity potential, $\eta = \eta(x,y,t)$ is the surface elevation, g is the acceleration of gravity, and h is the constant depth of the fluid when at a state of rest. In this paper, we ignore the effects of surface tension. We require periodic boundary conditions with period L_1 in the x-direction and L_2 in the y-direction. This provides additional boundary conditions on the surface elevation as well as on the velocity potential once the induced mean flow has been eliminated. We have assumed finite depth in the above formulation. The formulation is also valid in infinite depth.

As discussed in Ref. [**DO**], Euler's equations as written above are challenging to work with directly: they are a free-boundary problem with nonlinear boundary conditions specified at the unknown boundary. For the two-dimensional problem (i.e., one-dimensional surface), there are several reformulations that reduce these complications. However for the three-dimensional problem (with two-dimensional surface), there are fewer alternative formulations. One of the commonly used reformulations is that due to Zakharov, Craig and Sulem[**Zak68, CS93**]. A minor disadvantage to the Zakharov-Craig-Sulem formulation is that for numerical implementations, one must truncate the series expansion of the Dirichlet-to-Neumann operator. The nonlocal formulation presented below does not require such a truncation.

Ablowitz, Fokas and Musslimani (AFM) [**AFM06**] introduced a new, nonlocal reformulation of the Euler equations, valid for surface waves localized on the whole line or the whole plane. It is essentially trivial to extend this formulation to periodic boundary conditions. In a traveling coordinate frame where $\mathbf{x} = (x,y)$ is moving with speed $\bar{c} = (c_x, c_y)$, the equations of motion can be written as

$$\begin{aligned}\int_0^{L_2}\int_0^{L_1} & e^{-i\bar{k}\cdot\mathbf{x}}\Big(i(\eta_t - \bar{c}\cdot\nabla\eta)\cosh(\kappa(h+\eta)) \\ & - \frac{\bar{k}\cdot\nabla q}{\kappa}\sinh(\kappa(h+\eta))\Big)dx\,dy = 0,\end{aligned} \tag{2.5}$$

$$q_t - \bar{c}\cdot\nabla q + \frac{1}{2}\left|\nabla q\right|^2 + g\eta - \frac{1}{2}\cdot\frac{(\eta_t + \nabla q\cdot\nabla\eta)^2}{1+\left|\nabla\eta\right|^2} = 0, \tag{2.6}$$

where $\nabla = (\partial_x, \partial_y)$, $q(x,y,t) = \phi(x,y,\eta(x,y,t),t)$ is the velocity potential evaluated at the surface, $\bar{k} = (k_x, k_y)$, $\kappa = \sqrt{k_x^2 + k_y^2}$ and

$$\Lambda = \left\{\bar{k}\in\mathbb{R}^2 \,\middle|\, k_x = \frac{2m\pi}{L_1}, k_y = \frac{2n\pi}{L_2},\ (m,n)\in\mathbb{Z}^2 - \{(0,0)\}\right\}.$$

For the remainder of this paper, we use this nonlocal formulation of the water-wave problem. We refer the reader to Refs. [**DO, AFM06, AH08**] for additional information regarding this exact reformulation of the water-wave problem.

Since our goal is to determine the stability of one-dimensional periodic traveling waves with respect to two-dimensional perturbations, we must use the full equations for the two-dimensional surface gravity wave to examine stability. Since we only consider the stability of one-dimensional solutions, we reduce the equations of motion by eliminating the y-dependence in (2.5) and (2.6) to determine the traveling wave solutions moving in the x-direction with speed c. In Ref. [**DO**], we demonstrated that periodic one-dimensional traveling wave solutions are solutions of

$$\int_0^L e^{-ikx}\sqrt{(1+\eta_x^2)(c^2-2g\eta)}\sinh\left(k(\eta+h)\right)dx = 0, \tag{2.7}$$

$\forall\, k \in \frac{2\pi}{L}\mathbb{Z}_0$, for the surface elevation $\eta(x)$. This equation provides solutions to the fully nonlinear problem, and is easily solved numerically. To determine the velocity potential $q(x)$ at the surface (which is needed when examining the stability of traveling waves), we use the relationship

$$q_x = c \pm \sqrt{(1+\eta_x^2)\,(c^2-2g\eta)}. \tag{2.8}$$

where we work with the negative sign in order to enforce that $u-c<0$ within the fluid bulk as well as along the free surface for all but the limiting wave. One-dimensional solutions are computed using an iterative pseudo-continuation approach as outlined in Ref. [**DO**]. All solutions were calculated using at least 64 Fourier Modes and a residual error $< 10^{-14}$.

3. Stability Formulation

Before we proceed to investigate the stability of the traveling wave profiles, we discuss what perturbations we wish to allow. In this paper, we plan to investigate the effects of two-dimensional pertubations. It may appear natural to consider disturbances of the same period as the underlying stationary wave in the x direction, and vary the period of the perturbation in the y direction. However, we wish to work with a more general class of disturbances, namely those that are bounded on the whole real plane. Specifically, for some disturbance $U(x,y)$, we use the notation $||U(x,y)|| < \infty$ to mean that $\sup_{\mathbb{R}^2}|U(x,y)| < \infty$ and that $U(x,y)$ is continuous for all $(x,y)\in\mathbb{R}^2$. It is important to realize that this class is the largest class of perturbations allowed by the physical problem at hand. Indeed, disturbances should be bounded and continuous functions, but there is no physical reason to restrict their spatial dependence to be periodic.

In order to investigate the stability of the traveling wave profiles with respect to such perturbations, it is necessary to reformulate the governing equations. Equation (2.6) is a local statement and does not require modification. However, the current incarnation of the nonlocal equation (2.5) applies specifically to waves of period L_1 in the x-direction and L_2 in the y-direction. Let $\langle\langle f\rangle\rangle$ represent the average value of a function $f(\mathbf{x})$, $f:\mathbb{R}^2\to\mathbb{R}$ by

$$\langle\langle f\rangle\rangle = \lim_{|A|\to\infty}\frac{1}{|A|}\iint_A f(x,y)dA,$$

where A is a rectangle in $\mathbb{R}^2$ and $|A|$ denotes its area.

It is clear that if $f(x,y)$ is periodic in both x and y with period L_1 and L_2 respectively, then $\langle\langle f\rangle\rangle$ is well defined. However, this two-dimensional spatial average is defined for the larger class of so-called almost-periodic functions, which contains the set of quasi-periodic functions[**Boh47**]. This leads us to replace (2.5) with the more general nonlocal equation

$$\left\langle\!\!\left\langle e^{-i\bar{k}\cdot\mathbf{x}}\left(i(\eta_t-\bar{c}\cdot\nabla\eta)\cosh(\kappa(h+\eta))\right.\right.\right. \left.\left.\left.-\frac{\bar{k}\cdot\nabla q}{\kappa}\sinh(\kappa(h+\eta))\right)\right\rangle\!\!\right\rangle=0, \tag{3.1}$$

which is valid for all $\bar{k}\in\mathbb{R}_0^2=\mathbb{R}^2-\{(0,0)\}$. In fact, if we were to consider perturbations of increasingly larger period, the set of $\bar{k}$ values to be considered in Equation (2.5) would approach a dense subset of the real plane.

Having generalized the dynamical equations to accommodate the perturbations we wish to consider, we briefly discuss the definition of spectral stability. A stationary solution of a nonlinear problem is spectrally stable if there are no exponentially growing modes of the corresponding linearized problem. To determine the spectral stability of the periodic traveling wave solutions, we start by considering a one-dimensional traveling wave solution $(\eta_0(x-ct),q_0(x-ct),c)$, which solves equation (2.7). In the same traveling coordinate frame, we add a small perturbation in both the x and y direction of the form

$$\begin{aligned} q(x-ct,y,t) &= q_0(x-ct)+\epsilon q_1(x-ct,y,t)+\mathcal{O}(\epsilon^2),\\ \eta(x-ct,y,t) &= \eta_0(x-ct)+\epsilon\eta_1(x-ct,y,t)+\mathcal{O}(\epsilon^2),\end{aligned}$$

where ϵ is a small parameter. The perturbation is moving in a traveling reference frame with speed $\bar{c}=(c,0)$. Our goal is to determine the time dependence of the perturbation in order to determine how the deviation from the unperturbed solution evolves.

We begin with the local equation (2.6). Substituting the above into Equation (2.6) and keeping only terms to order ϵ, we obtain

$$q_{1,t}-cq_{1,x}+q_{0,x}q_{1,x}+g\eta_1-f\cdot\left(\eta_{1,t}+\eta_{1,x}(q_{0,x}-c)+\eta_{0,x}q_{1,x}-f\cdot\eta_{0,x}\eta_{1,x}\right)=0, \tag{3.2}$$

where

$$f=\frac{\eta_{0,x}(q_{0,x}-c)}{1+\eta_{0,x}^2}.$$

Since the problem does not depend on t or y explicitly (recall that the functions η_0 and q_0 are both independent of y and t, and depend only on x), we can use separation of variables to decompose $\eta_1(x,y,t)$ and $q_1(x,y,t)$ as

$$\eta_1(x,y,t)=e^{\lambda t}e^{i\rho y}\tilde{\eta}_1(x)+c.c., \tag{3.3a}$$

$$q_1(x,y,t)=e^{\lambda t}e^{i\rho y}\tilde{q}_1(x)+c.c., \tag{3.3b}$$

without loss of generality. Here ρ is the transverse wave number of the perturbation. Substituting into (3.2), we have

$$\lambda\tilde{q}_1-c\tilde{q}_{1,x}+q_{0,x}\tilde{q}_{1,x}+g\tilde{\eta}_1-f\cdot\left(\lambda\tilde{\eta}_1+\tilde{\eta}_{1,x}(q_{0,x}-c)+\eta_{0,x}\tilde{q}_{1,x}-\eta_{0,x}f\cdot\tilde{\eta}_{1,x}\right)=0, \tag{3.4}$$

which is the same linearization as that obtained in (Ref. [**DO**]) for the local equation with respect to one-dimensional perturbations.

Now we turn our attention to the nonlocal equation. For the nonlocal equation, we substitute the perturbation expansion into (3.1). This gives

$$\left\langle\!\!\left\langle e^{-i\bar{k}\cdot\mathbf{x}}\left(i\eta_{1,t}\mathcal{C}_\kappa-ic\eta_{0,x}\kappa\eta_1\mathcal{S}_\kappa-ic\eta_{1,x}\mathcal{C}_\kappa-\frac{1}{\kappa}\left(k_xq_{1,x}+k_yq_{1,y}\right)\mathcal{S}_\kappa-k_xq_{0,x}\eta_1\mathcal{C}_\kappa\right)\right\rangle\!\!\right\rangle=0,$$

where

$$\mathcal{S}_\kappa = \sinh(\kappa(\eta_0 + h), \ \mathcal{C}_\kappa = \cosh(\kappa(\eta_0 + h)).$$

As before, since the problem does not depend on y or t explicitly, we use separation of variables to decompose $\eta_1(x,y,t)$ and $q_1(x,y,t)$ as in (3.3). Using this decomposition, we obtain

$$\left\langle\!\!\left\langle e^{-i\bar{k}\cdot\mathbf{x}} e^{i\rho y} \quad \Big(i\tilde{\eta}_{1,t}\mathcal{C}_\kappa - ic\eta_{0,x}\kappa\tilde{\eta}_1\mathcal{S}_\kappa - ic\tilde{\eta}_{1,x}\mathcal{C}_\kappa \right.\right.$$
$$\left.\left. - \frac{1}{\kappa}\left(k_x\tilde{q}_{1,x} + ik_y\rho\tilde{q}_1\right)\mathcal{S}_\kappa - k_x q_{0,x}\tilde{\eta}_1\mathcal{C}_\kappa \Big) \right\rangle\!\!\right\rangle = 0,$$

Since our new integrand depends on y only through $e^{i(\rho-k_y)y}$, we eliminate the integration with respect to y completely. Calculating the average value along the y direction results in

$$\lim_{M\to\infty} \frac{1}{M}\int_{-M/2}^{M/2} e^{-i(k_y-\rho)y} dy = \lim_{M\to\infty} \frac{2}{M(k_y-\rho)} \sin\left(\frac{M(k_y-\rho)}{2}\right).$$

Taking the limit as $M\to\infty$, the contribution is identically zero if $k_y \neq \rho$. Otherwise, if $k_y = \rho$, the contribution from the integral is unity. Thus, the linearization of the nonlocal equation simplifies to

$$\left\langle e^{-ik_x x}\Big(i\lambda\tilde{\eta}_1\mathcal{C}_\kappa - ic\tilde{\eta}_{1,x}\mathcal{C}_\kappa - \eta_{0,x}c\kappa\tilde{\eta}_1\mathcal{S}_\kappa \right.$$
$$\left. - \frac{1}{\kappa}\left(k_x\tilde{q}_{1,x} + i\rho^2\tilde{q}_1\right)\mathcal{S}_\kappa - kq_{0,x}\tilde{\eta}_1\mathcal{C}_\kappa \right\rangle = 0, \tag{3.5}$$

where now $\kappa = \sqrt{k_x^2 + \rho^2}$ and

$$\langle f \rangle = \lim_{M\to\infty} \frac{1}{M}\int_{-M/2}^{M/2} f(x)\, dx.$$

It should be noted that if we equate $\rho = 0$, the above equation reduces to the expression for the one-dimensional surface case presented elsewhere[**DO**].

Having linearized the equations about the traveling wave solution, we have the following generalized eigenvalue problem:

$$\lambda\tilde{q}_1 - c\tilde{q}_{1,x} + q_{0,x}\tilde{q}_{1,x} + g\tilde{\eta}_1 - f\cdot\Big(\lambda\tilde{\eta}_1$$
$$+ \tilde{\eta}_{1,x}(q_{0,x} - c) + \eta_{0,x}\tilde{q}_{1,x} - \eta_{0,x}f\cdot\tilde{\eta}_{1,x}\Big) = 0, \tag{3.6}$$

$$\lambda\left\langle e^{-ik_x x}\left(i\tilde{\eta}_1\mathcal{C}_\kappa\right)\right\rangle = \left\langle e^{-ik_x x}\Big(ic\tilde{\eta}_{1,x}\mathcal{C}_\kappa + i\eta_{0,x}c\kappa\tilde{\eta}_1\mathcal{S}_\kappa \right.$$
$$\left. + \frac{1}{\kappa}\left(k_x\tilde{q}_{1,x} + i\rho^2\tilde{q}_1\right)\mathcal{S}_\kappa + k_x q_{0,x}\tilde{\eta}_1\mathcal{C}_\kappa\Big)\right\rangle. \tag{3.7}$$

Equations (3.6) and (3.7) are valid for $(k_x, \rho) \in \mathbb{R}^2 - \{(0,0)\}$, and can be rewritten compactly as

$$\lambda \mathcal{L}_1 U(x) = \mathcal{L}_2 U(x) \tag{3.8}$$

where $\mathcal{L}_1$ and $\mathcal{L}_1$ are 2×2 matrices of linear operators that depend on ρ, and whose entries are easily read off from (3.6) and (3.7).

Since the time dependence of the perturbation depends exponentially on λ, we can determine information about the stability of the underlying traveling wave by determining the solutions of this generalized eigenvalue problem. If any solutions $U(x)$ exist for which the corresponding λ has a positive real part, the linear approximation of the solution grows exponentially in time and the perturbed solution will diverge from the stationary solution in the linear approximation. With this intuition in mind, the concept of spectral stability, as applicable to the generalized eigenvalue problem (3.8) is defined as

DEFINITION 3.1 (Spectral Stability). A one-dimensional periodic traveling wave solution $(\eta_0(x), q_0(x), c)$ of Equations 2.7 and 2.8 (and thus of the Euler equations) is spectrally stable with respect to two-dimensional perturbations parameterized by ρ if no element of the spectrum for the generalized eigenvalue problem 3.8 has strictly positive real part.

In order to use this definition, we recall the definition of the spectrum of the generalized eigenvalue problem.

DEFINITION 3.2 (Spectrum of the Generalized Eigenvalue Problem). The spectrum of the generalized Eigenvalue problem $\lambda \mathcal{L}_1 U(x) = \mathcal{L}_2 U(x)$ is given by

$$\sigma(\mathcal{L}_2, \mathcal{L}_1) = \{\lambda \in \mathbb{C} : ||U(x)|| < \infty\}, \tag{3.9}$$

where $U(x)$ solves $\lambda \mathcal{L}_1 U(x) = \mathcal{L}_2 U(x)$.

In order to examine spectral stability or instability, we need to locate the spectrum of the generalized eigenvalue problem 3.8. By definition of the spectrum, we determine those values of $\lambda \in \mathbb{C}$ for which Equation 3.8 has bounded solutions. Since the coefficient functions of Equation (3.8) are periodic in x with period L_1, we decompose the perturbations further using Floquet's Theorem, see for instance Ref. [**DK06**]. Thus, both $\tilde{q}_1(x)$ and $\tilde{\eta}_1(x)$ can be written as

$$\tilde{q}_1(x) = e^{i\mu x}\bar{q}_1(x), \text{and } \tilde{\eta}_1(x) = e^{i\mu x}\bar{\eta}_1(x),$$

where $\mu \in [-\pi/L_1, \pi/L_1)$ and $\bar{q}_1$ and $\bar{\eta}_1$ are 2π periodic functions.

Substituting the Floquet decomposition into the eigenvalue problem yields

$$\begin{aligned} \lambda\left(\tilde{q}_1 - f\tilde{\eta}_1\right) &= c(i\mu+\partial_x)\bar{q}_1 - q_{0,x}(i\mu+\partial_x)\bar{q}_1 - g\bar{\eta}_1 \\ &\quad + f\Big((q_{0,x}-c)(i\mu+\partial_x)\bar{\eta}_1 + \eta_{0,x}(i\mu+\partial_x)\bar{q}_1 - \eta_{0,x} f(i\mu+\partial_x)\bar{\eta}_1\Big), \end{aligned} \tag{3.10}$$

$$\begin{aligned} \lambda\left\langle e^{i(\mu-k_x)x}\left(i\bar{\eta}_1\mathcal{C}_\kappa\right)\right\rangle &= \Big\langle (e^{i(\mu-k_x)x}\Big(ic\mathcal{C}_\kappa(i\mu+\partial_x)\bar{\eta}_1 + i\eta_{0,x}c\kappa\bar{\eta}_1\mathcal{S}_\kappa \\ &\quad + \frac{1}{\kappa}\left(k_x(i\mu+\partial_x)+i\rho^2\right)\bar{q}_1\mathcal{S}_\kappa + k_x q_{0,x}\bar{\eta}_1\mathcal{C}_\kappa\Big)\Big\rangle = 0. \end{aligned} \tag{3.11}$$

We represent the 2π-periodic portion of the eigenfunctions $\bar{q}_1$ and $\bar{\eta}_1$ by their Fourier series. In other words,

$$\bar{q}_1(x) = \sum_{m=-\infty}^{\infty} \hat{Q}_m e^{imx}, \; \bar{\eta}_1(x) = \sum_{m=-\infty}^{\infty} \hat{N}_m e^{imx}.$$

Additionally, we represent the periodic variable coefficient functions by their Fourier series as well. Substituting the series representations into the Floquet decomposition, we are able to determine a bi-infinite matrix eigenvalue problem, where the eigenvectors are the Fourier coefficients $\hat{Q}_j$ and $\hat{N}_j$ for $j \in \{0, \pm 1, \pm 2, \ldots\}$.

The Fourier decomposition proceeds in a straightforward manner for the local equation. Substituting the Fourier expansions into the local equation, we obtain

$$\begin{aligned}\lambda\left(\sum_{m=-\infty}^{\infty} \hat{F}_{3,n-m}\hat{N}_m - \hat{Q}_n\right) = -g\hat{N}_n &+ \sum_{m=-\infty}^{\infty} (i\mu + im)\,\hat{F}_{1,n-m}\hat{N}_m \\ &+ \sum_{m=-\infty}^{\infty} \hat{F}_{2,n-m}\,(i\mu + im)\,\hat{Q}_m,\end{aligned} \tag{3.12}$$

for $n \in \mathbb{Z} - \{0\}$, where the coefficients $\hat{F}_{j,n}$ are given by the following expressions

$$\begin{aligned}\hat{F}_{1,n} &= \int_0^{L_1} e^{-\frac{2in\pi x}{L_1}} f(\eta_0, q_0)\,(q_{0,x} - c - \eta_{0,x} f(\eta_0, q_0))\,dx, \\ \hat{F}_{2,n} &= \int_0^{L_1} e^{-\frac{2in\pi x}{L_1}} (q_{0,x} - c - \eta_{0,x} f(\eta_0, q_0))\,dx, \\ \hat{F}_{3,n} &= \int_0^{L_1} e^{-\frac{2in\pi x}{L_1}} f(\eta_0, q_0) dx.\end{aligned}$$

For the nonlocal equation, we employ a similar strategy as used elsewhere[**DO**]. There are two slight differences. First, the hyperbolic sine and cosine terms depend on $\kappa = \sqrt{k_x^2 + \rho^2}$ instead of k_x directly. Second, there is an additional term depending on the perturbation of the velocity potential and on ρ.

Substituting the Fourier series into the nonlocal equation and representing all variable coefficients by their Fourier series representation, we have

$$\begin{aligned}&\Big[\sum_{j=-\infty}^{\infty}\sum_{m=-\infty}^{\infty} \left(\hat{G}_{1,j}^{(k_x)} + ic(i\mu + im)\hat{G}_{2,j}^{(k_x)}\right)\hat{N}_m \\ &\quad + \sum_{j=-\infty}^{\infty}\sum_{m=-\infty}^{\infty} \left(\frac{k_x}{\kappa}(i\mu + im) + \frac{i\rho^2}{\kappa}\right)\hat{G}_{3,j}^{(k_x)}\hat{Q}_m\Big]\langle E_{jm}\rangle \\ &\quad = i\lambda \sum_{j=-\infty}^{\infty}\sum_{m=-\infty}^{\infty} \hat{G}_{2,j}^{(k_x)}\hat{N}_m\,\langle E_{jm}\rangle\end{aligned} \tag{3.13}$$

where

$$\begin{aligned}
\hat{G}_{1,n}^{(k_x)} &= \int_0^{L_1} e^{-\frac{2in\pi x}{L_1}} \left(i\kappa c\eta_{0,x}\mathcal{S}_\kappa + k_x q_0 \mathcal{C}_\kappa\right) \, dx, \\
\hat{G}_{2,n}^{(k_x)} &= \int_0^{L_1} e^{-\frac{2in\pi x}{L_1}} \mathcal{C}_\kappa \, dx, \\
\hat{G}_{3,n}^{(k_x)} &= \int_0^{L_1} e^{-\frac{2in\pi x}{L_1}} \mathcal{S}_\kappa \, dx.
\end{aligned}$$

Each term in the double infinite sum contains the average value of

$$E_{jm} = \exp\left[ix\left(\mu - k_x + m + j\right)\right].$$

This average value is identically zero unless the quantity $\mu - k_x + m + j$ is zero so that $\langle E_{jm} \rangle = 1$. In other words the contribution from the integral is identically zero unless

$$k_x = \mu + n,$$

for some integer n. Thus, the only contribution in the integral equation occurs when k_x is a shift of the original dual lattice. This allows us to collapse the double infinite sum to a single infinite sum given by

$$\begin{aligned}
&\sum_{m=-\infty}^{\infty} \left(\hat{G}_{1,n-m}^{(k(\mu))} + ic(i\mu + im)\hat{G}_{2,n-m}^{(k(\mu))}\right) \hat{N}_m \\
&\quad + \sum_{j=-\infty}^{\infty} \sum_{m=-\infty}^{\infty} (i\mu + im)\hat{G}^{(k(\mu))}(\mu + n)_{3,n-m}\hat{Q}_m\Big], \qquad (3.14)\\
&= i\lambda \sum_{m=-\infty}^{\infty} \hat{G}_{2,n-m}^{(k(\mu))} \hat{N}_m
\end{aligned}$$

where $k(\mu) = \mu + n$. The above equation holds for all $n \in \mathbb{Z}$.

Combining the above equation with (3.12) yields a generalized bi-infinite eigenvalue problem for the spectrum of the linearized operator about the stationary traveling wave solutions. We define the quantity $\hat{\mathbf{X}}$ as the bi-infinite vector of Fourier coefficients $\hat{Q}_j$ and $\hat{N}_j$ for all $j \in \mathbb{Z}$: $\hat{\mathbf{X}} = \left[\ldots, \hat{N}_{-1}, \hat{N}_0, \hat{N}_1, \ldots \hat{Q}_{-1}, \hat{Q}_0, \hat{Q}_1, \ldots\right]^T$. With this notation, we rewrite the generalized eigenvalue problem as

$$\begin{bmatrix} A(\mu) & B(\mu) \\ C(\mu,\rho) & D(\mu,\rho) \end{bmatrix} \hat{\mathbf{X}} = \lambda \begin{bmatrix} -I & S(\mu) \\ 0 & V(\mu,\rho) \end{bmatrix} \hat{\mathbf{X}},$$

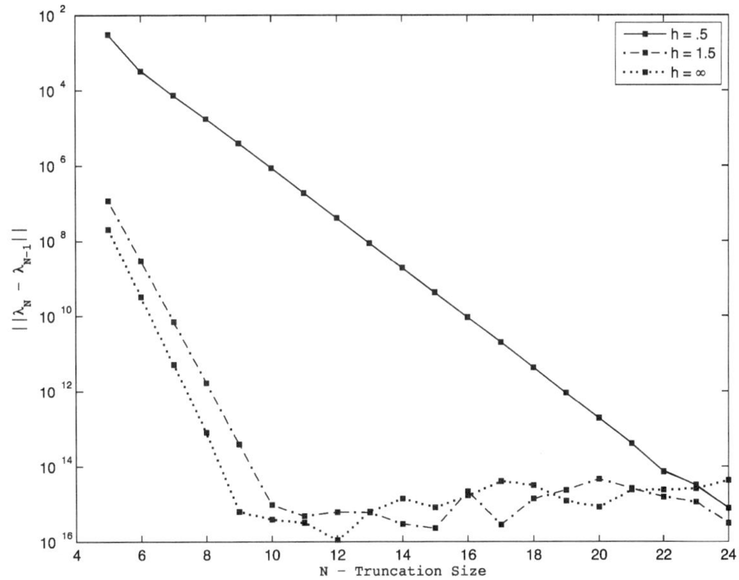

FIGURE 2. Cauchy error for the maximal eigenvalue corresponding to a solution with amplitude $a = .01$ in water of various depths.

where

$$
\begin{aligned}
A(\mu)_{n,m} &= -g\delta(n-m) + (i\mu + im)\,\hat{F}_{1,n-m}, \\
B(\mu)_{n,m} &= \hat{F}_{2,n-m}\,(i\mu + im)\,, \\
C(\mu)_{n,m} &= \hat{G}^{(k(\mu))}_{1,n-m} + ic(i\mu + im)\hat{G}^{(k(\mu))}_{2,n-m}, \\
D(\mu)_{n,m} &= (i\mu + im)\hat{G}^{(k(\mu))}_{3,n-m}, \\
S(\mu)_{n,m} &= \hat{F}_{3,n-m}, \quad V(\mu)_{n,m} = i\hat{G}^{(k(\mu))}_{2,n-m}, \qquad \text{and} \\
\delta(n-m) &= \begin{cases} 1 & \text{if } m = n \\ 0 & \text{if } m \neq n \end{cases}.
\end{aligned}
$$

We solve this generalized eigenvalue problem numerically by truncating the Fourier series representation from $\mathbb{Z}$ to $\{0, \pm 1, \pm 2, \ldots, \pm N\}$ for a discrete sampling for all possible μ and ρ values. We note that due to the underlying symmetries in the problem, we may further restrict the μ interval from $[-\pi/L_1, \pi/L_1]$ to $[0, \pi/L_1]$. For traveling wave solutions of period 2π, $\mu \in [0, .5]$, while the parameter ρ can be any number in $[0, \infty)$.

The convergence and reliability of this truncation was investigated by Curtis & Deconinck[**CD10**] as well as Johnson & Zumbrun [**JZ12**]. Those results do not formally apply to our setting, but can be easily generalized once we note that

$S(\mu, \rho)$ is invertible. It follows that we can trust the numerical results obtained. With this truncation, we obtain $2(2N+1)$ equations for $2(2N+1)$ unknown Fourier coefficients $\hat{Q}_n$ and $\hat{N}_n$.

As seen in Figure 2, the adjustment to the maximal eigenvalue decays to ϵ_{mach} as N increases. It should be noted that the decay rate for $h = .5$ is significantly slower than the decay rate for the other depths $h = 1.5$ and $h = \infty$ with the same amplitude. This is not surprising: $a = .1$ corresponds to a significantly more nonlinear wave for $h = .5$ than for the other two depths.

We further establish confidence in our numerical approach by investigating its Cauchy convergence. Let λ_N represent a particular eigenvalue corresponding to a truncation size N (which corresponds to $2N + 1$ Fourier coefficients). We define the Cauchy error as

$$e_N = |\lambda_N - \lambda_{N-1}|.$$

As we increase the size of our truncated matrix, the calculated eigenvalues should be found to converge numerically. We track the approximation of the eigenvalues with the largest real part as a function of truncation size for three different depths corresponding to traveling wave solutions with a dimensionless amplitude $ak = 0.1$. The Cauchy errors are given in Table 1.

4. Results

4.1. Flat Water. Since we use a continuation approach, we start by carefully considering the stability of the trivial solution $\eta(x) = 0$, $q(x) = 0$, corresponding to a traveling wave of zero amplitude. We can analytically determine the spectrum of this trivial traveling wave. We choose the wave speed c corresponding to the starting point of the bifurcation branch of periodic solutions to be

$$c = \sqrt{\frac{g \tanh(hk)}{k}},$$

where we choose $k = 1$ for 2π periodic solutions. Other bifurcation branches start elsewhere, but those branches correspond to mere scalings of the branch we consider. Their stability properties are identical, and no information is lost by considering only the one branch.

TABLE 1. Cauchy error for the calculation of the eigenvalue with largest real part for $h = .5$, $h = 1.5$, and $h = \infty$ with $a = .1$.

N	$h = .5$	$h = 1.5$	$h = \infty$	N	$h = .5$	$h = 1.5$	$h = \infty$
5	3.1e-03	1.2e-07	2.0e-08	13	8.9e-09	5.1e-16	1.3e-15
6	3.2e-04	3.0e-09	3.3e-10	14	2.0e-09	5.7e-16	2.1e-15
7	7.5e-05	7.2e-11	5.2e-12	15	4.3e-10	2.4e-15	2.8e-15
8	1.8e-05	1.7e-12	7.9e-14	16	9.3e-11	3.5e-15	3.0e-15
9	4.0e-06	4.0e-14	6.7e-16	17	2.0e-11	7.0e-16	2.2e-15
10	8.7e-07	1.2e-17	6.8e-16	18	4.3e-12	3.4e-15	4.0e-15
11	1.9e-07	4.0e-16	1.9e-16	19	9.2e-13	6.6e-16	3.1e-15
12	4.1e-08	6.4e-16	1.2e-15	20	1.9e-13	1.2e-15	8.1e-17

Since the linear stability problem for this solution has constant coefficients, we easily determine the spectrum analytically. Repeating the method outlined in the previous section, we consider perturbations of the form

$$\eta(x) = \epsilon e^{i\mu x + i\rho y + \lambda t} \sum_{j=-\infty}^{\infty} \hat{\eta}_j e^{ijx}, \tag{4.1}$$

$$q(x) = \epsilon e^{i\mu x + i\rho y + \lambda t} \sum_{j=-\infty}^{\infty} \hat{q}_j e^{ijx}, \tag{4.2}$$

where ϵ is a small parameter.

Substituting this solution form into the equations of motion and simplifying as before, we find that the eigenvalues λ are purely imaginary, as expected. They are given by

$$\lambda_m^s = -ic(\mu + m) + is\sqrt{g\kappa \tanh(\kappa h)} \tag{4.3}$$

where $s = \pm 1$, $m \in \mathbb{Z}$, $\kappa = \sqrt{(\mu + m)^2 + \rho^2}$, and $\kappa \neq 0$. Thus, for each perturbation (parameterized by μ and ρ), we can determine the growth-rates λ_m^s for all $m \in \mathbb{Z}$.

Since all eigenvalues are purely imaginary, it follows that flat water is spectrally stable. Using the results of MacKay & Saffman[**MS86**], we know that for an instability to arise in the water wave system, it is necessary for two eigenvalues with opposite signatures to collide, where the signature of a particular eigenvalue is given by

$$\text{sig}(\lambda_m^s) = -\text{Im}\,(s\lambda_m^s)\,.$$

Using the signature and the form of the eigenvalues λ, we see that a necessary condition for an instability to arise is that

$$\lambda_m^{s_1} = \lambda_n^{s_2}, \ -\text{Im}\,(s_1\lambda_m^{s_1}) = \text{Im}\,(s_2\lambda_n^{s_2})\,.$$

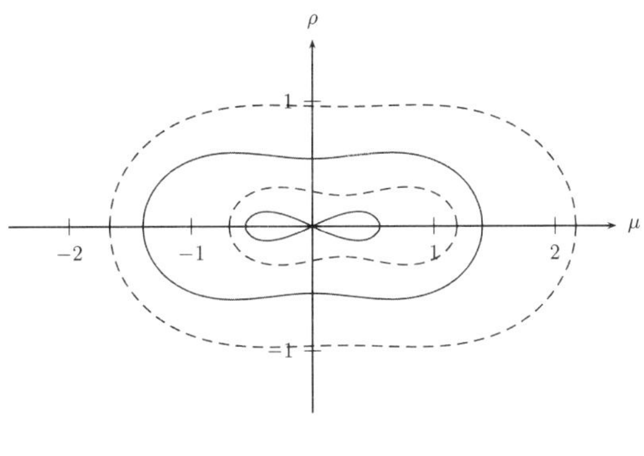

FIGURE 3. Resonant Curves for $h = .5$. The solid lines are associated with the Class I instabilities for $m = 1$ and $m = 2$, and the dashed lines with the Class II instabilities for $m = 1$, and $m = 2$.

Solving these two equations, it follows that $s_1 = -s_2$. This implies we can reduce the above two conditions to the single condition

$$\lambda_m^s = \lambda_n^{-s}$$

Following the work of McLean [**McL82a, McL82b**] and Francius & Kharif [**FK06**], there are two main conditions that we need to investigate for collisions of eigenvalues with opposite signatures:

Class I: $n = -m$ (the case of complex conjugates colliding at the origin),
Class II: $n = -m - 1$ (collisions on the imaginary axis, away from the origin).

There are additional collisions of interest (such as when $n = -m - 2$, etc). However, recalling that the perturbations are periodic in μ with period 1, we can omit these additional collisions from consideration. These conditions give rise to the well-known resonant curves that are found throughout the literature (see for example (Ref. [**McL82a, McL82b, FK06**]). They are shown in Figure 3 for $h = .5$. These curves connect the values of μ and ρ for which eigenvalues with opposite signatures collide.

4.2. General Results. As mentioned in the introduction, much is known about the transverse stability of one-dimensional solutions of Euler's equations. Our intent is to demonstrate that the use of the AFM method reproduces these known results for the same parameter values as those used most recently by Francius & Kharif [**FK06**], with less computational effort, or equivalently, with greater accuracy for comparable computational effort. The comparison is not straightforward. In [**FK06**], the authors used solutions that were allowed to have non-zero average value for a fixed depth whereas with our method, we strictly inforce zero-average values solutions.

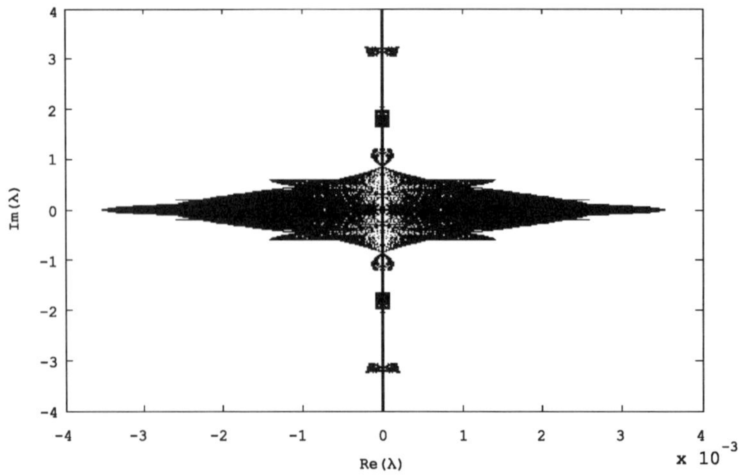

FIGURE 4. The spectrum $\{\lambda\}$ (real vs. imaginary part) for $h = .5$ and amplitude $a = .1$. Computed with $N = 16$ Fourier modes, 1000 μ values, and 300 ρ values.

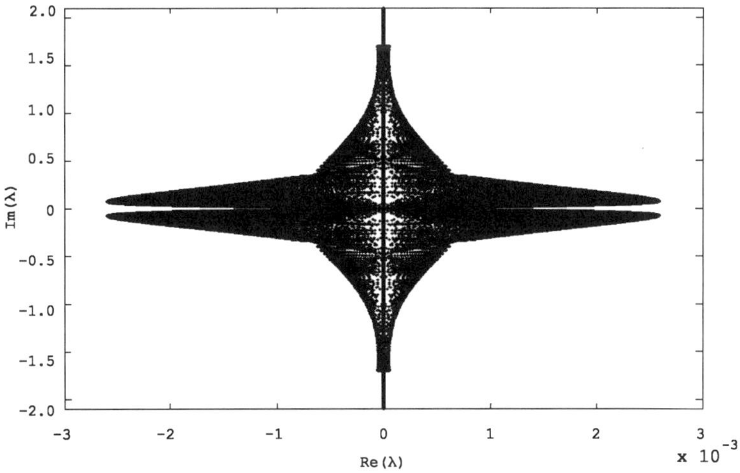

FIGURE 5. The spectrum $\{\lambda\}$ (real vs. imaginary part) for $h = 1.5$ and amplitude $a = .1$. Computed with $N = 16$ Fourier modes, 1000 μ values, and 300 ρ values.

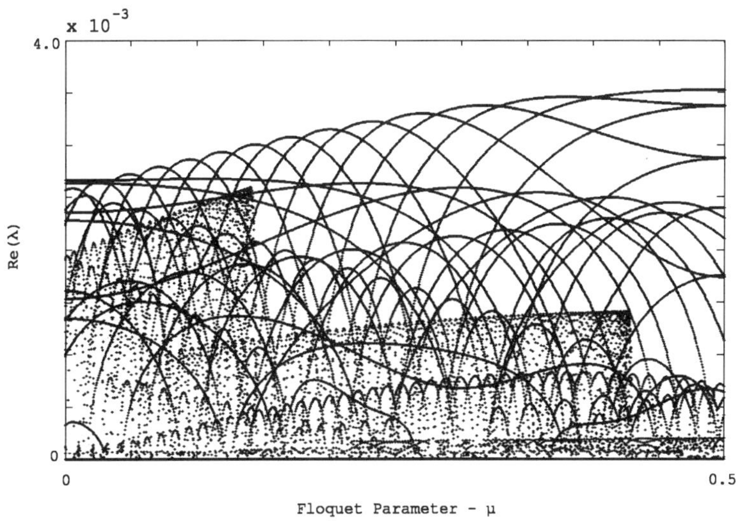

FIGURE 6. The real part of the spectrum as a function of the Floquet parameter μ for depth $h = .5$ and amplitude $a = .1$. Computed with $N = 16$ Fourier modes, 1000 μ values, and 300 ρ values.

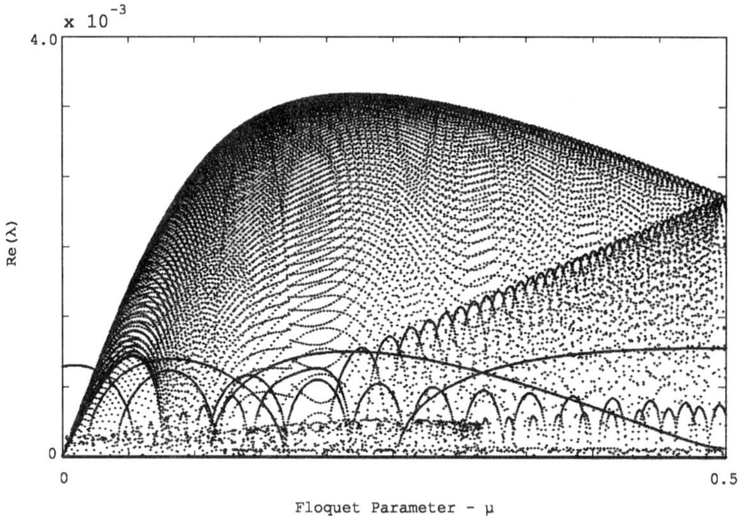

FIGURE 7. The real part of the spectrum as a function of the Floquet parameter μ for depth $h = 1.5$ and amplitude $a = .1$. Computed with $N = 16$ Fourier modes, 1000 μ values, and 300 ρ values.

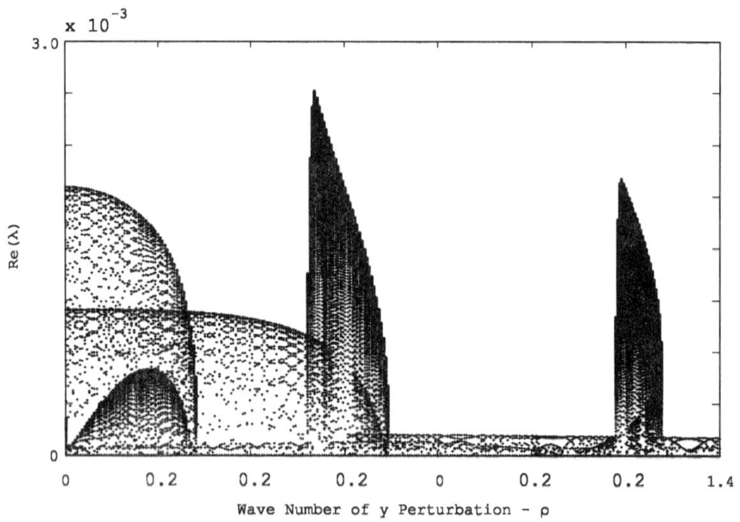

FIGURE 8. The real part of the spectrum as a function of the transverse wave number ρ for depth $h = 1.5$ and amplitude $a = .1$. Computed with $N = 16$ Fourier modes, 1000 μ values, and 300 ρ values.

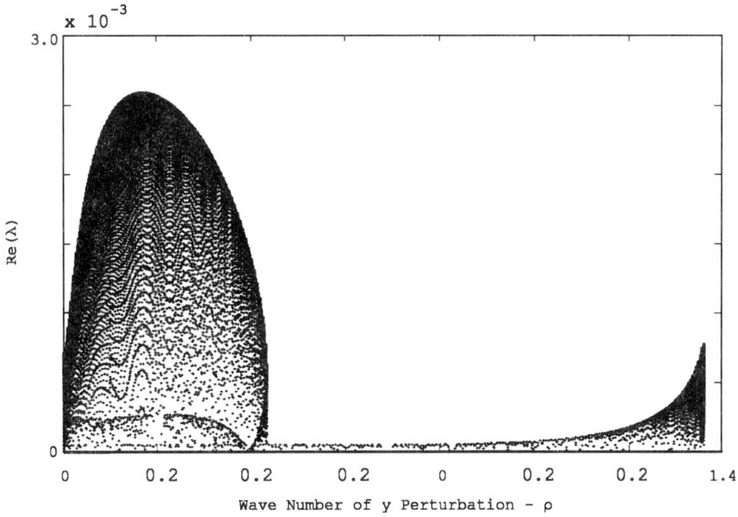

FIGURE 9. The real part of the spectrum as a function of the transverse wave number ρ for depth $h = 1.5$ and amplitude $a = .1$. Computed with $N = 16$ Fourier modes, 1000 μ values, and 300 ρ values.

We begin our investigations in shallow water ($h = .5$) with a solution of moderate amplitude ($a = .1$). We know that for this particular choice of wave height $ak = .1$ and depth $h = .5$ the wave is unstable with respect to one-dimensional perturbations[**DO**]. Further, as expected, we find that this wave is unstable with respect to two-dimensional perturbations as demonstrated in the full spectrum given in Figure 4. Likewise, the full spectrum for $h = 1.5$ is shown in Figure 5. It is important to note that the largest real part of the spectrum in both cases is much larger than the largest real spectral elements when only one-dimensional perturbations are considered (see Ref. [**DO**] for comparison). This demonstrates that the dominant instabilities for these waves are the instabilities that arise from transverse or fully two-dimensional perturbations. However, this is not always the case. For $h = \infty$ and $a = .3$, we found that the dominant instabilities were the result of one-dimensional perturbations.

4.2.1. *Maximum Growth Rates as a Function of μ and ρ.* The full spectra demonstrated in Figures 4 and 5 provide a good visualization of the growth rates and frequencies present in the instabilities. We may also examine the unstable parts of the spectrum as a function of μ and ρ. First, we consider the spectrum as a function of μ. In Figures 6 and 7, we display the real part of the spectra for $h = .5$ and $h = 1$. Additionally, we show the same spectra as a function of ρ in Figures 8 and 9.

It is important to note that as the values of the parameter ρ are increased, bands of large instabilities appear (Figure 8 near $\rho = .6$ and $\rho = 1.25$). As ρ increases, these bands continue to appear. However, the magnitude of the dominant instability in these bands decreases with each successive band. Each of these bands corresponds to a specific resonant curve for the trivial solution, demonstrated in Figure 3.

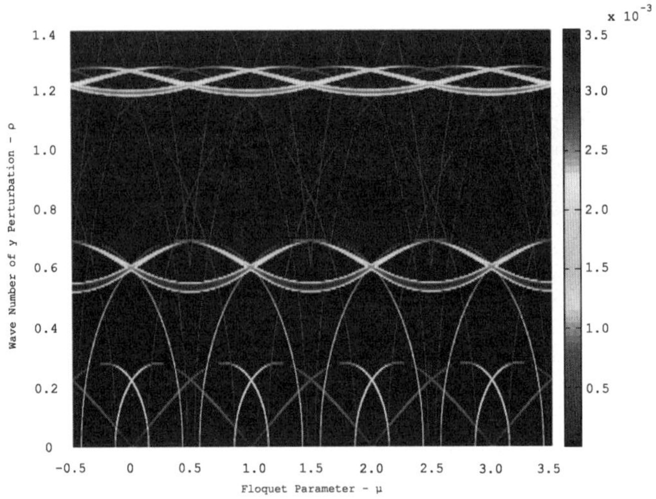

FIGURE 10. The maximum real part of the spectrum as a function of the Floquet parameter μ and the transverse wave number ρ for depth $h = .5$ and amplitude $a = .1$. Computed with $N = 16$ Fourier modes, 1000 μ values, and 300 ρ values.

In addition, we visualize the instabilities in the (μ, ρ) plane where we use color to represent the magnitude of the maximum real part of the spectrum calculated for each perturbation. Figures 10 and 11 show the bands of instability. The periodicity with respect to the Floquet parameter μ is obvious in this representation. The instability bands given in Figure 10 again correspond to the resonant curves for the zero-amplitude solution. We notice the same trends as those found elsewhere

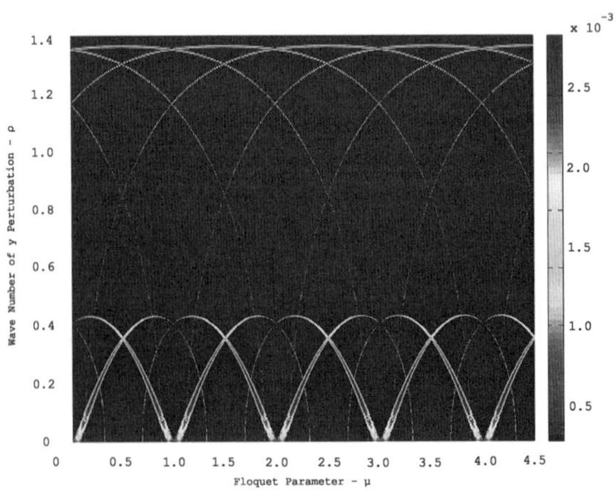

FIGURE 11. The maximum real part of the spectrum as a function of the Floquet parameter μ and the transverse wave number ρ for depth $h = 1.5$ and amplitude $a = .1$. Computed with $N = 16$ Fourier modes, 1000 μ values, and 300 ρ values.

in the literature[**FK06, McL82a, McL82b**]; as the amplitude of the underlying traveling wave increases, the instability bands (which start as curves) widen to unstable bands.

4.2.2. *Eigenfunction Corresponding to the Most Unstable Eigenvalue.* Yet another interesting investigation is to examine the eigenfunctions corresponding to the most unstable perturbations. These eigenfunctions provide a glimpse at the dynamics following the onset of instability, and allow one to predict the patterns that might be observed. For depths $h = .5$ and $h = 1.5$, we calculate the sum of the eigenfunctions corresponding to the quadruplet of eigenvalues $(\lambda, -\lambda, \lambda^*, -\lambda^*)$, where λ is the eigenvalue with maximal real part, to incorporate the Hamiltonian structure of the stability problem. Combining the appropriate eigenfunctions and normalizing the amplitude to one, we obtain the level sets shown in Figures 12 and 13.

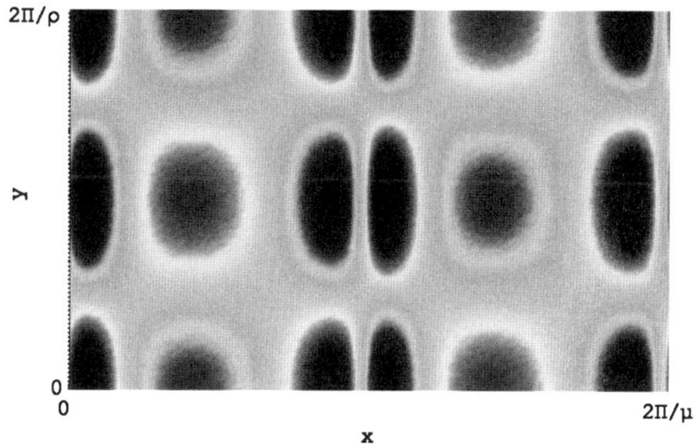

FIGURE 12. The combined eigenfunctions corresponding to the most unstable eigenvalue and its opposite, and their complex conjugates for depth $h = .5$ and amplitude $a = .1$.

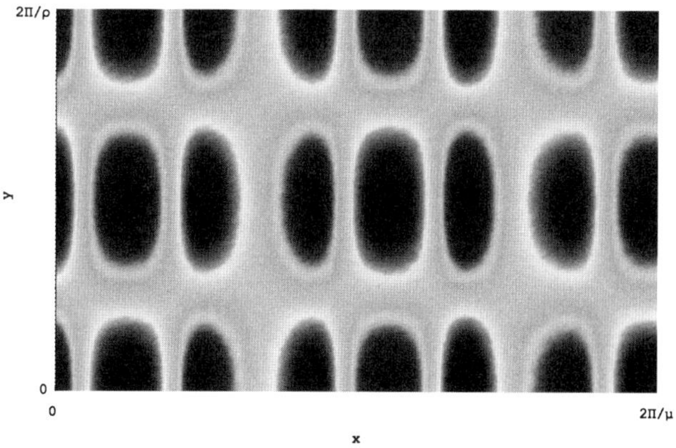

FIGURE 13. The combined eigenfunctions corresponding to the most unstable eigenvalue and its oppose, and their complex conjugates for depth $h = 1.5$ and amplitude $a = .1$.

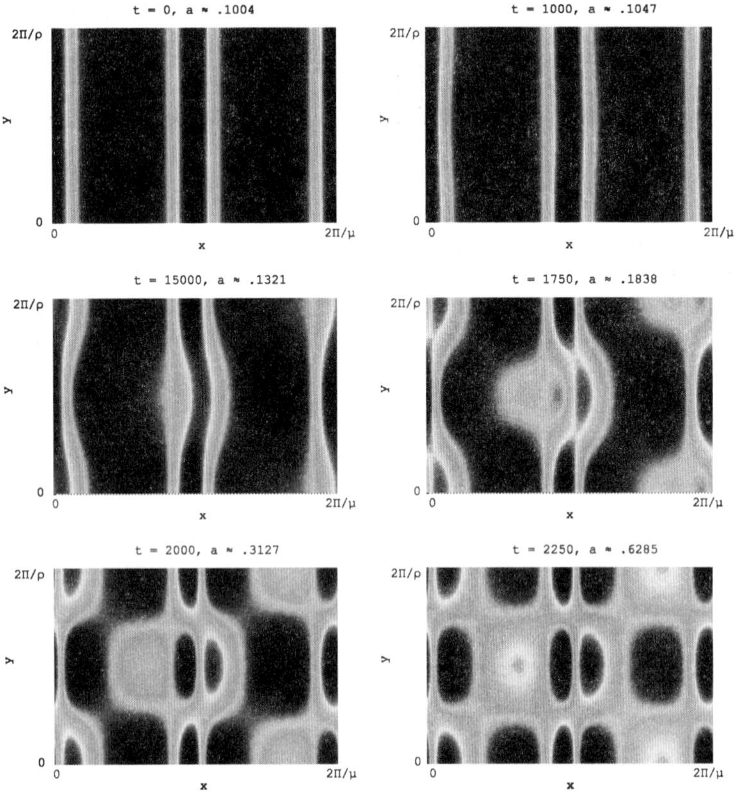

FIGURE 14. The solution to the linear problem for depth $h = .5$ and amplitude $a = .1$ corresponding to the most unstable perturbation. The dynamics displays the so-called snake instability, where wave crests are deformed but the amplitude is mostly unchanged along these deformed crests, at least initially.

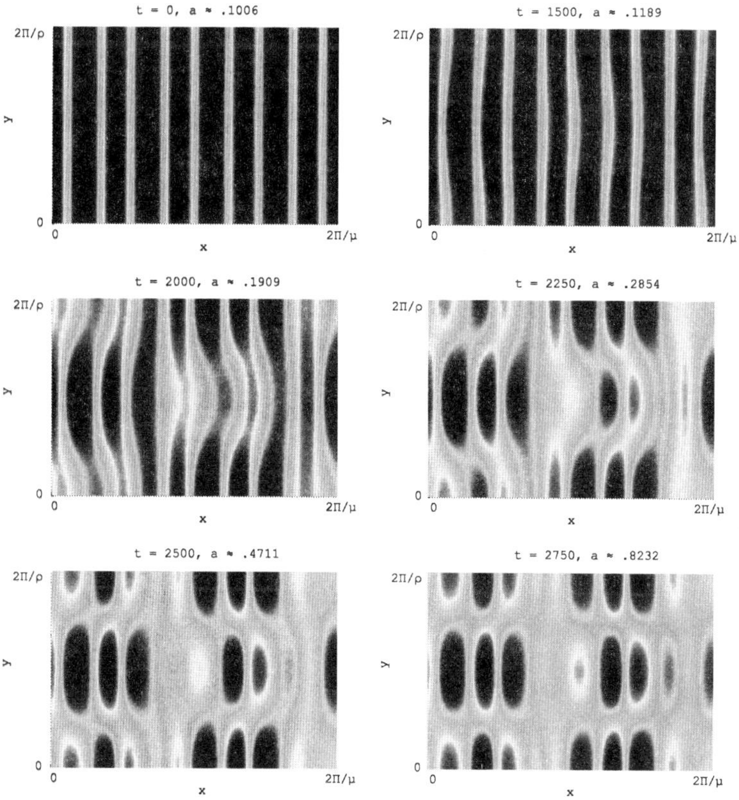

FIGURE 15. The solution to the linear problem for depth $h = 1.5$ and amplitude $a = .1$ corresponding to the most unstable perturbation. The dynamics displays the so-called neck instability, where the amplitude along wave crests oscillates (most notably in the third frame), followed by a secondary snake instability.

Recall that the perturbation of the one-dimensional solution is of the form

$$\eta(x,y,t) = \eta_0(x) + \epsilon e^{\lambda t} e^{i\rho y} e^{i\mu x} \eta_1(x) + c.c. + \mathcal{O}(\epsilon^2)$$

In addition to the mode corresponding to the complex conjugate eigenvalue, we also add those corresponding to the opposite eigenvalue and its complex conjugate. We use this to examine the linear time evolution of the traveling wave. For short times, before nonlinear effects become important, this demonstrates the dynamics of the perturbed wave pattern. Here "short time" should be considered relative to the size of the eigenvalue: the important measure is the product λt. The time steps are shown in Figures 14 and 15. Since the perturbations are unstable, the amplitude of the solutions grow as demonstrated in each of the figures.

5. Conclusions

We have presented a small sampling of the instabilities of one-dimensional traveling wave solutions with respect to two-dimensional perturbations using the AFM formulation. By using the nonlocal formulation in combination with Hill's method, we have determined various spectra numerically with high degree of accuracy. Additionally, as demonstrated in Figure 2, the error of the approximation of the most unstable eigenvalue converges rapidly as a function of the number of Fourier modes used in the truncation.

When comparing our results to the method presented by Francius & Kharif [**FK06**], our method converges for similar solutions using a smaller number of Fourier modes. To quantify this, where 80 Fourier modes are needed in [**FK06**] to obtain eigenvalues with Cauchy error on the order of 10^{-8}, we only require 48 modes to obtain a comparable Cauchy error for the most unstable eigenvalue when $h = 0.3$ and $a \approx 0.10$ (data obtained from the appendix of Ref. [**FK06**], available online only).

TABLE 2. Comparisons of Results with Francius & Kharif for $h = 0.3$ for convergence to six digits of accuracy. For the results from [**FK06**], $a = .1$. However, for our results, $a = .1003$.

		Francius & Kharif ($a = .1$)	
Class Instability	Parameter Values	Modes	$\mathcal{R}(\lambda)_{max}$
Class I ($n = 4$)	$\mu = 0,\ \rho = 0.954$	60	.034650
Class I ($n = 6$)	$\mu = 0,\ \rho = 0.954$	70	.041305
Class II ($n = 3$)	$\mu = 0,\ \rho = 0.954$	70	.021491
Class II ($n = 5$)	$\mu = 0,\ \rho = 0.954$	70	.040446

		Nonlocal ($a = .1003$)	
Class Instability	Parameter Values	Modes	$\mathcal{R}(\lambda)_{max}$
Class I ($n = 4$)	$\mu = 0,\ \rho = 0.954$	48	.033527
Class I ($n = 6$)	$\mu = 0,\ \rho = 0.954$	52	.042458
Class II ($n = 3$)	$\mu = 0,\ \rho = 0.954$	50	.020000
Class II ($n = 5$)	$\mu = 0,\ \rho = 0.954$	60	.040528

TABLE 3. Comparisons of Results with Francius & Kharif for $h = 0.5$ for convergence to six digits of accuracy. For the results from [**FK06**], $a = .17$. However, for our results, $a = .1696$.

		Francius & Kharif		Nonlocal Formulation	
Class Instability	Parameter Values	Modes	$\mathcal{R}(\lambda)_{max}$	Modes	$\mathcal{R}(\lambda)_{max}$
Class I ($n = 4$)	$\mu = 0,\ \rho = 1.102$	70	.068502	48	.073054
Class I ($n = 6$)	$\mu = 0,\ \rho = 3.169$	70	.056946	52	.061713
Class II ($n = 3$)	$\mu = 0.5,\ \rho = 0.475$	80	.051826	48	.053728
Class II ($n = 5$)	$\mu = 0.5,\ \rho = 1.973$	80	.067256	58	.073044

In addition to the convergence rates, we also compared calculated growth rates with previously known results. Tables 2 & 3, show comparisons of the calculated growth rates with the results from Francius & Kharif[**FK06**] when $h = .3$ and $h = .5$. While there are some minor discrepencies between the final calculated growth rates for the chosen values of (μ, ρ), these discrepencies are easily explained since (1) the amplitudes of the traveling waves are slightly different, and (2) the average value of the traveling waves in Francius & Kharif [**FK06**] are nonzero, whereas the average value of the traveling waves using the nonlocal formulation are set to be zero. Furthermore, the results demonstrate the same patters of instabilities which were found by Francius & Kharif[**FK06**]. For example, they showed that for $a = 0.1$ and $h = 0.3$ they dominant instability corresponded to a Class I instability (when $n = 6$). Similarly, we found that the same trend exists for our calculations as well (see Table 2).

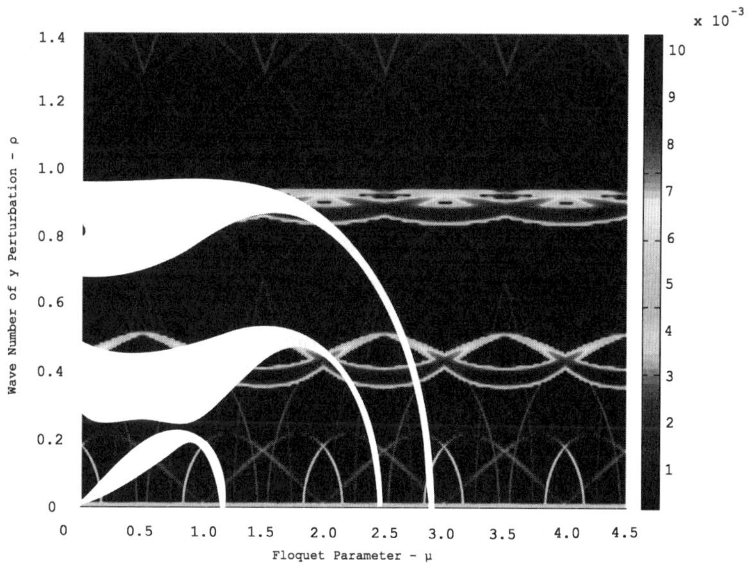

FIGURE 16. The maximum real part of the spectra as a function of Floquet parameter μ and wave number ρ for $h = .5$ and amplitude $a = .160$ compared with the results obtained by Francius and Kharif[**FK06**] (outlined region).

We have obtained results consistent with those of Francius & Kharif. However, we have noticed that our instability regions (regions in the (μ, ρ) plane for which the spectrum intersects the right-half plane) appear to be smaller than those presented in [**FK06**]. For example, in Figure 16 we demonstrate the stability regions obtained in [**FK06**] along with our own computed for $h = .5$ and $a = .160$. As shown in Figure 16, the general shape is comparable, but our instability regions appear to be smaller than those presented by Francius & Kharif.

There are several possible explanations for the discrepancy in the size of the stability region. Perhaps the most logical is that we are able to calculate the underlying traveling wave solution to a high degree of accuracy using far fewer Fourier modes than those used in Francius & Kharif[**FK06**]. This is a benefit of the reduced complexity of the nonlocal formulation for the one-dimensional traveling wave solutions, and perhaps its amenability to spectral numerical methods. In addition, It is important to note that since we incorporated the periodicity with respect to μ in our computations of the spectrum, our results are essentially tiled versions of the results presented in Francius & Kharif[**FK06**].

Summarizing, we have confirmed that for small amplitude waves, the dominant instability is two-dimensional (occurring when $\rho \neq 0$) with the magnitude of the dominant two-dimensional instability was sometimes multiple orders of magnitude greater than the dominant one-dimensional instability. The results were found earlier by Bryant [**Bry78**]. For $h = .5$, he found that for four-wave interactions, the dominant instability was fully two-dimensional. However, as the perturbations were time-evolved in his truncated equations, the two-dimensional instabilities dominated. As mentioned in Francius & Kharif[**FK06**], it would be interesting to explore the time evolution of a traveling wave perturbed with both of the dominant one- and two-dimensional instabilities to determine how each affects the long-time dynamics. This is an interesting possibility for future work since little is know about the time propagation of perturbations in shallow water using the fully nonlinear equations. Another interesting results is that for $h = .5$, the dominant instability corresponds to $\mu = 1/2$, as also noted in the previous literature[**FK06**]. This implies that the dominant instability is phase-locked with the original solution expanded over twice the original period.

References

[AFM06] M. J. Ablowitz, A. S. Fokas, and Z. H. Musslimani, *On a new non-local formulation of water waves*, Journal of Fluid Mechanics **562** (2006), 313–343.

[AH08] M. J. Ablowitz and T. S. Haut, *Spectral formulation of the two fluid Euler equations with a free interface and long wave reduction.*, Analysis and Applications **6** (2008), 323–348.

[Ben67] T.B. Benjamin, *Instability of periodic wave trains in nonlinear dispersive systems.*, Proceedings, Royal Society of London, A **299** (1967), 59–79.

[BF67] T.B. Benjamin and J.E. Feir, *The disintegration of wave trains on deep water. part i. theory.*, Journal of Fluid Mechanics **27** (1967), 417–430.

[Boh47] H. Bohr, *Almost periodic functions*, Chelsea Publishing Company, New York, N.Y., 1947.

[Bry78] P. J. Bryant, *Oblique instability of periodic waves in shallow water*, Journal of Fluid Mechanics **86** (1978), 783–792.

[CD10] C. W. Curtis and B. Deconinck, *On the convergence of Hill's method*, Mathematics of Computation **79** (2010), 169–187.

[CS93] W. Craig and C. Sulem, *Numerical simulation of gravity waves*, J. Comput. Phys. **108** (1993), no. 1, 73–83.

[DK06] B. Deconinck and J. N. Kutz, *Computing spectra of linear operators using Hill's method*, Journal of Computational Physics **219** (2006), 296–321.

[DO] B. Deconinck and K. Oliveras, *The instability of periodic surface gravity waves*, J. Fluid Mech. **Submitted**.

[FK06] M. Francius and C. Kharif, *Three-dimensional instabilities of periodic gravity waves in shallow water*, Journal of Fluid Mechanics **561** (2006), 417–437.

[JZ12] Mathew A Johnson and Kevin Zumbrun, *Convergence of hill's method for nonselfadjoint operators*, SIAM Journal on Numerical Analysis **50** (2012), no. 1, 64–78.

[KR90] C. Kharif and A. Ramamonjiarisoa, *On the stability of gravity waves on deep water*, Journal of Fluid Mechanics **218** (1990), 163–170.

[LH78a] M. S. Longuet-Higgins, *The instabilities of gravity waves of finite amplitude in deep water II. subharmonics*, Proceedings, Royal Society of London, A **360** (1978), 489–505.

[LH78b] M.S. Longuet-Higgins, *The instabilities of gravity waves of finite amplitude in deep water I. superharmonics*, Proceedings, Royal Society of London, A **360** (1978), 471–488.

[McL82a] J. W. McLean, *Instabilities of finite-amplitude gravity waves on water of finite depth*, Journal of Fluid Mechanics **114** (1982), 331–341.

[McL82b] J. W. McLean, *Instabilities of finite-amplitude water waves.*, Journal of Fluid Mechanics **114** (1982), 315–330.

[MMM+81] J. W. McLean, Y.C. Ma, D. U. Martin, P. G. Saffman, and H. C. Yuen, *Three-dimensional instability of finite-amplitude water waves*, Physical Review Letters **46** (1981), no. 13, 817–820.

[MS86] R. S. MacKay and P.G. Saffman, *Stability of water waves*, Proceedings, Royal Society of London, A **406** (1986), 115–125.

[Nic09] D. P. Nicholls, *Spectral data for travelling water waves: singularities and stability*, Journal of Fluid Mechanics **625** (2009), 339–360.

[Whi67] G. B. Whitham, *Non-linear dispersion of water waves*, Journal of Fluid Mechanics **27** (1967), 399–412.

[Zak68] V. E. Zakharov, *Stability of periodic waves of finite amplitude on the surface of a deep fluid.*, J. Appl. Mech. Tech. Phys. **9** (1968), 190–194.

Mathematics Department, Seattle University, Seattle, Washington 98122
E-mail address: oliveras@seattleu.edu

Department of Applied Mathematics, University of Washington, Seattle, Washington 98195
E-mail address: bernard@amath.washington.edu

Contemporary Mathematics
Volume **635**, 2015
http://dx.doi.org/10.1090/conm/635/12715

Relationships between the pressure and the free surface independent of the wave-speed

Katie Oliveras and Vishal Vasan

ABSTRACT. The focus of this paper is the derivation of a direct relationship between the surface of an inviscid traveling gravity-wave in two dimensions, and the pressure at any point in the fluid. We obtain this relationship without approximation and without knowledge of the traveling-wave speed. Using this relationship, we numerically generate the pressure at arbitrary depths beneath a traveling wave. We also demonstrate that this relationship can be used to determine the correct surface profile from pressure measurements inside the fluid domain.

1. Introduction

In field experiments, the surface elevation is often determined indirectly by measuring the pressure along the bottom of the fluid. Using this data, one reconstructs the surface elevation employing the classical relationship between the pressure $p(x)$, constant depth of the fluid h, and the elevation of the surface from a zero-average state $\eta(x)$ given by

$$p = \rho g(h + \eta), \tag{1}$$

where ρ is density, and g is the acceleration due to gravity. This relationship is obtained by linearizing the equations of fluid motion about the trivial solution *i.e.* a flat free surface with no fluid motion. While this simplified relationship is accurate on some scales, this model (which we will refer to as the linear model) fails to reconstruct the surface elevation accurately under the conditions of most interest: nonlinear waves. As discussed in [**3**] and demonstrated in [**14**], for waves in shallow-water, errors in the relationship between p and η given by (1) can easily exceed 20% in terms of the peak amplitude. Thus we must consider nonlinear corrections to equation (1).

Recently several nonlinear relationships between the surface elevation profile of a water wave, $\eta(x)$, and the pressure at the bottom of a fluid $(p(x, -h))$ have been discovered [**4, 6, 7, 11, 14**]. While these new models are a significant improvement over the linear theory, they do require that the pressure be measured at the bottom of the fluid domain and preclude any direct relationship between the pressure in the bulk of the fluid, and the free-surface. Recently, in [**20**], the authors proposed a

2010 *Mathematics Subject Classification.* Primary 76B15; Secondary 35R35, 76B07, 35Q80.
Key words and phrases. Nonlinear waves, Euler's Equations, Surface Gravity Waves.
The first author was supported in part by NSF Grant #DMS-1313049.

method to relate the free surface η with the pressure at any point (x, z) inside the fluid domain for a traveling wave with constant vorticity γ. The method proposed is equally valid in the irrotational case. Further, the relationship between the pressure and the free surface may be inverted for the irrotational case ($\gamma = 0$) to reconstruct the surface elevation profile given the fluid pressure at some vertical height within the fluid [**14**].

Even the fully nonlinear relationships given in [**4**, **14**] between the pressure in the fluid bulk and the free surface suffers from the same limitation that other models suffer from; they all require knowledge of the wave-speed c (or equivalently, the Bernoulli constant). In practice, this is a difficult quantity to measure. If one were to attempt to measure the wave-speed in field experiments, an array of sensors all directly aligned in the direction of the waves propagation would be needed.

For certain physical regimes, knowledge of the wave-speed c is not an absolute prerequisite for relating p and η. Evidently, relation (1) states reconstruction is possible without the wave-speed for linear waves. As shown in [**14**] certain nonlinear corrections do not depend on the wave-speed and can provide remarkably accurate surface-profile reconstructions. Additionally, in the shallow water regime, the wave-speed is well approximated by the quantity $\sqrt{gh}$ and is sufficient to reconstruct the wave profile from pressure measurements [**11**, **14**]. However, as the nonlinearity of the wave-profile is increased, these approximations introduce error into the reconstruction even in shallow water. If the speed c is eliminated from the relationship, then such approximations (i.e. $c = \sqrt{gh}$) would not be needed. Thus, errors in the wave-speed would not results in errors for the reconstruction of highly nonlinear waves.

Demonstrating that the wave-speed c could be completely removed from the relationship between p and η would yield an interesting improvement on current formulae. Here, we demonstrate that this is possible. The new formulation obtained does not yield the most efficient reconstruction method in practice (especially when compared to the heuristic formulation as given in [**14**]). However, it does explain the lack of sensitivity to the wave-speed c as seen in previous numerical results [**11**, **14**]. In addition to eliminating the wave-speed c, we furthermore demonstrate that it is indeed possible to find a relationship between the pressure p at any point inside the fluid bulk and the free-surface η that does not involve the wave-speed c.

The content of this paper is outlined as follows. In Section 2, we present the relevant equations of fluid motion for an incompressible, irrotational inviscid water-wave. In Section 3, the relationship between the pressure at any point in the fluid is derived. This follows directly from the work in [**20**] and yields an implicit relationship between η and p that depends on the wave-speed c. In the following section, we show the wave-speed may be eliminated by employing an operator that maps the normal derivative to the tangential derivative. Finally in Section 5, we present numerical reconstructions of the free surface using pressure measurements made internal to the fluid domain.

2. Equations of motion

We begin by considering the Euler equations for a traveling wave in an ideal, irrotational two-dimensional fluid. The one-dimensional surface profile moves with speed c. These equations are given by

$$\phi_{xx} + \phi_{zz} = 0, \qquad (x, z) \in D, \tag{2}$$

$$\phi_z = 0, \qquad z = -h, \tag{3}$$

$$\eta_x (\phi_x - c) = \phi_z, \qquad z = \eta(x), \tag{4}$$

$$-c\phi_x + \frac{1}{2}\left(\phi_x^2 + \phi_z^2\right) + g\eta = 0, \qquad z = \eta(x), \tag{5}$$

where $\phi(x, z)$ represents the velocity potential of the fluid, $\eta(x)$ represents the surface elevation, and we have ignored the effects of surface tension. In the formulation given by (2-5), the pressure at the free surface of the fluid has been normalized to zero.

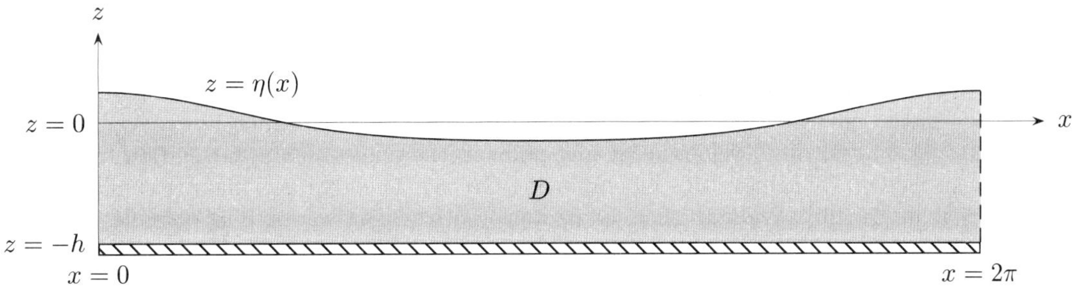

FIGURE 1. The fluid domain.

The primary goal of this paper is to relate the pressure $p(x, z)$ at any point interior to the fluid with the elevation of the fluid surface $\eta(x)$ subject to periodic boundary conditions in x (without loss of generality, we assume that the period is 2π). While the pressure does not show up explicitly in the above formulation, we know the Bernoulli equation is valid throughout the fluid domain for an irrotational fluid. For a traveling wave moving with speed c, we have

$$-c\phi_x + \frac{1}{2}\left(\phi_x^2 + \phi_z^2\right) + p(x, z) = B, \tag{6}$$

where B represents the Bernoulli constant, and

$$p(x, z) = gz + \frac{P(x, z)}{\rho},$$

and represents the non-static portion of the pressure. Thus, if we can find a relationship between $\phi(x, z)$ inside the fluid domain and the surface elevation η, we can directly connect the two quantities of interest. Equation (6) will serve as the foundation for relating the pressure at any point in the fluid with the free surface.

REMARK 1. *In the following we set $B = 0$. This is equivalent to suitably redefining the speed of the wave and hence introducing a horizontal current relative to the frame of reference* [**18**].

3. Relationship between pressure and the free surface

In order to relate the pressure at any point inside the fluid to the free-surface η, we first express the Bernoulli condition at $z = \eta$ in terms of surface variables alone. Let $q(x) = \phi(x, \eta(x))$ represent the velocity potential at the surface. Equations (4) and (5) can be combined to obtain

$$-cq_x + \frac{1}{2}q_x^2 + g\eta - \frac{1}{2}\frac{\left(\eta_x(q_x - c)\right)^2}{1+\eta_x^2} = 0. \tag{7}$$

As shown in [**10**], (7) can be solved explicitly for q_x to find

$$q_x - c = -\sqrt{(c^2 - 2g\eta)(1+\eta_x^2)}, \tag{8}$$

where we have chosen the negative square root to preserve $\phi_x - c \leq 0$ throughout the fluid and at the free surface.

3.1. Interior to the fluid domain.

In order to obtain the pressure at any point in the fluid in terms of the free surface, we need to determine how $\phi_x(x, z)$ and $\phi_z(x, z)$, the fluid velocities at any point in the fluid domain, depend on the free surface η. We achieve this by employing Green's Theorem in a manner similar to that in [**19, 20**]. We note that for irrotational traveling waves, it is possible to extend the fluid domain to a rectangle of the form $(x, z) \in [0, 2\pi] \times [-h, \max(\eta)]$. As noted by [**4**] pg 468, this is possible due to the specific decay rate of the Fourier coefficients of the traveling wave profile [**15**].

Having established the extended harmonic domain we proceed to derive a relation for the fluid velocities at any point in the fluid domain. For this we adapt the procedure in [**1, 19**] and consider a function E which is given by

$$E = e^{-ikx+lz}.$$

For $l = \pm k$, E is a harmonic function of x and z. It is then straightforward to show

$$\left(\phi_x E_z + \phi_z E_x\right)_x + \left(-\phi_x E_x + \phi_z E_z\right)_z = 0, \tag{9}$$

which can easily be verified by expanding the terms in parentheses [**1**]. As the above equation is identically zero throughout the fluid domain and since our goal is to relate information at some depth $z = z_0$ with the free surface at $z = \eta$, we choose to integrate (9) over the domain $(x, z) \in [0, 2\pi] \times [z_0, \eta(x)]$. Figure 2 (a)-(b) depicts two possible domains over which we may integrate (9). Hence we obtain

$$\int_0^{2\pi}\int_{z_0}^{\eta(x)} \left(\phi_x E_z + \phi_z E_x\right)_x - \left(\phi_x E_x - \phi_z E_z\right)_z \, dz\, dx = 0, \tag{10}$$

which is an integral in divergence form. Employing Green's Theorem in the plane, equation (10) becomes

$$\int_{\partial D} \left(E\left(-ik\phi_x - l\phi_z\right)\right) dx + \left(E\left(l\phi_x - ik\phi_z\right)\right) dz = 0. \tag{11}$$

If we restrict k to $\mathbb{Z}\backslash\{0\}$, the integral contributions from $x = 0$ and $x = 2\pi$ cancel by periodicity. Thus we find the relationship given by

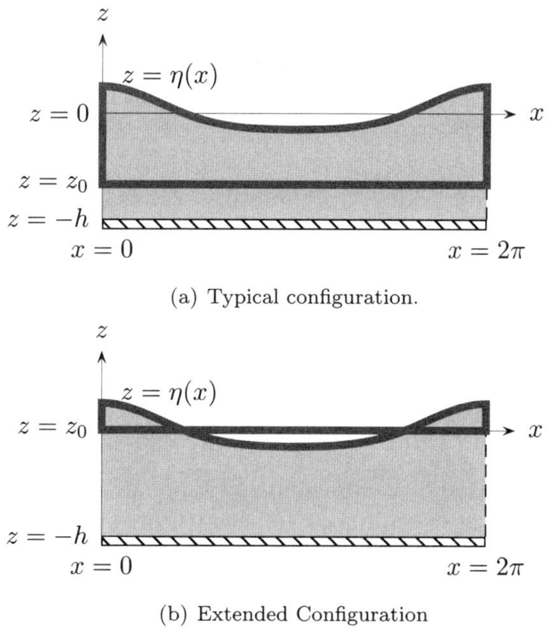

FIGURE 2. Integration paths for various values of $z = z_0$.

$$\int_0^{2\pi} e^{-ikx} e^{lz_0} \left(-ik\phi_x(x, z_0) - l\phi_z(x, z_0)\right) dx$$
$$= \int_0^{2\pi} e^{-ikx} e^{l\eta} \left(-ik\left(\phi_x + \eta_x\phi_z\right) - l\left(\phi_z - \phi_x\eta_x\right)\right)_{z=\eta} dx.$$

As a consequence of $q = \phi(x, \eta(x))$ and (4), we can rewrite the above equation as

$$\int_0^{2\pi} e^{-ikx} e^{lz_0} \left(-ik\phi_x(x, z_0) - l\phi_z(x, z_0)\right) dx = -ik\int_0^{2\pi} e^{-ikx} e^{l\eta} \left(q_x - c\right) dx, \tag{12}$$

where we have integrated the right-hand-side by parts as a matter of convenience.

Following the work of [**1**, **19**], we can consider the separate cases where $l = k$ and $l = -k$. Adding the two equations we find

$$\int_0^{2\pi} e^{-ikx} \left(\phi_x(x, z_0)\right) dx = \int_0^{2\pi} e^{-ikx} \left((q_x - c)\cosh(k(\eta - z_0))\right) dx, \tag{13}$$

whereas on subtracting the two equations we obtain

$$\int_0^{2\pi} e^{-ikx} \left(\phi_z(x, z_0)\right) dx = \int_0^{2\pi} e^{-ikx} \left(i\left(q_x - c\right)\sinh(k(\eta - z_0))\right) dx. \tag{14}$$

Equations (13) - (14) form a system of two equations which can be used to relate the two unknowns ϕ_x and ϕ_z along the horizontal line $z = z_0$. Specifically, since both ϕ_x and ϕ_z are 2π-periodic, (13) and (14) represent the Fourier coefficients of the x- and z-derivatives of the velocity potential ϕ respectively. Thus

$$\phi_x(x,z_0) = \sum_{k\neq 0} e^{ikx}\left(\int_0^{2\pi} e^{-iky}\left((q_x-c)\cosh(k(\eta-z_0))\right)dy\right), \tag{15}$$

$$\phi_z(x,z_0) = \sum_{k\neq 0} e^{ikx}\left(\int_0^{2\pi} e^{-iky}\left(i\,(q_x-c)\sinh(k(\eta-z_0))\right)dy\right). \tag{16}$$

Further, since $q_x - c$ may be expressed directly in terms of c and η using (8), $\phi_x(x,z_0)$ and $\phi_z(x,z_0)$ can be expressed explicitly in terms of the free surface η. Thus upon substituting (15 -16) into (6), we find a direct relationship between the free surface η and the pressure at any point inside the fluid domain $p(x,z_0)$.

4. Eliminating the wave-speed

The expressions derived in the previous section relate the pressure at any point (x,z) inside the fluid to the free surface η by substituting (15 -16) into (6). However, the resulting expression for the pressure requires knowledge of the wave-speed c. Specifically, on substituting (15 -16) into (6) (and using (8)), equation (6) represents a single equation in terms of three unknown quantities: η, p, and c. A special case of this equation is obtained as we limit to the free surface where $p = 0$. Thus we have two equations: one involving η, p and c valid in the interioir of the fluid domain and another involving only η and c valid at the free surface. Given these two relations, we show it is possible to eliminate c and obtain a relationship between p and η alone.

The elimination of the wave-speed is possible because the fluid velocities (15 - 16) are proportional to c. This is a consequence of the tangential velocity at the free surface q_x itself being proportional to the wave-speed c. In the following, using the methods outlined in [**13**], we present a relationship between q_x and η to explicitly show this fact by employing the operator $\mathcal{H}(\eta)$ that maps the normal derivative at the free surface to the derivative of the Dirichlet condition at $z=\eta$.

4.1. Determining the operator $\mathcal{H}(\eta)$. In order to characterize $\mathcal{H}(\eta)$, we consider the solution of the following boundary-value problem

$$\phi_{xx}+\phi_{zz}=0, \qquad (x,z)\in D \tag{17}$$

$$\phi_z-\eta_x\phi_x=f(x), \qquad z=\eta(x) \tag{18}$$

$$\phi_z=0, \qquad z=-h, \tag{19}$$

where $f(x)$ is 2π-periodic and is suitably smooth. As before, we impose 2π-periodic boundary conditions on the gradient of the velocity potential. Note that (18) prescribes that the normal derivative of the velocity potential along the free surface is given by the function $f(x)$.

We introduce the operator $\mathcal{H}(\eta)$ such that $\mathcal{H}(\eta)$ maps the normal derivative of ϕ at the free-surface to the tangential derivative of ϕ at the free-surface. In other words,

$$\mathcal{H}(\eta)\{\phi_z-\eta_x\phi_x\}=\frac{d}{dx}\phi(x,\eta(x)).$$

Following [**1, 2, 13**], it can be shown that $\mathcal{H}(\eta)\{f(x)\}$ satisfies the relationship

$$\int_0^{2\pi} e^{-ikx}\left[i\cosh(k(\eta+h))f(x)-\sinh(k(\eta+h))\mathcal{H}(\eta,D)\{f(x)\}\right]dx=0, \tag{20}$$

where $k \in \mathbb{Z}$.

The operator $\mathcal{H}(\eta)\{f\}$ may either be numerically determined from the above expression or expressed as a Taylor series expansion about $\eta = 0$ (see [**9**, **13**] for more details). Alternately, one may use standard solvers for Laplace's equation to determine $\mathcal{H}(\eta)\{f\}$. Here we briefly outline the Taylor series of the operator expanded about the zero-amplitude solution $\eta = 0$.

To determine a Taylor series expression for $\mathcal{H}(\eta, D)$, we assume that $\mathcal{H}(\eta, D)$ has a series representation in η of the form

$$\mathcal{H}(\eta,D)\{f\}=\sum_{j=0}^{\infty}\mathcal{H}_j(\eta,D)\{f\},$$

where each $\mathcal{H}_j(\eta, D)$ is homogeneous of order j in η, *i.e.* $\mathcal{H}_j(\lambda\eta, D) = \lambda^j\mathcal{H}_j(\eta, D)$. A calculation similar to the one presented in [**2**] allows us to determine the following recursive relationship for $\mathcal{H}_j(\eta, D)$ in terms of lower-order terms:

$$\begin{aligned}\int_0^{2\pi} e^{-ikx}\mathcal{H}_j(\eta,D)\{f\}\,dx &= i\int_0^{2\pi} e^{-ikx}\frac{(k\eta)^j}{j!}\begin{bmatrix}\coth(kh); & j \text{ even}\\ 1; & j \text{ odd}\end{bmatrix} f\,dx\\ &-\int_0^{2\pi} e^{-ikx}\sum_{m=1}^{j}\left(\mathcal{H}_{j-m}(\eta)\{f\}\frac{(k\eta)^m}{m!}\begin{bmatrix}1; & m \text{ even}\\ \coth(kh); & m \text{ odd}\end{bmatrix}\right)dx.\end{aligned} \tag{21}$$

In the above, we have used the brackets [ ] as a conditional multiplier at the appropriate index of summation. Thus, for a suitable function $f(x)$, we can determine $\mathcal{H}(\eta)\{f\}$ through either (20) or (21).

REMARK 2. *From the relationship given by (*21*), it is clear that $\mathcal{H}(\eta)$ is a linear operator acting on $f(x)$. Specifically, $\mathcal{H}(\eta)\{f+g\} = \mathcal{H}(\eta)\{f\} + \mathcal{H}(\eta)\{g\}$, and $\mathcal{H}(\eta)\{\alpha f\} = \alpha\mathcal{H}(\eta)\{f\}$ for any scalar α. Of course, the operator $\mathcal{H}(\eta)$ may be defined abstractly through the solution of the boundary-value problem (*17-19*). Using standard techniques it is possible to show this operator acts linearly on the Neumann condition, much as the classical Dirichlet-to-Neumann operator is proved a bounded linear operator* [**5**].

4.2. Relating the surface and the wave-speed. Returning to the problem of interest, assume that (q_x, η, c) is a solution set to Equations (2-5) and the quantity $-c\eta_x$ is the normal derivative of the potential due to (4). Recalling that q_x is the tangential derivative of the velocity potential at the free-surface, we have

$$\mathcal{H}(\eta)\{-c\eta_x\}=q_x. \tag{22}$$

Substituting $q_x = \mathcal{H}(\eta)\{-c\eta_x\}$ into the Bernoulli equation at the free-surface given by (7), we find

$$-c\mathcal{H}(\eta)\{-c\eta_x\}+\frac{1}{2}\left(\mathcal{H}(\eta)\{-c\eta_x\}\right)^2+g\eta-\frac{1}{2}\frac{\eta_x^2\left(\mathcal{H}(\eta)\{-c\eta_x\}-c\right)^2}{1+\eta_x^2}=0. \tag{23}$$

Since $\mathcal{H}(\eta)\{f\}$ is a linear operator, $\mathcal{H}(\eta)\{-c\eta_x\} = -c\mathcal{H}(\eta)\{\eta_x\}$. Thus, (23) can be explicitly solved for the parameter c^2 to find

$$c^2 = \frac{-2g\eta\left(1+\eta_x^2\right)}{\left(\mathcal{H}(\eta)\{\eta_x\}+1\right)^2-\left(1+\eta_x^2\right)}, \tag{24}$$

where we note that the right-hand side of (24) does not depend on the parameter c. A remark concerning the denominator on the right-hand side of equation (24): for a sufficiently smooth η that satisfies the traveling-wave problem (2-5), equation (23) is satisfied point-wise and any singularity on the right-hand side of (24) is removable. Of course, existence of traveling-wave solutions to (2-5) has been established with appropriate regularity [**8**, **12**, **16**, **17**] and hence we eliminate any possible singularity in the right-hand side of (24).

4.3. Relating the surface and interior. Expression (24) is a consequence of the Bernoulli condition evaluated at the free surface. Assuming there is an analogous statement within the bulk of the fluid in terms of bulk fluid velocities, then as c is a uniform constant for the flow, we may eliminate the wave speed. However, as seen in (15) & (16), the expressions for $\phi_x(x,z_0)$ and $\phi_z(x,z_0)$ depended not only on η, but also on the quantity c. Consequently, we redefine $\phi_x(x,z_0)$ and $\phi_z(x,z_0)$ as

$$\begin{aligned}\phi_x(x,z_0) &= -c\,\mathcal{U}\{\eta\},\\ \phi_z(x,z_0) &= -c\,\mathcal{V}\{\eta\},\end{aligned}$$

where we have introduced the operators $\mathcal{U}\{\eta\}$ and $\mathcal{V}\{\eta\}$. These operators are similar to those given in (15) and (16), with q_x replaced by $-c\mathcal{H}(\eta)\{\eta_x\}$. It is straightforward to show that $\mathcal{U}\{\eta\}$ and $\mathcal{V}\{\eta\}$ are given by

$$\mathcal{U}\{\eta\} = \sum_{k\neq 0} e^{ikx}\left(\int_0^{2\pi} e^{-iky}\left(\left(\mathcal{H}(\eta)\{\eta_x\}+1\right)\cosh(k(\eta-z_0))\right)dy\right), \tag{25}$$

$$\mathcal{V}\{\eta\} = \sum_{k\neq 0} e^{ikx}\left(\int_0^{2\pi} e^{-iky}\left(i\left(\mathcal{H}(\eta)\{\eta_x\}+1\right)\sinh(k(\eta-z_0))\right)dy\right). \tag{26}$$

Substituting the expressions for $\mathcal{U}\{\eta\}$ and $\mathcal{V}\{\eta\}$ into the Bernoulli equation valid inside the bulk of the fluid (6), we find

$$c^2\,\mathcal{U}\{\eta\} + \frac{1}{2}c^2\left(\mathcal{U}\{\eta\}\right)^2 + \frac{1}{2}c^2\left(\mathcal{V}\{\eta\}\right)^2 + p(x,z_0) = 0. \tag{27}$$

Once again, we note the operators $\mathcal{H}(\eta)\{\eta_x\}$, $\mathcal{U}\{\eta\}$, and $\mathcal{V}\{\eta\}$ are independent of the wave speed c. Thus we solve equation (27) for the parameter c^2. Combining this with equation (24) we find

$$\begin{aligned}p(x,z_0)&\left(\left(\mathcal{H}(\eta)\{\eta_x\}+1\right)^2-\left(1+\eta_x^2\right)\right)\\ &= g\eta\left(1+\eta_x^2\right)\left(2\,\mathcal{U}\{\eta\}+\left(\mathcal{U}\{\eta\}\right)^2+\left(\mathcal{V}\{\eta\}\right)^2\right).\end{aligned} \tag{28}$$

Equation (28) relates the pressure at any point in the fluid $p(x,z_0)$ and the free surface η without knowledge of the traveling wave speed c. Notice the operators $\mathcal{U}\{\eta\}$ and $\mathcal{V}\{\eta\}$ may also be defined through the solution of a boundary-value problem precisely as $\mathcal{H}(\eta)\{\eta_x\}$ is defined. Thus in the above expression, the exact representation used for $\mathcal{H}(\eta)\{\eta_x\}$, $\mathcal{U}\{\eta\}$ and $\mathcal{V}\{\eta\}$ is a matter of convenience. One may equally well employ a boundary-integral approach or conformal map to specify these operators. These alternate approaches do not require the fluid extension used here. In the present work we use the global relation approach of [**1**] which affords considerable facility in obtaining asymptotic reductions of the full expression. Such asymptotic expressions will be investigated in future work.

Lastly, we remark that it is immediate that given the surface elevation profile, the pressure in the bulk of the flow is uniquely determined from (28). The problem of surface reconstruction considers the inverse map. In the following section we present numerical evidence of a map from the bottom pressure to the surface elevation profile, with no knowledge of the wave-speed c.

5. Numerics

In this section, we numerically test the relationship between the pressure and the free-surface for numerically computed traveling wave solutions. We consider two separate problems: the forward problem (given $\eta(x)$, find $p(x, z_0)$), and the inverse problem (given $p(x, z_0)$, find $\eta(x)$). In the following sections, we first consider this forward problem where we assume that η is known. Using the pressure data generated by this forward problem, we demonstrate the inverse relationship can be numerically solved. That is, given this pressure data $p(x)$, we can numerically solve (28) in order to recover the original free surface η.

5.1. The Forward Problem. As mentioned in the previous section, the forward problem is direct. In particular, given η, (20) provides a direct map from $\eta \to \mathcal{H}(\eta)\{\eta_x\}$ which then allows us to directly map to p. In previous work, the wave-speed c must be known in order to compute the fully nonlinear relationship between the surface elevation $\eta(x)$ and the pressure $p(x)$. The ideas presented in the previous sections work both with and without knowledge of the wave speed. Here, we outline the numerical procedures in both scenarios: (1) with knowledge of the wave speed c, and (2) without knowledge of c.

5.1.1. *With knowledge of the wave-speed c.* We begin by considering the forward problem where both the wave-speed c and the free surface are known quantities. In terms of the forward problem, given a periodic traveling wave solution set $\eta(x)$ *and* c, one can determine the pressure along various locations inside the fluid domain using both (6) and (28). For example, if we are given $\eta(x)$ with the measured wave-speed c, we compute the pressure at any point in the fluid directly via (6) through the following steps.

First, using both $\eta(x)$ and c, we can determine the quantity $q_x - c$ through (8). This can be achieved using a simple pseudo-spectral method where nonlinear operations are computed in the physical domain and derivatives are taken in the Fourier domain. Next, using the calculated quantity $q_x - c$, ϕ_x and ϕ_z can be determined at the desired depth $z = z_0$ through (15) and (16). Finally, we can substitute the expressions for ϕ_x and ϕ_z directly into (6). This gives the pressure pressure $p(x, z_0)$ at any point in the fluid.

5.1.2. *Without knowledge of the wave-speed c.* Similarly, given $\eta(x)$ without knowledge of c, we numerically compute the map from the free-surface η to $p(x, z_0)$ via (28). As a first step, using $\eta(x)$, we must first determine the operator $\mathcal{H}(\eta)\{\eta_x\}$. This can be achieved one of two different ways.

One option is numerically solve (20) for the quantity $\mathcal{H}(\eta)\{\eta_x\}$. Since (20) is linear in the quantity $\mathcal{H}(\eta)\{\eta_x\}$, solving for this quantity involves inverting a linear

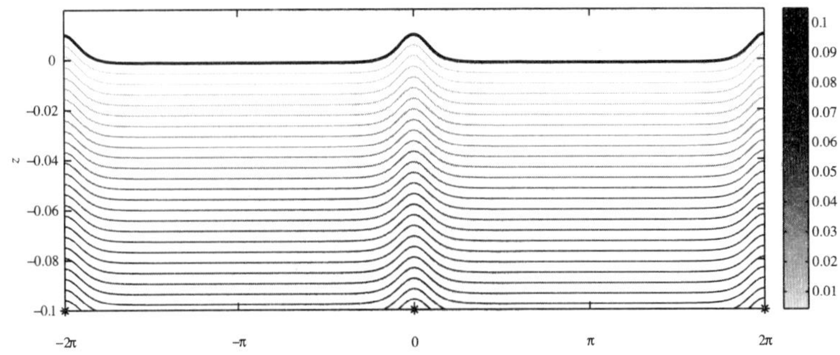

FIGURE 3. Pressure contours for periodic traveling waves with $h = 0.1$, $g = 1$, $L = 2\pi$, and $||\eta||_\infty = 0.01$ calculated using (6) which requires knowledge of the wave-speed c. The asterisk * indicates the point in the fluid domain with the maximum pressure.

system. Specifically, if we represent $\mathcal{H}(\eta)\{\eta_x\}$ by a Fourier series with unknown coefficients of the form

$$\mathcal{H}(\eta)\{\eta_x\} = \sum_{m=-\infty, m\neq 0}^{\infty} e^{imx}\hat{H}_m,$$

we can solve for the coefficients H_m by rewriting (20) in the form

$$\int_0^{2\pi} e^{ikx}\left[i\cosh(k(\eta+h))\eta_x - \sinh(k(\eta+h))\sum_{m=-\infty, m\neq 0}^{\infty} e^{imx}\hat{H}_m\right]dx = 0,$$

for all $k \in \mathbb{Z}/\{0\}$. For numerical purposes, we truncate the admissible values of k such that $k = -N, \dots -1, 1, \dots N$. Similarly, we truncate the infinite series for $\mathcal{H}(\eta)\{\eta_x\}$ generating $2N$ unknown coefficient values. Thus, (5.1.2) generates a system of $2N$ *linear* equations for the $2N$ unknown values of $\hat{H}_m$ which can be easily calculated numerically.

Alternatively, one can use the Taylor series expansion of the operator $\mathcal{H}(\eta)\{\eta_x\}$ as described in [**13**]. We numerically tested both methods (using 10 terms in the Taylor series expansion) and found that for the solutions tested, the results obtained using both methods are comparable.

Using $\mathcal{H}(\eta)\{\eta_x\}$, we can then proceed as before by determining $\mathcal{U}\{\eta\}$ and $\mathcal{V}\{\eta\}$ at the desired depth $z = z_0$ via (25) and (26). Once these quantities have been determined, they can be directly substituted into (28) to determine the pressure $p(x, z_0)$ at any point in the fluid.

Using the parameter values $h = 0.1$, $g = 1$, $\rho = 1$ and $L = 2\pi$, we calculate the relationship between pressure $p(x, z_0)$ and $\eta(x)$ for various solution amplitudes and speeds using either of the above outlined methods. For example, given η, Figure 3 shows lines of constant pressure throughout the fluid domain as calculated via (28).

Similarly, Figure 4 shows various maps from the free-surface $\eta(x)$ to the pressure $p(x, z_0)$ for $z_0 = -h$, $z_0 = -\frac{1}{2}h$, $z_0 = \min(\eta(x))$, and $z_0 = 0$, the last of which extends outside of the fluid domain D. As Figure 4 demonstrates, both methods produce the same pressure profile (consistent up to 10^{-15}) at the desired z_0 values in the fluid. As expected and confirmed, Figure 4 (b) (evaluated at $z_0 = \min(\eta)$) obtains its minimum pressure value at $x = \pi$, precisely where $\eta(x)$ obtains its

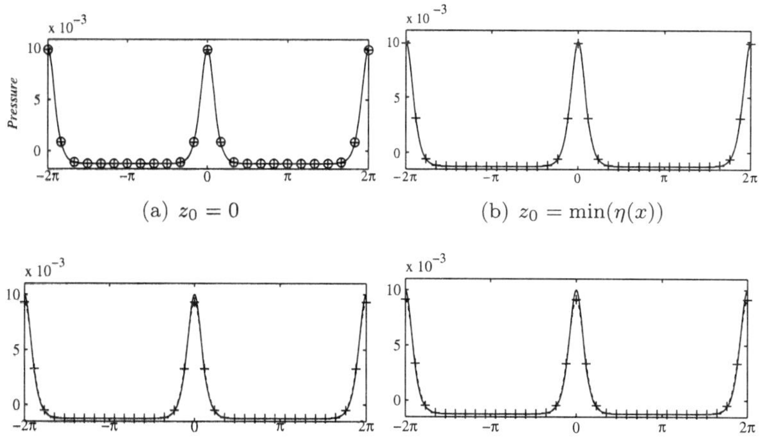

(a) $z_0 = 0$ (b) $z_0 = \min(\eta(x))$

(c) $z_0 = -\frac{1}{2}h$ (d) $z_0 = -h$

FIGURE 4. Pressure calculated at various depths and evaluated for a periodic traveling waves with $h = 0.1$, $g = 1$, $L = 2\pi$, and $||\eta||_\infty = 0.01$. The solid line represents pressure calculated using the methods outlined in [**20**]. The symbol 'o' represents pressures calculated using (6) while '+' represents pressure calculations made using (28).

minimum value. At this point, the pressure is exactly the prescribed value given by the boundary condition $p(x, \eta) = g\eta$ (up to 10^{-15}). Furthermore, Figure 4 (a) examines the pressure along the line $z_0 = 0$. In this case, we have only drawn the pressure for x values such that $(x, 0)$ lies inside or on the boundary of the fluid domain D (see Figure 2(b) for reference).

5.2. The Inverse Problem. We now consider the inverse problem, *given $p(x, z)$ measured at some height $z = z_0$, can we determine $\eta(x)$*. We proceed to only use (28) for this purpose as the wave-speed c is completely eliminated from this formulation. Specifically, using the pressure measured at various depths obtained in the previous section, we numerically solve (28) for the free surface η using a Newton method with an error tolerance of 10^{-14}. As before, we use a pseudo-spectral method with differentiation carried out in Fourier space and multiplication is carried out in physical space.

As an initial guess for our Newton method, we use the zero averaged portion of given pressure which yields a hydrostatic approximation for the free-surface η based on the depth at which the pressure is measured. We find that this initial guess underestimates the peak height of the traveling wave as demonstrated in [**14**] and seen again in Figures 5(a-c) & 7(a-c). As expected, the error between the initial guess (the hydrostatic approximation for the free surface) and the true free-surface η decreases as z_0 increases from the bottom of the fluid ($z = -h$) to the free-surface $z_0 = \eta(x)$.

Specifically, when pressure measurements are made near the free surface for small amplitude waves, the hydrostatic approximation provides a very close estimate to the free-surface η (see for example Figures 6(a) and 8(a)when $z_0 =$

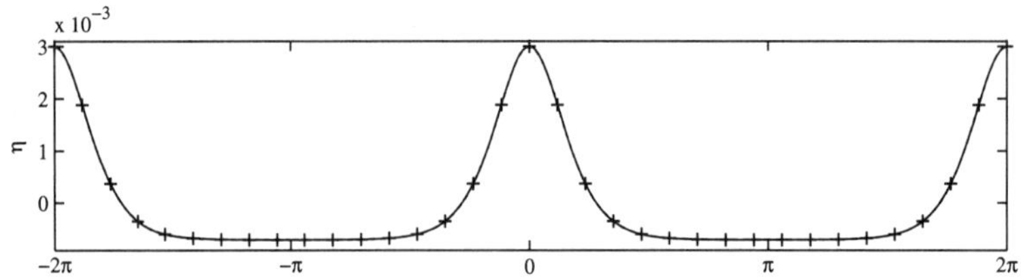

(a) $z_0 = \min(\eta(x))$

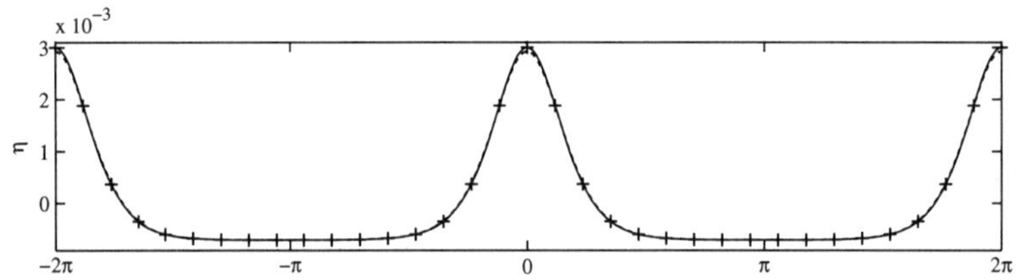

(b) $z_0 = -\frac{1}{2}h$

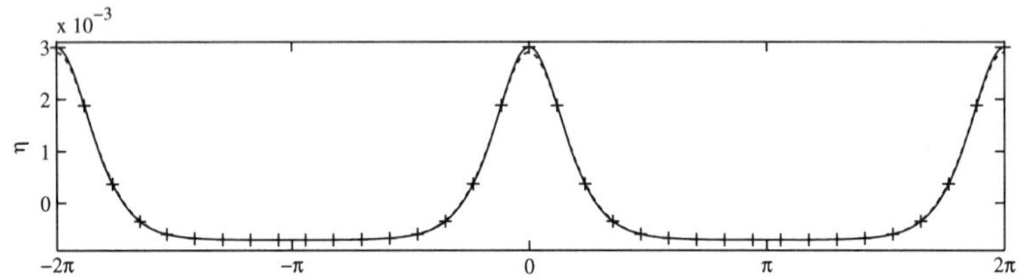

(c) $z_0 = -h$

FIGURE 5. Pressure calculated at various depths and evaluated for a periodic traveling waves with $h = 0.1$, $g = 1$, $L = 2\pi$, and $||\eta||_\infty = 0.003$. The solid line represents the true value of the free surface $\eta(x)$. The dashed line represents the hydrostatic approximation to the free surface, and the '+' represents the reconstruction to the free surface found via (28).

$\min(\eta(x))$). However, as pressure measurements are made at the bottom of the fluid for highly nonlinear waves, this approximation can be off by over 20% (see Figure 8(c)).

While the solutions shown throughout this section are measured in shallow water, the non-dimensional amplitudes are well beyond the linear / KdV regime (see [**14**] for more details). For these nonlinear waves, as seen in Figure (8)(a-c), we are able to use (28) to determine the correct surface profile from pressure measured from multiple depths z_0 up to 10^{-14} without knowledge of the wave speed c.

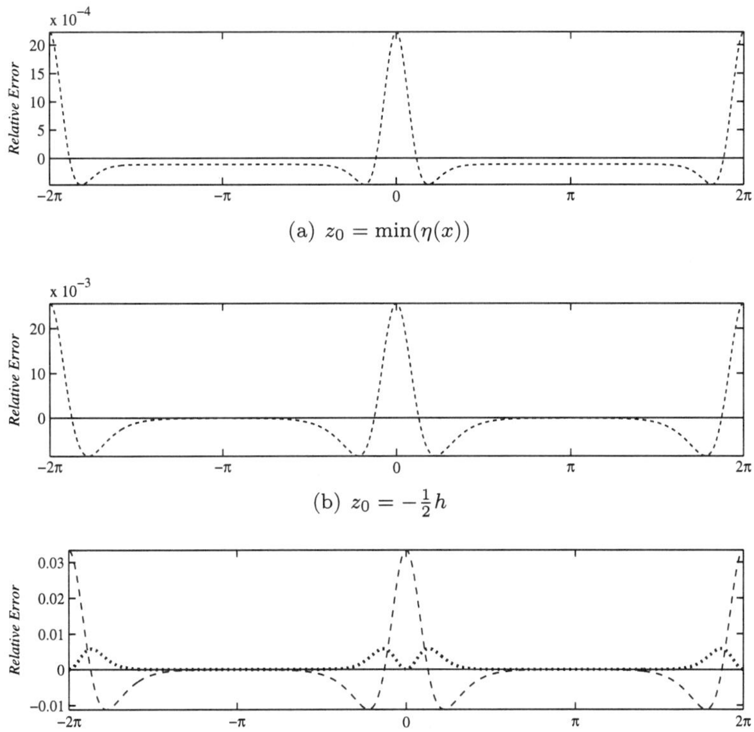

(a) $z_0 = \min(\eta(x))$

(b) $z_0 = -\frac{1}{2}h$

(c) $z_0 = -h$

FIGURE 6. The absolute error between the true value of the free-surface η and the reconstructed free-surface based on pressure measurements at various depths for $h = 0.1$, $g = 1$, $L = 2\pi$, and $||\eta||_\infty = 0.003$. The solid line represents the $|\eta - \eta_r|$ where η_r is the reconstruction from (28). The dashed lined represents the error between the true surface and the hydrostatic approximation, and the dotted line (panel (c) only) represents the error based on a KdV reconstruction (see [**14**] for details).

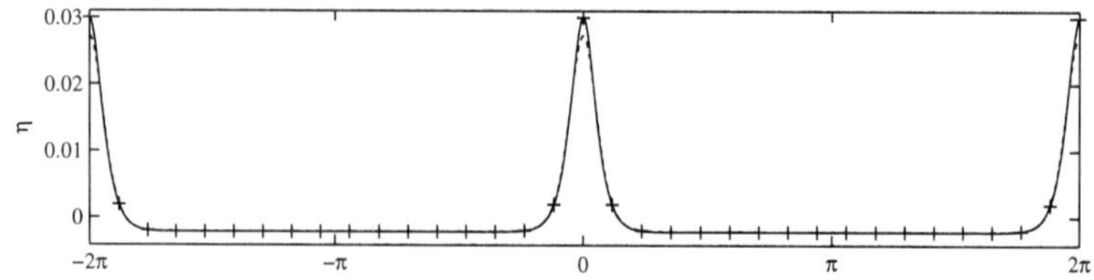

(a) $z_0 = \min(\eta(x))$

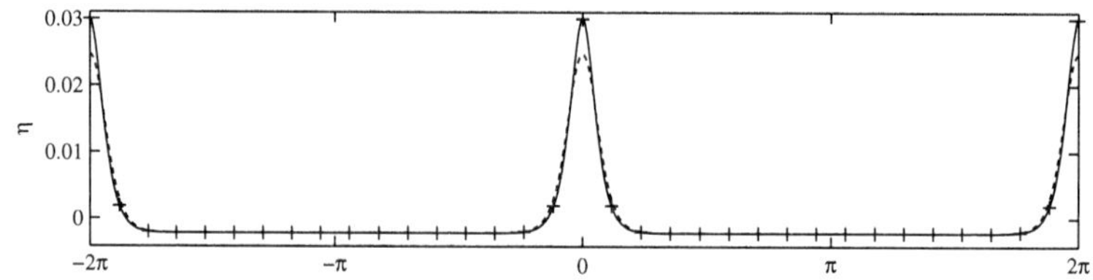

(b) $z_0 = -\frac{1}{2}h$

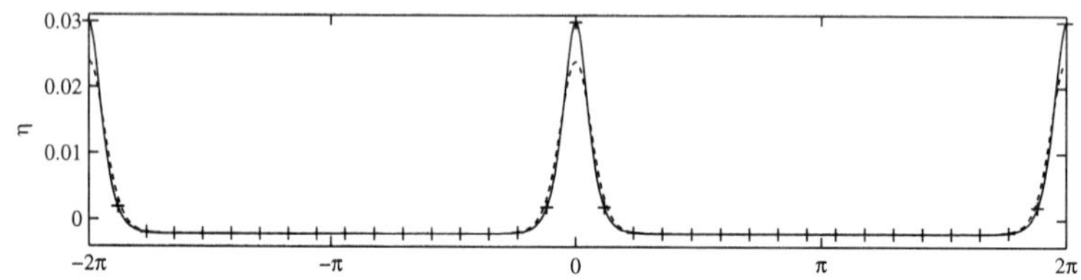

(c) $z_0 = -h$

FIGURE 7. The free-surface η calculated from pressure data measured at various depths and evaluated for a periodic traveling waves with $h = 0.1$, $g = 1$, $L = 2\pi$, and $||\eta||_\infty = 0.03$. The solid line represents the true value of the free surface $\eta(x)$. The dashed line represents the hydrostatic approximation to the free surface, and the '+' represents the reconstruction to the free surface found via (28).

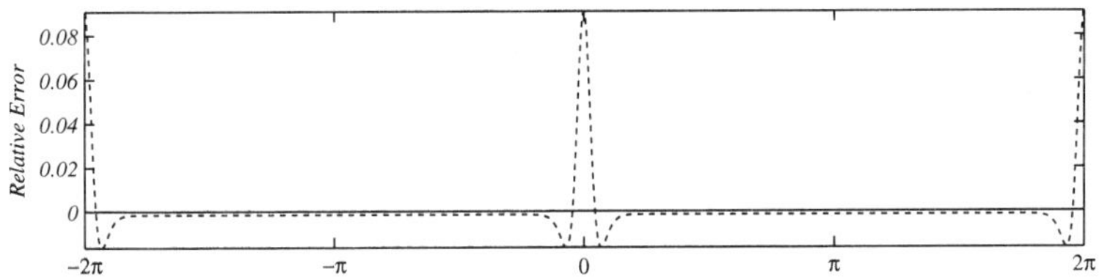

(a) $z_0 = \min(\eta(x))$

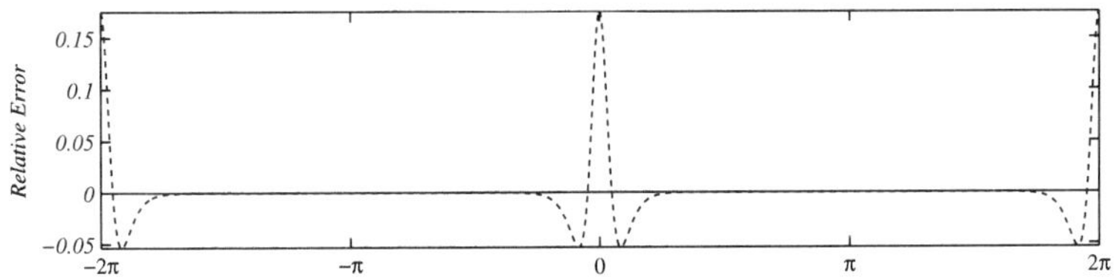

(b) $z_0 = -\frac{1}{2}h$

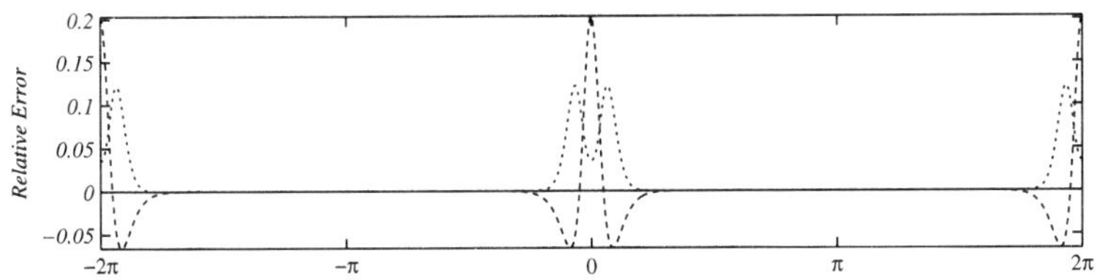

(c) $z_0 = -h$

FIGURE 8. The absolute error between the true value of the free-surface η and the reconstructed free-surface based on pressure measurements at various depths for $h = 0.1$, $g = 1$, $L = 2\pi$, and $||\eta||_\infty = 0.03$. The solid line represents the $|\eta - \eta_r|$ where η_r is the reconstruction from (28). The dashed lined represents the error between the true surface and the hydrostatic approximation, and the dotted line (panel (c) only) represents the error based on a KdV reconstruction (see [**14**] for details).

References

[1] M. J. Ablowitz, A. S. Fokas, and Z. H. Musslimani, *On a new non-local formulation of water waves*, J. Fluid Mech. **562** (2006), 313–343, DOI 10.1017/S0022112006001091. MR2263547 (2007k:76013)

[2] M. J. Ablowitz and T. S. Haut, *Spectral formulation of the two fluid Euler equations with a free interface and long wave reductions*, Anal. Appl. (Singap.) **6** (2008), no. 4, 323–348, DOI 10.1142/S0219530508001213. MR2459114 (2010a:76016)

[3] C. T. Bishop and M. A. Donelan. Measuring waves with pressure transducers. *Coastal Engineering*, 11:309–328, 1987.

[4] D. Clamond and A. Constantin, *Recovery of steady periodic wave profiles from pressure measurements at the bed*, J. Fluid Mech. **714** (2013), 463–475, DOI 10.1017/jfm.2012.490. MR3008068

[5] R. R. Coifman and Yves Meyer, *Nonlinear harmonic analysis and analytic dependence*, Pseudodifferential operators and applications (Notre Dame, Ind., 1984), Proc. Sympos. Pure Math. vol. 43, Amer. Math. Soc., Providence, RI, 1985, pp. 71–78, DOI 10.1090/pspum/043/812284. MR812284

[6] A. Constantin, *On the recovery of solitary wave profiles from pressure measurements*, J. Fluid Mech. **699** (2012), 376–384, DOI 10.1017/jfm.2012.114. MR2923684

[7] Adrian Constantin, Joachim Escher, and Hung-Chu Hsu, *Pressure beneath a solitary water wave: mathematical theory and experiments*, Arch. Ration. Mech. Anal. **201** (2011), no. 1, 251–269, DOI 10.1007/s00205-011-0396-0. MR2807138 (2012f:76032)

[8] Walter Craig and David P. Nicholls, *Travelling two and three dimensional capillary gravity water waves*, SIAM J. Math. Anal. **32** (2000), no. 2, 323–359, DOI 10.1137/S0036141099354181. MR1781220 (2002c:76016)

[9] W. Craig and C. Sulem, *Numerical simulation of gravity waves*, J. Comput. Phys. **108** (1993), no. 1, 73–83, DOI 10.1006/jcph.1993.1164. MR1239970 (94h:76064)

[10] Bernard Deconinck and Katie Oliveras, *The instability of periodic surface gravity waves*, J. Fluid Mech. **675** (2011), 141–167, DOI 10.1017/S0022112011000073. MR2801039 (2012j:76057)

[11] B. Deconinck, K. L. Oliveras, and V. Vasan, *Relating the bottom pressure and the surface elevation in the water wave problem*, J. Nonlinear Math. Phys. **19** (2012), no. suppl. 1, 1240014, 11, DOI 10.1142/S1402925112400141. MR2999408

[12] T. Levi-Civita, *Détermination rigoureuse des ondes permanentes d'ampleur finie* (French), Math. Ann. **93** (1925), no. 1, 264–314, DOI 10.1007/BF01449965. MR1512238

[13] Katie Oliveras and Vishal Vasan, *A new equation describing travelling water waves*, J. Fluid Mech. **717** (2013), 514–522, DOI 10.1017/jfm.2012.591. MR3018618

[14] K. L. Oliveras, V. Vasan, B. Deconinck, and D. Henderson, *Recovering the water-wave profile from pressure measurements*, SIAM J. Appl. Math. **72** (2012), no. 3, 897–918, DOI 10.1137/110853285. MR2968755

[15] Pavel I. Plotnikov and John F. Toland, *The Fourier coefficients of Stokes' waves*, Nonlinear problems in mathematical physics and related topics, I, Int. Math. Ser. (N. Y.) vol. 1, Kluwer/Plenum, New York, 2002, pp. 303–315, DOI 10.1007/978-1-4615-0777-2_18. MR1970618 (2004c:76023)

[16] D. J. Struik, *Détermination rigoureuse des ondes irrotationelles périodiques dans un canal à profondeur finie* (French), Math. Ann. **95** (1926), no. 1, 595–634, DOI 10.1007/BF01206629. MR1512296

[17] J. F. Toland, *Stokes waves*, Topol. Methods Nonlinear Anal. **7** (1996), no. 1, 1–48. MR1422004 (97j:35130)

[18] Vishal Vasan and Bernard Deconinck, *The Bernoulli boundary condition for traveling water waves*, Appl. Math. Lett. **26** (2013), no. 4, 515–519, DOI 10.1016/j.aml.2012.12.008. MR3019985

[19] Vishal Vasan and Bernard Deconinck, *The inverse water wave problem of bathymetry detection*, J. Fluid Mech. **714** (2013), 562–590, DOI 10.1017/jfm.2012.497. MR3008073

[20] V. Vasan and K. L. Oliveras. Pressure beneath a traveling wave with constant vorticity. *submitted for publication*, 2013.

Mathematics Department, Seattle University, Seattle, Washington 98102
E-mail address: oliveras@seattleu.edu

Department of Mathematics, Pennsylvania State University, University Park, Pennsylvania 16802
E-mail address: vasan@math.psu.edu

Contemporary Mathematics
Volume **635**, 2015
http://dx.doi.org/10.1090/conm/635/12713

Comparison of Five Methods of Computing the Dirichlet–Neumann Operator for the Water Wave Problem

Jon Wilkening and Vishal Vasan

ABSTRACT. We compare the effectiveness of solving Dirichlet–Neumann problems via the Craig–Sulem (CS) expansion, the Ablowitz-Fokas-Musslimani (AFM) implicit formulation, the dual AFM formulation (AFM*), a boundary integral collocation method (BIM), and the transformed field expansion (TFE) method. The first three methods involve highly ill-conditioned intermediate calculations that we show can be overcome using multiple-precision arithmetic. The latter two methods avoid catastrophic cancellation of digits in intermediate results, and are much better suited to numerical computation.

For the Craig–Sulem expansion, we explore the cancellation of terms at each order (up to 150th) for three types of wave profiles, namely band-limited, real-analytic, or smooth. For the AFM and AFM* methods, we present an example in which representing the Dirichlet or Neumann data as a series using the AFM basis functions is impossible, causing the methods to fail. The example involves band-limited wave profiles of arbitrarily small amplitude, with analytic Dirichlet data. We then show how to regularize the AFM and AFM* methods by over-sampling the basis functions and using the singular value decomposition or QR-factorization to orthogonalize them. Two additional examples are used to compare all five methods in the context of water waves, namely a large-amplitude standing wave in deep water, and a pair of interacting traveling waves in finite depth.

1. Introduction

The water wave equations, also known as Euler's equations for inviscid, irrotational waves, describe the motion of the free surface of an ideal fluid. For periodic waves in two dimensions over a flat surface in the absence of surface tension, they are given as follows [**9**]:

$$\begin{aligned} \phi_{xx} + \phi_{yy} &= 0, & -h < y < \eta, \\ \eta_t + \phi_x \eta_x &= \phi_y, & y = \eta(x,t), \\ \phi_t + \frac{1}{2}\phi_x^2 + \frac{1}{2}\phi_y^2 + g\eta &= 0, & y = \eta(x,t), \\ \phi_y &= 0, & y = -h. \end{aligned}$$

Here ϕ is the velocity potential (related to the fluid velocity by $\mathbf{u} = \nabla\phi$), η is the graph of the free surface of the water, g is the acceleration of gravity, h is the

2010 *Mathematics Subject Classification.* Primary 76B07, 76B15, 31C20, 45B05, 47H14.

depth of the undisturbed fluid, and subscripts denote partial derivatives. Since we are interested in periodic waves, we consider periodic boundary conditions in the horizontal direction. We note that the above equations represent a free-boundary value problem for Laplace's equation, *i.e.* both η and ϕ are unknowns. These equations readily generalize to three-dimensional fluids (with two-dimensional surfaces); however, in the present work, we limit ourselves to one-dimensional surfaces.

The water-wave equations may be reformulated in terms of only the surface variables $\eta(x,t)$ and $q = \phi(x,\eta,t)$ as [**11**, **33**]

$$\begin{aligned} \frac{\partial \eta}{\partial t} &= G(\eta)q, \\ \frac{\partial q}{\partial t} &= -g\eta - \frac{q_x^2}{2} + \frac{(G(\eta)q + \eta_x q_x)^2}{2(1+\eta_x^2)}, \end{aligned}$$

where $G(\eta)$ represents the Dirichlet-Neumann operator (DNO) defined as

$$G(\eta)q = \phi_y(x,\eta) - \eta_x \phi_x(x,\eta),$$

where ϕ is the solution to the following boundary-value problem

$$\begin{aligned} \phi_{xx} + \phi_{yy} &= 0, & -h < y < \eta, \\ \phi(x,y) &= q(x), & y = \eta, \\ \phi_y &= 0, & y = -h. \end{aligned}$$

Thus $G(\eta)$ maps the given Dirichlet data q to the associated Neumann data at the free surface. Consequently, to evolve the surface variables in time using some numerical scheme, we require the solution to Laplace's equation at every time step. Numerically solving Laplace's equation at every time step is expensive, particularly in three dimensions.

A characterization of $G(\eta)$ that avoids the expensive numerical solution of Laplace's equation is appealing. In the current work we discuss four such characterizations developed in the context of water waves. These four methods are the operator expansion method of Craig & Sulem (CS) [**11**], the transformed field expansion method (TFE) of Bruno & Reitich [**7**] and Nicholls & Reitich [**23**, **24**], the nonlocal implicit formulation of the DNO given by Ablowitz, Fokas & Musslimani (AFM) [**1**], and a dual version to the AFM method, derived by Ablowitz & Haut [**2**], which we denote by AFM*. Each of these methods has had remarkable theoretical utility in deriving reduced models for water waves in various physical regimes [**10**], in deriving conserved quantities [**1**], and also in providing the theoretical framework to pose some inverse problems [**25**, **28**]. Additionally, each of these methods readily generalizes to the case of both varying bottom boundaries and three dimensional fluids.

Of course, many traditional methods to numerically solve Laplace's equation exist, including the boundary integral method, conformal mapping techniques, and the finite element method. In the present work we chose to compare the CS and TFE operator expansions and the AFM/AFM* nonlocal formulations with the boundary integral method. We do not consider conformal mapping [**14**], as it does not extend to three dimensional problems, nor traditional finite elements [**26**], as the trade-off between high-order elements and sparsity of the stiffness matrix makes them expensive when high accuracy is desired. Though it is not usually described in this way, we regard the TFE method as a spectrally accurate variant of the finite element method. The boundary integral method is particularly efficient in two dimensions for periodic problems as the lattice sums involved have an explicit analytical representation. However, in three dimensions, boundary integral methods are considerably more difficult to implement.

The overall goal of the present work is to understand the relative effectiveness and accuracy of each method for two-dimensional fluids. The CS, AFM and AFM* methods involve highly ill-conditioned intermediate calculations, so much of the paper focuses on whether accurate results can be obtained if multiple-precision arithmetic is employed in these intermediate calculations. In particular, for timestepping the water wave, it is important that the methods work if the input data is only known with limited accuracy — additional precision in intermediate calculations is permissible as long as the output is roughly as accurate as the input. Given the care required to obtain high accuracy in the present work, one must be cautious about using these methods in double-precision without carefully monitoring condition numbers and cancellation of digits.

An outline of the paper is as follows. In § 2, we briefly introduce the CS operator expansion, the AFM and AFM* nonlocal formulations, and the boundary integral and TFE representations of the DNO. In §3 we present comparisons of the first four of these methods for specific choices of the free surface η. Here we seek to quantify how accurate the CS, AFM and AFM* methods are. We assume the exact DNO is obtained from the boundary integral method, and, where suitable, perform computations using higher precision. Section 3.1 discusses the subtle cancellation properties associated with the series representation of Craig & Sulem. In §3.2, we present a specific example where the AFM* method (in its usual interpretation as a system of equations for the coefficients of a certain series) is guaranteed to fail, and other instances where AFM and AFM* successfully converge to the correct Neumann data. The subtle cancellation of the CS expansion is mirrored by the rapidly decaying singular values associated with the AFM/AFM* methods. This leads us to consider regularized versions of AFM/AFM* that involve oversampling the AFM basis functions to accurately approximate a Gram-Schmidt orthogonalization procedure via QR factorization or the singular value decomposition. In the SVD approach, we also investigate the use of a pseudo-inverse cutoff threshold. We find that the AFM basis functions can be more efficient at representing solutions than a Fourier basis, but with the drawback of poor conditioning. Finally, in §3.3, we discuss the behavior of all the methods on examples relevant to water waves.

2. Representations of the Dirichlet–Neumann operator

As mentioned in the introduction, the Dirichlet–Neumann operator plays an important role in the mathematical formulation of the motion of surface gravity waves. To efficiently compute the time-dependent motion, we require a fast and efficient means to solve Laplace's equation, or alternatively, a direct method to compute the Dirichlet–Neumann operator. In this section, we outline five commonly used approaches. The first, due to Craig & Sulem [**11**], involves expanding the DNO in a Taylor series. The second, due to Ablowitz, Fokas and Musslimani [**1**], involves deriving a global relation between the Dirichlet and Neumann data that can be used as an integral equation to solve for the Neumann data. The third [**2, 12, 25, 28**] is a dual variant of the second, formulated more directly. The fourth is a boundary integral collocation method [**5, 6, 16, 19, 20, 31, 32**]. And the fifth is the transformed field expansion method of Bruno & Reitich [**7**] and Nicholls & Reitich [**23, 24**].

2.1. Power Series Expansion of the DNO. Consider Laplace's equation

$$\phi_{xx} + \phi_{yy} = 0,$$

posed on the domain $\Omega = \{(x, y) \in \mathbb{R}^2 : 0 < x < L, -h < y < \eta(x)\}$, where η is a smooth periodic function with period L. Further assume that ϕ is periodic in the horizontal variable x with period L, and

$$\phi_y(x, -h) = 0.$$

Thus we restrict ourselves to a flat bottom boundary at $y = -h$. Let $\mathscr{D}(x)$ and $\mathscr{N}(x)$ be the Dirichlet and Neumann values of the function ϕ at $y = \eta(x)$. If either $\mathscr{D}(x)$ or $\mathscr{N}(x)$ is given (in appropriate function spaces), the problem of determining ϕ in Ω is well-posed in the Hadamard sense. As we require the map from the Dirichlet to the Neumann data, assume we are given a Dirichlet condition at the boundary $y = \eta(x)$. The associated Neumann condition at the boundary $y = \eta(x)$ is given in terms of the solution to the following boundary-value problem:

$$\begin{aligned} \phi_{xx} + \phi_{yy} &= 0, & -h < y < \eta(x), \\ \phi(x + L, y) &= \phi(x, y), & -h < y < \eta(x), \\ \eta(x + L) &= \eta(x), & \\ \phi(x, \eta(x)) &= \mathscr{D}(x), & \\ \phi_y(x, -h) &= 0. & \end{aligned}$$

In abstract terms, the Dirichlet–Neumann operator G is given by

$$G(\eta)\mathscr{D} = \phi_y - \eta_x \phi_x,$$

where ϕ satisfies the above boundary-value problem. Note that a function of the form

$$\varphi = \exp(ikx) \cosh(k(y + h)),$$

satisfies Laplace's equation, periodicity and the boundary condition at $y = -h$ for $k = 2\pi n/L$, $n \in \mathbb{Z}$. Hence

$$G(\eta)\varphi(x, \eta) = ke^{ikx} \sinh(k(\eta + h)) - ik\eta_x e^{ik} \cosh(k(\eta + h)). \tag{2.1}$$

It is well-known (see [**11, 23**] and references therein) that for a Lipschitz domain, the DNO is an analytic function of the domain shape. Thus, G has a power series expansion in η. Writing

$$G(\eta) = \sum_{j=0}^{\infty} G_j(\eta),$$

where $G_j(\lambda\eta) = \lambda^j G_j(\eta)$ for $\lambda \in \mathbb{R}$, we obtain an explicit representation for $G_j(\eta)$ from (2.1) by expanding the hyperbolic terms in their respective Taylor series and identifying terms of the same degree in η. To lowest order we obtain

$$G_0 e^{ikx} = k \tanh(kh) e^{ikx}.$$

By decomposing the given Dirichlet condition in a Fourier series, we obtain the following representation of the lowest order term of the DNO

$$\mathcal{F}\big[G_0\mathscr{D}\big] = k \tanh(kh) \mathcal{F}\left[\mathscr{D}\right].$$

Similarly, proceeding to higher order, we obtain further terms in the expansion of the DNO. For instance,

$$G_1 = D\eta D - G_0\eta G_0,$$
$$G_2 = -\frac{1}{2}\left(G_0\eta^2 D^2 - 2G_0\eta G_0\eta G_0 + D^2\eta^2 G_0\right),$$

where $D = -i\partial_x$ and $G_0 = D\tanh(hD)$. Terms such as $\tanh(hD)$ are understood as pseudo differential operators, *i.e.* they are defined through associated Fourier multipliers. We remark that higher order terms of the Taylor expansion of G involve increasingly higher order derivatives. Although the Taylor series of G exists for η with a limited (even finite) degree of smoothness, the formulas for $G_j(\eta)$ are not valid in such cases, or must be interpreted very carefully, as a whole, rather than as a sum of individual operators. Even when η is real analytic, there is a delicate balance existing among the terms that leads to a high degree of cancellation [**23**]. This causes numerical difficulties in finite precision arithmetic. We explore the extent of these cancellations in arbitrary precision arithmetic in §3. Of course, many of these problems can be avoided by flattening the domain through a change of variables [**7**, **17**, **23**], or using boundary integral methods to compute the DNO [**31**, **32**]. However, the original Craig–Sulem expansion remains of theoretical interest, and is the one most closely related to the AFM approach, discussed next.

2.2. AFM Implicit Representation. Following [**1**, **2**], we now derive the global relation of the Ablowitz–Fokas–Musslimani reformulation of the water-wave problem. The distinguishing feature of this reformulation is the implicit nonlocal characterization of the Dirichlet–Neumann operator. Solving the resulting integral equation gives the full DNO, effectively summing all the terms in the expansion of Craig & Sulem without having to compute them order by order.

As above, the functions

$$\psi = \exp(ikx)\cosh(k(y+h)) \tag{2.2}$$

play a role, but now as dual functions rather than basis functions. From Green's second identity, we have

$$\begin{aligned}
0 &= \int_D \left(\psi(\phi_{xx}+\phi_{yy}) - \phi(\psi_{xx}+\psi_{yy})\right) dx\, dy,\\
&= \int_{\partial D}\left(\psi\frac{\partial\phi}{\partial n} - \phi\frac{\partial\psi}{\partial n}\right) dS,\\
&= \int_0^L \psi(x,\eta)\left[\phi_y(x,\eta) - \eta_x\phi_x(x,\eta)\right] dx - \int_0^L \phi(x,\eta)\left[\psi_y(x,\eta) - \eta_x\psi_x(x,\eta)\right] dx,\\
&= \int_0^L \psi(x,\eta)\mathscr{N}(x)dx - \int_0^L \mathscr{D}(x)\left[\psi_y(x,\eta) - \eta_x\psi_x(x,\eta)\right] dx,
\end{aligned} \tag{2.3}$$

where $\partial/\partial n$ is the normal derivative to the surface. Using the definition of ψ and noting that

$$e^{ikx}\left(k\sinh(k(\eta+h)) - ik\eta_x\cosh(k(\eta+h))\right) = -i\partial_x\left(e^{ikx}\sinh(k(\eta+h))\right),$$

we obtain

$$\int_0^L e^{ikx}\cosh(k(\eta+h))\mathscr{N}(x)\,dx = \int_0^L ie^{ikx}\sinh(k(\eta+h))\partial_x\mathscr{D}(x)\,dx, \tag{2.4}$$

which is the Ablowitz-Fokas-Musslimani (AFM) global relation [**1**] for Laplace's equation. Note that the global relation thus obtained is but a rephrasing of Green's identity. In [**1**], the authors obtain this expression in a different, but equivalent, manner. By a spectral collocation technique, this yields an algorithm in which approximate values $\mathscr{N}(x_j)$ are obtained on a grid $\{x_j\}_{j=0}^{M-1}$ by solving a linear system. In that case, (2.4) is enforced for wave numbers $|k| \le (2\pi/L)M/2$.

A third approach, due to Ablowitz & Haut [**2**], can be derived from (2.4) through a type of inverse Fourier transform. The resulting DNO algorithm boils down to solving

$$\sum_k \hat{\Psi}_k e^{ikx}\cosh(k(\eta+h)) = \mathscr{D}(x), \tag{2.5}$$

for the expansion coefficients $\hat{\Psi}_k$, and then computing

$$\mathscr{N}(x) = -i\partial_x\left(\sum_k \hat{\Psi}_k e^{ikx}\sinh(k(\eta+h))\right). \tag{2.6}$$

For those functions $\mathscr{D}(x)$ that permit an expansion of the form (2.5), the normal derivative is expected (from term by term differentiation) to be of the form (2.6).

The operation of solving the integral equation (2.4) is the formal adjoint of solving the system (2.5), (2.6). Indeed, if the (conjugate of the) former is written $\mathscr{N} = (A^*)^{-1}B^*\partial_x\mathscr{D}$, then the latter becomes $\mathscr{N} = -\partial_x BA^{-1}\mathscr{D}$, which are consistent since $G(\eta)$ is self-adjoint. Here

$$\begin{aligned} Ac &= \sum_k c_k e^{ikx}\frac{\cosh(k(\eta(x)+h))}{w_k}, \qquad (A^*f)_k = \int_0^L e^{-ikx}\frac{\cosh(k(\eta(x)+h))}{w_k}f(x)\,dx, \\ Bc &= \sum_k c_k ie^{ikx}\frac{\sinh(k(\eta(x)+h))}{w_k}, \qquad (B^*f)_k = \int_0^L e^{-ikx}\frac{\sinh(k(\eta(x)+h))}{iw_k}f(x)\,dx, \end{aligned} \tag{2.7}$$

and the weights w_k are chosen to make A and B bounded from $\ell^2(\mathbb{Z})$ to $L^2(0,L)$. Formally, these weights cancel internally in the products $(A^*)^{-1}B^*$ and BA^{-1}. Indeed, the results are formally unchanged if both sides of (2.4) are multiplied by w_k^{-1}, or if the weights are absorbed into $\hat{\Psi}_k$ in (2.5), (2.6). These statements are only formal since A and A^* are not invertible.

2.3. Boundary Integral Method. Whereas the AFM* method represents ϕ in the fluid as a superposition of basic solutions of the Laplace equation of the form $e^{ky}e^{ikx}$, the boundary integral method represents ϕ as a superposition of dipoles distributed along the surface,

$$\phi(z) = \int_{-\infty}^{\infty} -\frac{\partial N}{\partial n_\zeta}(z,\zeta(\alpha))\mu(\zeta(\alpha))|\zeta'(\alpha)|\,d\alpha, \qquad N(z,\zeta) = \frac{1}{2\pi}\log|z-\zeta|.$$

Here $\zeta(\alpha) = \alpha + i\eta(\alpha)$ is a parametrization of the free surface and $z = x + iy$ is a field point in the fluid. We then use

$$-\frac{\partial N}{\partial n_\zeta}\,ds = \operatorname{Im}\left\{\frac{\zeta'(\alpha)}{z-\zeta(\alpha)}\right\}d\alpha, \qquad \frac{1}{2}\cot\frac{z}{2} = PV\sum_k \frac{1}{z+2\pi k}$$

to reduce the integral to a period cell, and obtain

$$\phi(z) = \frac{1}{2\pi}\int_0^{2\pi}\widetilde{A}(z,\alpha)\mu(\alpha)\,d\alpha, \qquad \widetilde{A}(z,\alpha) = \operatorname{Im}\left\{\frac{\zeta'(\alpha)}{2}\cot\left(\frac{z-\zeta(\alpha)}{2}\right)\right\}.$$

Using the Plemelj formula [**22**], we take the limit as the field point approaches the boundary from below to obtain a second kind Fredholm integral equation for μ:

$$\text{(2.8)}\qquad \begin{aligned}&\frac{\mu(\alpha)}{2} + \frac{1}{2\pi}\int_0^{2\pi} A(\alpha,\beta)\mu(\beta)\,d\beta = \mathscr{D}(\alpha),\\ &A(\alpha,\beta) = \operatorname{Im}\left\{\frac{\zeta'(\beta)}{2}\cot\left(\frac{\zeta(\alpha)-\zeta(\beta)}{2}\right) - \frac{1}{2}\cot\left(\frac{\alpha-\beta}{2}\right)\right\}.\end{aligned}$$

Including $\frac{1}{2}\cot\left(\frac{\alpha-\beta}{2}\right)$ in the formula has no effect on $A(\alpha,\beta)$, but shows that A is in fact a smooth function when η is smooth. Indeed, as $\beta \to \alpha$, the $(\alpha-\beta)^{-1}$ singularities of the terms in braces cancel, yielding $A(\alpha,\alpha) = \operatorname{Im}\{-\zeta''(\alpha)/[2\zeta'(\alpha)]\}$. Once μ is known, the Neumann data is readily shown to satisfy

$$\text{(2.9)}\qquad \begin{aligned}&\mathscr{N}(\alpha) = \frac{1}{2}H[\mu'](\alpha) + \frac{1}{2\pi}\int_0^{2\pi} B(\alpha,\beta)\mu'(\beta)\,d\beta,\\ &B(\alpha,\beta) = \operatorname{Re}\left\{\frac{\zeta'(\alpha)}{2}\cot\left(\frac{\zeta(\alpha)-\zeta(\beta)}{2}\right) - \frac{1}{2}\cot\left(\frac{\alpha-\beta}{2}\right)\right\},\end{aligned}$$

where H is the Hilbert transform, with symbol $\hat{H}_k = -i\,\mathrm{sgn}(k)$. To carry this out numerically, M collocation points are used to turn the integral equation (2.8) into an $M \times M$ matrix equation, where integrals are approximated by the trapezoidal rule. The derivative and Hilbert transform in (2.9) are easily computed using the FFT. The work involved in setting up and solving these integral equations is very similar to that of the AFM and AFM* methods. However, the condition number is much better in the BIM approach since the underlying infinite dimensional system is a second-kind Fredholm integral equation. This makes a big difference in practice since intermediate calculations need only be done in double-precision to achieve double-precision results, and iterative methods such as GMRES can be employed to reduce the work of solving the equations from $O(M^3)$ to $O(M^2)$. See [**4–6, 16, 19–21, 27, 32**] for similar boundary integral methods, including formulations that incorporate a bottom boundary and allow the interface to overturn.

2.4. Transformed Field Expansion method. The aim of this approach [**7, 23**] is to compute successive terms in the Craig–Sulem expansion via formulas that do not suffer from catastrophic cancellation of digits in floating point arithmetic. The price we pay for this improvement is that the bulk fluid must be discretized. For simplicity, we consider only the finite depth case in two dimensions. The three-dimensional case is considered in [**23, 24**], while infinite depth is treated in [**24**] by introducing a fictitious interface coupling the unbounded problem on a half-space to the finite-depth problem with a curved upper boundary and a flat lower boundary.

Instead of deriving the perturbation expansion for $G(\eta)$ using the ill-conditioned basis functions $e^{ikx}e^{\pm ky}$, as was done in (2.1) above, we perform a boundary-flattening change of variables:

$$u(x,y) = \phi\big(x, (1+h^{-1}\eta)y + \eta\big), \qquad 0 \le x < L, \qquad -h < y < 0.$$

A straightforward calculation reveals that

$$u_{xx} = \phi_{xx} + \big(1+h^{-1}y\big)\big(1+h^{-1}\eta\big)^{-1}\eta_x\partial_y[u_x - \circledast] + \partial_x[\circledast],$$
$$u_{yy} = \phi_{yy} + \Big[1 - \big(1+h^{-1}\eta\big)^{-2}\Big]u_{yy},$$

where $\circledast = (1+h^{-1}y)(1+h^{-1}\eta)^{-1}\eta_x u_y$. Since $\Delta\phi = 0$, we find that

$$\Delta u = \partial_x\big(F_1\big) + \partial_y\big(F_2\big) + F_3,$$

where

$$F_1 = \big(1+h^{-1}y\big)E\eta_x u_y, \qquad E = \big(1+h^{-1}\eta\big)^{-1},$$
$$F_2 = \big(1+h^{-1}y\big)E\eta_x u_x - \big(1+h^{-1}y\big)^2E^2\eta_x^2u_y + \big(1-E^2\big)u_y,$$
$$F_3 = -h^{-1}E\eta_x u_x + h^{-1}(1+h^{-1}y)E^2\eta_x^2 u_y.$$

Next, we write $\eta(x) = \varepsilon f(x)$ and expand E and E^2 in powers of ε to conclude that the terms of the series $u(x,y) = \sum_{n=0}^\infty \varepsilon^n u_n(x,y)$ satisfy

$$\Delta u_0 = 0, \qquad u_0(x,0) = \mathscr{D}(x), \tag{2.10}$$

$$\Delta u_n = \partial_x\big(F_1^n\big) + \partial_y\big(F_2^n\big) + F_3^n, \qquad u_n(x,0) = 0, \tag{2.11}$$

as well as $u_{n,y}(x,-h) = 0$ and $u_n(x+L,y) = u_n(x,y)$. Here

$$F_1^n = (1+h^{-1}y)f_x\sum_{m=0}^{n-1}(-h^{-1}f)^m u_{n-1-m,y},$$

$$F_2^n = (1+h^{-1}y)(F_4 - F_5) + \frac{f}{h}\sum_{m=0}^{n-1}(m+2)(-h^{-1}f)^m u_{n-1-m,y},$$

$$F_3^n = h^{-1}(F_5 - F_4), \qquad F_4^n = f_x\sum_{m=0}^{n-1}(-h^{-1}f)^m u_{n-1-m,x},$$

$$F_5^n = (1+h^{-1}y)f_x^2\sum_{m=0}^{n-2}(l+1)(-h^{-1}f)^m u_{n-2-m,y}.$$

Finally, we use

$$G(\eta)\mathscr{D} = \mathbf{n}\cdot\big[\nabla\phi\big]_{y=0} = \Big[-\eta_x u_x + \big(1+\eta_x^2\big)(1+h^{-1}\eta)^{-1}u_y\Big]_{y=0},$$
$$\mathbf{n} = (-\eta_x, 1), \qquad \nabla\phi = \Big(u_x - (1+h^{-1}y)(1+h^{-1}\eta)^{-1}\eta_x u_y\,,\,(1+h^{-1}\eta)^{-1}u_y\Big)$$

to conclude that $G(\varepsilon f) = \sum_{n=0}^\infty \varepsilon^n G_n(f)$ with

$$G_n(f)\mathscr{D} = -f_x u_{n-1,x} + \sum_{m=0}^{n}(-h^{-1}f)^m u_{n-m,y} + f_x^2\sum_{m=0}^{n-2}(-h^{-1}f)^m u_{n-2-m,y},$$

where empty sums (with upper index smaller than lower index) are zero.

In our code, $u_{n,x}(x,y)$ and $u_{n,y}(x,y)$ are stored on a rectilinear grid with M uniformly spaced mesh points in the x-direction and $N+1$ Chebyshev-Lobatto nodes in the y-direction, mapped by an affine transformation to obtain $y_0 = -h$ and $y_N = 0$. Functions on the grid are stored as matrices with entries in a column indexed by x, holding y fixed. The formulas for $F_j^n(x,y)$ are evaluated pointwise on the grid from the known values of $f(x)$, $f_x(x)$, $u_{m,x}(x,y)$ and $u_{m,y}(x,y)$ for $m = 0,\dots,n-1$. The FFT of the zeroth order term $u_0(x,y)$ is computed from (2.10) by expanding $\mathscr{D}(x)$ in a Fourier series and evaluating $\hat{u}_0(k,y) = \hat{\mathscr{D}}_k \cosh(k(y+h))\operatorname{sech}(kh)$ at the grid points y_j. To obtain $u_{0,x}(x,y)$, we multiply $\hat{u}_0(k,y)$ by ik and take the inverse FFT. Similarly, $u_{0,y}(x,y)$ is obtained by taking the iFFT of $\partial_y \hat{u}_0(k,y)$. This latter function is computed by transforming the rows of $\hat{u}_0$ to their Chebyshev coefficients (also using an FFT), differentiating the Chebyshev polynomials, and evaluating the result on the grid using the Clenshaw recurrence formula. The differentiation procedure amounts to determining the coefficients of $\partial_y \hat{u}_0$ from those of $\hat{u}_0$:

$$\hat{u}_0(k,y) = \textstyle\sum_j \alpha_j(k) T_j(1+2h^{-1}y), \qquad \partial_y \hat{u}_0(k,y) = \textstyle\sum_j \beta_j(k) T_j(1+2h^{-1}y) 2h^{-1},$$
$$\beta_N = 0, \quad \beta_{N-1} = N\alpha_N, \quad \beta_j = (j+1)\alpha_{j+1} + \beta_{j+2}, \quad (j = N-2 : -1 : 0),$$

where $T_j(y)$ is the jth Chebyshev polynomial. Finally, (2.11) is solved using an FFT in the x-direction to convert the PDE into a collection of uncoupled boundary value problems in y. The kth BVP is

$$\partial_y^2 \hat{u}_n(k,y) - k^2 \hat{u}_n(k,y) = ik\hat{F}_1^n(k,y) + \partial_y \hat{F}_2^n(k,y) + \hat{F}_3^n(k,y),$$

subject to $\partial_y \hat{u}_n(k,-h) = 0$, $\hat{u}_n(k,0) = 0$. Instead of using the Chebyshev tau method [**8**, **24**] to solve this BVP, we multiply by a test function, integrate the first term by parts, and proceed as if implementing a finite element method using Chebyshev polynomials as the basis functions. Our implementation is similar to what was done in [**30**] to study the projected dynamics of kinetic diffusion equations in spaces of orthogonal polynomials. Once $\hat{u}_n(k,y)$ is known for each k, we compute $u_{n,x}$ and $u_{n,y}$ as described above for $u_{0,x}$ and $u_{0,y}$.

The key observation is that the derivatives on F_1^n and F_2^n in (2.11), and on u_n at the end of the procedure (to obtain $u_{n,x}$ and $u_{n,y}$), are balanced by the inverse Laplacian in (2.11). In other words, at the point in the algorithm where large numbers enter the computation due to applying derivatives in Fourier space, we divide by even larger numbers by applying the inverse Laplacian. By contrast, in the CS expansion, $G_n(f)$ is expressed as a sum of several terms, each involving n derivatives of products of $\mathscr{D}$ with powers of η. Each derivative amplifies roundoff error, making it difficult to extract the desired sum, which is often many orders of magnitude smaller than the individual terms, as shown below.

3. Comparison of the methods.

In this section we assess the merits and shortcomings of the CS method and the AFM/AFM* methods. In §3.1, we explore the convergence of the CS expansion and quantify the delicate cancellation of terms mentioned in [**23**]. In §3.2, we present an example illustrating a circumstance in which the AFM* method (in its usual interpretation) is guaranteed to fail. By oversampling the columns of the linear operators involved in the AFM/AFM* methods, we are once again able to compute the true normal derivative with spectral accuracy. We also report on the performance of the CS expansion method for the examples considered for the AFM method. Finally in §3.3 we consider the performance of these three methods, as well as the TFE and BIM methods, on two examples in which the Dirichlet data and free surface come from solutions of the water-wave equations. The first is a standing water wave and the second involves two interacting traveling water waves.

3.1. Cancellation properties of the DNO expansion. For simplicity, until §3.3, we restrict to the case of two-dimensional fluids of infinite depth. In that case, the DNO expansion takes the form [**23**]

$$G_0(f) = |D|, \tag{3.12}$$

$$G_n(f) = |D|^{n-1} D \frac{f^n}{n!} D - \sum_{s=0}^{n-1} |D|^{n-s} \frac{f^{n-s}}{(n-s)!} G_s(f), \quad n = 1, 2, 3, \dots, \tag{3.13}$$

where the symbols of D and $|D|$ are k and $|k|$, respectively. Note that $G_n(\eta) = \varepsilon^n G_n(f)$ when $\eta = \varepsilon f$. Computationally, it is convenient to absorb the $s = 0$ term into the first term:

$$G_n(f) = A_n(f) - \sum_{s=1}^{n-1} \frac{1}{(n-s)!} |D|^{n-s} f^{n-s} G_s(f), \quad n = 1, 2, 3, \dots, \tag{3.14}$$

where $A_n(f) = \frac{1}{n!}|D|^{n-1}\big(Df^nD - |D|f^n|D|\big)$ for $n \geq 1$. In Fourier space, $A_n(f)$ is an infinite matrix with the quadrants containing the main diagonal zeroed out:

$$A_n(f)^\wedge_{kj} = \begin{cases} -\dfrac{2|k|^n|j|}{n!}(f^n)^\wedge_{k-j}, & kj < 0, \\ 0 & kj \geq 0. \end{cases} \tag{3.15}$$

This already accomplishes a fair amount of cancellation since $|k-j| > |k|$ when $jk < 0$, so the rapid growth of $|k|^n/n!$ is balanced by decay of $(f^n)^\wedge_{k-j}$ when f is smooth. Indeed, if f is real analytic, one may show that there exist C and ρ such that $|(f^n)^\wedge_k| \leq C^n e^{-\rho|k|}$, which is enough to guarantee that $A_n(f)^\wedge$ maps l^2 sequences $\hat{\varphi}$ to exponentially decaying sequences:

$$\big|(A_n(f)\varphi)^\wedge_k\big| \leq \frac{|Ck|^n e^{-\rho|k|}}{n!} \sum_{j\in J(k)} 2|j|e^{-\rho|j|}\,|\hat{\varphi}_j| \leq \Big(\rho^{-3/2}e^{2\rho}\|\hat{\varphi}\|_{l^2}\Big)\frac{|Ck|^n}{n!}e^{-\rho|k|}.$$

Here $J(k)$ is the set of positive integers when k is negative and the set of negative integers when k is positive. Thus, if f is real analytic, (3.14) implies that $G_n(f)\varphi$ is real analytic for $n \geq 1$ when φ is merely L^2. Also, aside from $n = 0$, $G_n(f)$ is bounded on L^2 when f is real analytic.

Numerical experiments reveal that significant additional cancellations occur in (3.14), beyond combining Df^nD with $|D|f^n|D|$. Indeed, without these cancellations, even when f is real analytic, one would expect $(|D|f)^{n-1}A_1(f)$, which is one of the terms in (3.14) when the recursion is unrolled, to grow super-exponentially with n. In Figure 1, we plot the Frobenius norm of the operators $A_n(f)^\wedge$ and $G_n(f)^\wedge$ versus n for the functions

$$\text{Example 1:} \quad f(x) = \cos(x - \pi/6), \tag{3.16}$$

$$\text{Example 2:} \quad f(x) = \frac{\sinh(1)}{\cosh(1) - \cos(x)}, \tag{3.17}$$

$$\text{Example 3:} \quad f(x) = \sum_k \exp\left(-\frac{3}{2}|k|^{2/3}\right)e^{ikx}, \tag{3.18}$$

which are band-limited, real-analytic, and C^∞, respectively. The Frobenius norm of a matrix (in this case doubly-infinite) is the root sum of squares of the matrix

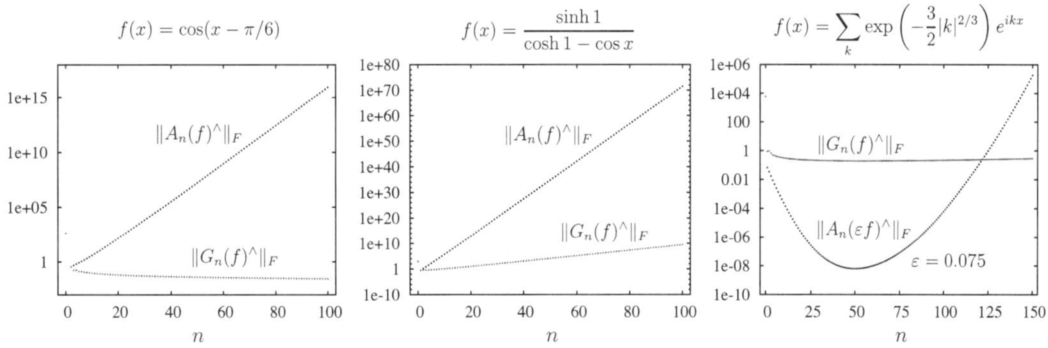

FIGURE 1. Illustration of the cancellations that occur in the recursion (3.14) for three types of functions: band-limited, real-analytic, and C^∞. In the third panel, the Frobenius norm of $A_n(f)^\wedge$ grows so much faster than that of $G_n(f)^\wedge$ that we had to re-scale it by ε^n, $\varepsilon = 0.075$. Note that $\varepsilon^{150} \approx 1.8 \times 10^{-169}$.

entries. We computed it in arbitrary precision arithmetic using (3.15) to evaluate $A_n(f)^\wedge$ and (3.14) to evaluate $G_n(f)$. The computations were done column by column, in parallel. We worked in Fourier space except when computing $f^{n-s}G_s(f)$, which was done by applying the inverse FFT to a column of $G_s(f)^\wedge$, then multiplying by f^{n-s} in real space, and finally applying the FFT again. The Frobenius norm of $A(f)^\wedge$ was computed from the indices in the range

$$k < 0, \qquad j > 0, \qquad |k - j| < M/2, \tag{3.19}$$

where M was chosen large enough that, for the range of n's considered, the terms $A_n(f)^\wedge_{kj}$ with $|k - j| \geq M/2$ are small compared to the largest of those satisfying $|k - j| < M/2$, and may be set to zero. As explained above, this is possible since the exponential decay of $(f^n)^\wedge_{k-j}$ dominates the polynomial growth of $|k|^n|j|$ in (3.15). This M was also used as the number of grid points in the FFT. We include a factor of 2 when summing the squares of the matrix entries to account for the entries $A_n(f)^\wedge_{-k,-j} = \overline{A_n(f)^\wedge_{kj}}$ in the opposite quadrant, $k > 0$, $j < 0$.

The other consideration for choosing M is that errors near the boundary propagate inward when computing $G_n(f)^\wedge_{kj}$. Thus, we choose a smaller integer K and compute the Frobenius norm of $G_n(f)^\wedge$ from the entries with indices

$$-K/2 < k < K/2, \qquad 0 < j < K/2. \tag{3.20}$$

The remaining columns of $G_n(f)^\wedge$ (with $K/2 \leq j < M/2$) are not computed, although the rows are computed out to $-M/2 < k < M/2$. We always zero out the Nyquist frequency, $|k| = M/2$. As before, $G_n(f)^\wedge_{-k,-j} = \overline{G_n(f)^\wedge_{kj}}$ is accounted for in Figure 1 with a factor of $\sqrt{2}$ in the root sum of squares.

The parameters used in these computations were

Example	M	K	$n_{\max}$	bits
1	256	128	100	300
2	2048	330	100	500
3	24576	768	150	1500

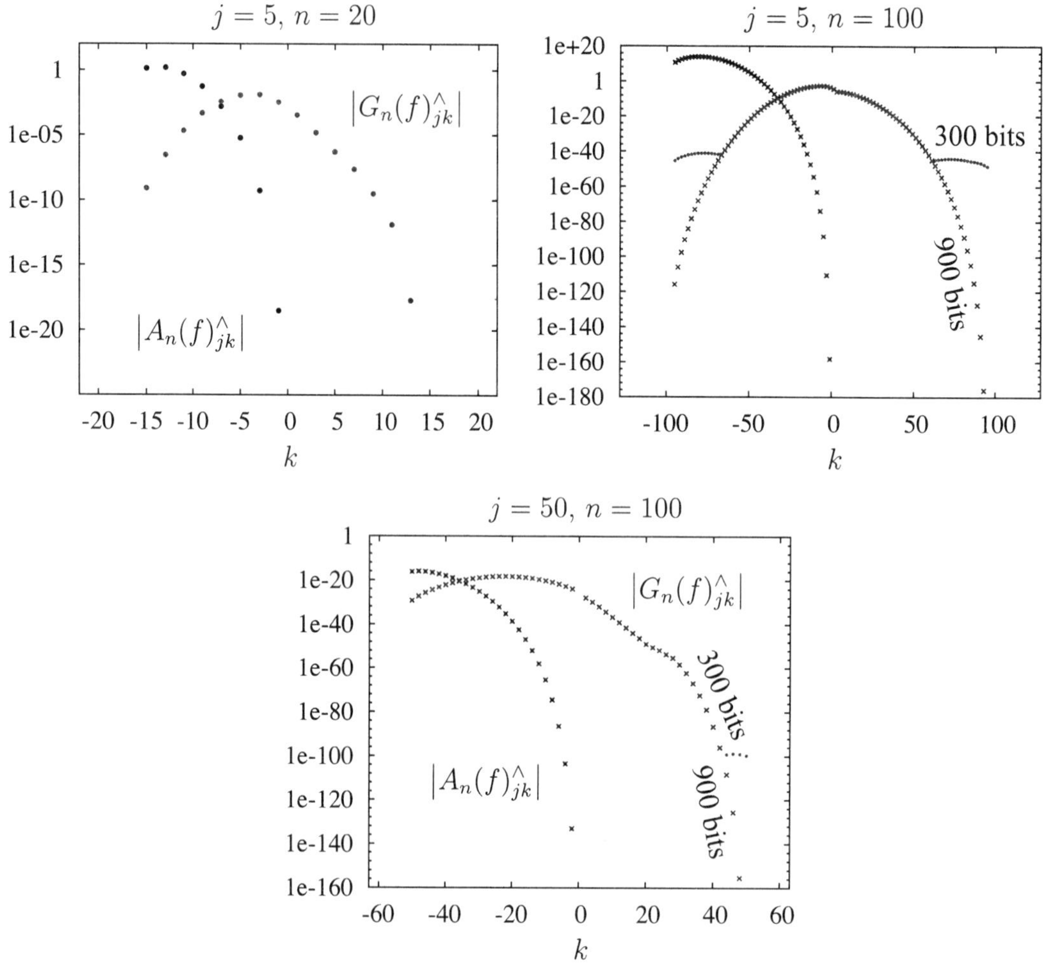

FIGURE 2. Plots of the magnitudes of the non-zero matrix entries in selected columns (indexed by j) of $A_n(f)^\wedge$ and $G_n(f)^\wedge$ for Example 1. The orange and black markers were computed with 300 bits of precision while the blue markers were computed with 900 bits of precision. The support of each column is finite since $(f^n)^\wedge_k = 0$ for $|k| > n$.

Here "bits" refers to the binary precision of the mantissa, where 53 would correspond to double-precision. We used MPFR [**15**] for the floating-point arithmetic, which provides IEEE-like arbitrary precision rounding behavior. We also used double-precision and quadruple-precision arithmetic (using the qd package) in some cases. The following table gives running times for multiplying two 1000×1000 matrices on a 3.33 GHz Intel Xeon X5680 system with 12 cores:

precision	double	quad	300 bits	500 bits	1500 bits
time (in seconds)	0.0157	1.66	12.5	17.1	63.5

The double-precision calculation is particularly fast due to the use of Intel's math kernel library. The higher-precision examples were parallelized using openMP, but do not employ block-matrix algorithms to re-use data that has been pulled from main memory to cache. Our general experience (excluding level 3 BLAS routines such as matrix-matrix multiplication) is that switching from double to quadruple to arbitrary precision slows down the calculation by a factor of 10 each.

In Figure 2, we plot the non-zero matrix entries of $A_n(f)^\wedge$ and $G_n(f)^\wedge$ for Example 1. Because $f(x) = \cos(x - \pi/6)$, the Fourier modes $(f^n)^\wedge_k$ are zero for $|k| > n$ or $k - n$ odd. As a result, $A_n(f)^\wedge$ and $G_n(f)^\wedge$ have only finitely many nonzero terms in this example:

$$\begin{aligned} A_n(f)^\wedge_{kj} &= 0 \quad \text{if} \quad kj \geq 0, \ |k| > n - |j|, \ \text{ or } \ k - j - n \ \text{ is odd,} \\ G_n(f)^\wedge_{kj} &= 0 \quad \text{if} \quad kj = 0, \ |k| > n - |j|, \ \text{ or } \ k - j - n \ \text{ is odd.} \end{aligned}$$

In Figure 2, we explicitly filtered the data to zero out matrix entries of $A_n(f)^\wedge$ and $G_n(f)^\wedge$ with indices in these ranges. If this is not done, roundoff error from the FFT is rapidly amplified by the recurrence (3.14), and requires additional precision to maintain accuracy. This is demonstrated in Figure 3, where we did not filter the data. Increasing the precision from 300 bits to 600 causes the correct values of $G_n(f)^\wedge$ to emerge from the roundoff noise. This is less of an issue for $A_n(f)^\wedge$, which involves errors from taking the FFT of f^n, amplified by $|k|^n/n!$, but no recurrence.

In Figures 4 and 5, we plot selected columns of $A_n(f)^\wedge$ and $G_n(f)^\wedge$ for Examples 2 and 3. The main change from the band-limited case of Example 1 is that the matrices are no longer of finite rank, and do not have compactly supported columns. For small n, the decay rate of each column (with respect to row index k) is still very fast, and the cancellations in obtaining $G_n(f)^\wedge$ from $A_n(f)^\wedge$ is fairly mild. However, as n increases, the cancellations become quite severe. The cancellations can be seen in the figures as the vertical difference from one curve to the other. Recall that $A_n(f)^\wedge$ is the first term in the formula (3.14) for $G_n(f)^\wedge$. The functions (3.17) and (3.18) have Fourier modes of the form

$$\text{Examples 2 and 3:} \qquad \hat{f}_k = e^{-\alpha|k|^\beta}, \tag{3.21}$$

where $\alpha = \beta = 1$ in Example 2 and $\alpha = \beta^{-1} = 3/2$ in Example 3. Example 2 is intended to represent a typical real-analytic function, while Example 3 was designed to check if super-exponential growth in the norms of the operators $A_n(f)$ might cause $G_n(f)$ to also grow super-exponentially with n. To see why $A_n(f)$ grows super-exponentially, note that iterated convolution of $\hat{f}$ with itself yields $(f^n)^\wedge_k \geq \hat{f}_k = e^{-\alpha|k|^\beta}$ for functions of the form (3.21). We then consider the $j = 1$ column of (3.15) and maximize

$$\max_k A_n(f)^\wedge_{k,1} \geq \max_k |k|^n e^{-\alpha|k|^\beta}/n!.$$

The maximum on the right occurs near $k^* = -(n/\alpha\beta)^{1/\beta}$, so we set $\alpha = \beta^{-1}$ and obtain, via Sterling's formula,

$$A_n(f)^\wedge_{k^*,1} \geq (\sqrt{2\pi n})^{\beta^{-1}} (n!)^{\beta^{-1}-1}.$$

We tried $\beta = 1/2$ and $\beta = 2/3$. The decay in the former case was too slow for the problem to be computationally tractable beyond $n = 50$. So we present the results in Figure 5 with $\beta = 2/3$, where we were able to compute terms out to $n = 150$. Returning to Figure 1, the right panel shows that the Frobenius norm of $A_n(f)^\wedge$

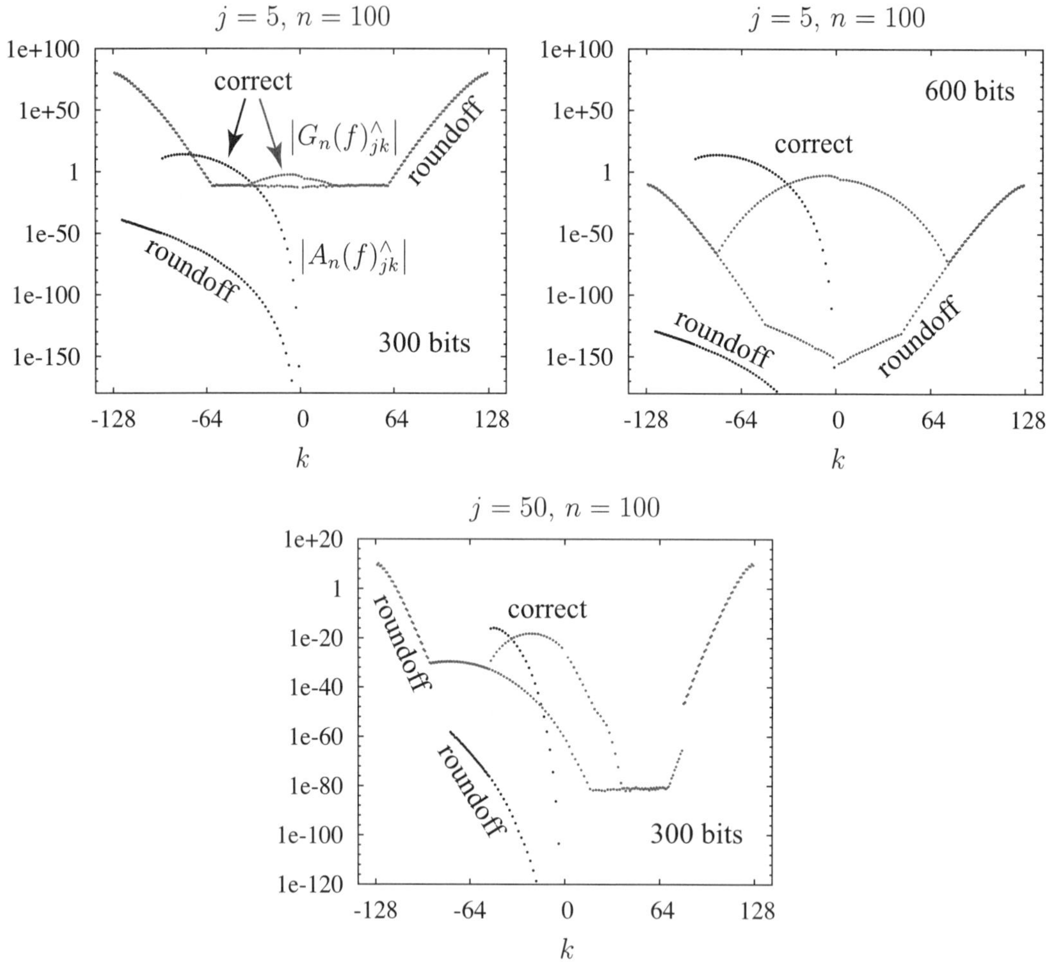

FIGURE 3. Repeat of the calculation of Figure 2 without imposing a zero-pattern filter on the matrices as they are constructed. The $n = 20$ solution has been replaced by a second instance of the $n = 100$ solution. (top left and bottom) With 300 bits of precision, $G_n(f)^\wedge$ is almost entirely corrupted with roundoff errors. (top right) With 600 bits of precision, roundoff error is suppressed enough to achieve an accurate result. Errors are largest near $k = \pm M/2$, and propagates inward as n increases.

does indeed grow super-exponentially, but the cancellations are strong enough that $G_n(f)^\wedge$ remains fairly flat. As a result, $\|G_n(\varepsilon f)^\wedge\|_F$ decays like ε^n for large n even though $\|A_n(\varepsilon f)^\wedge\|_F$ eventually stops decaying for any positive ε.

In addition to visual confirmation that roundoff error has not corrupted $G_n(f)^\wedge_{jk}$ for $|k|$ small, as demonstrated in Figures 3–5, we also validate the results by checking

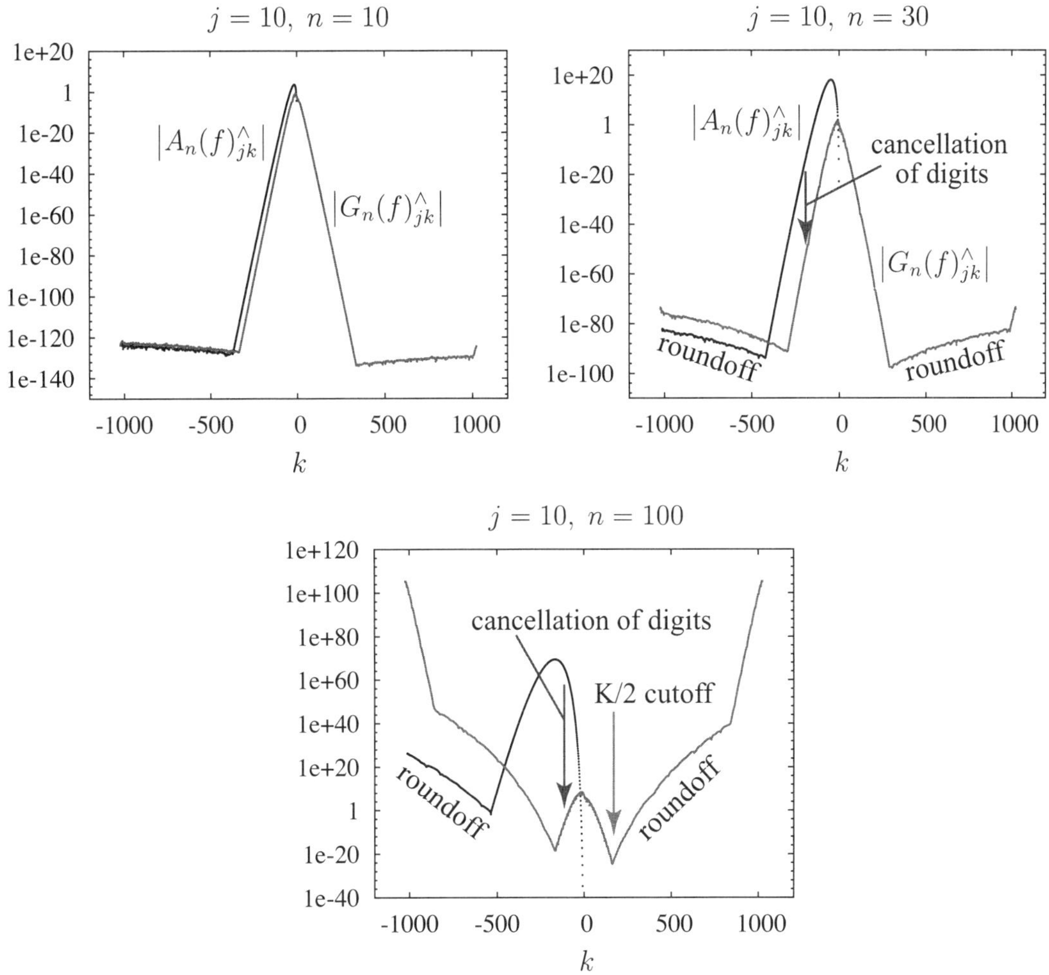

FIGURE 4. Plots of the magnitudes of the matrix entries in the $j = 10$th column of $A_n(f)^\wedge$ and $G_n(f)^\wedge$, with $n = 10, 30, 100$, for Example 2. For small n, the amplitude of $G_n(f)^\wedge_{kj}$ is similar to that of $A_n(f)^\wedge_{kj}$, indicating that little cancellation has occurred. By the time n reaches 100, the leading 90 digits of $A_n(f)^\wedge_{kj}$ have been eliminated in the recurrence (3.14) to obtain $G_n(f)^\wedge_{kj}$ for typical values of j, k.

self-adjointness of each $G_n(f)^\wedge$. This is done by measuring

$$r_n = \frac{\max_{j,k\in\mathbb{K}} \left| G_n(f)^\wedge_{kj} - \overline{G_n(f)^\wedge_{jk}} \right|}{\|G_n(f)^\wedge\|_F}, \qquad \mathbb{K} = \{k \,:\, |k| < K/2\}. \tag{3.22}$$

In Figure 6, we plot r_n versus n for each of the three examples. Note that 300 bits of precision was not sufficient in the unfiltered case of Example 1 to avoid $O(1)$ errors in the symmetry of $G_n(f)^\wedge$, whereas 600 bits gives at least 90 correct

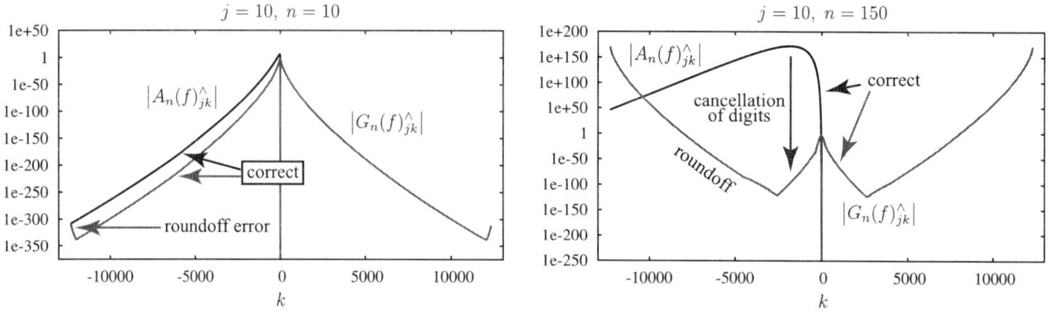

FIGURE 5. Plots of the magnitudes of the matrix entries in the 10th column of $A_n(f)^\wedge$ and $G_n(f)^\wedge$, with $n = 10$ and 150, for Example 3. M had to be chosen quite large ($M = 24576$) to prevent errors from propagating inward to $k = 0$ before $n = 150$. Note that more than 250 leading digits of $A_n(f)^\wedge$ cancel to form $G_n(f)^\wedge$ for this column when $n = 150$.

digits. This is consistent with the results of Figure 3, where the signal is barely distinguishable from the noise in the top left panel, but is many orders of magnitude larger in the top right panel. In hindsight, 1500 bits was overkill for Example 3 since the symmetry errors in $G_n(f)^\wedge$ are still below 10^{-200} when $n = 150$. Further validation of the correctness of the DNO expansion will be given in the following section, where it will be used to compute $G(\eta)\mathscr{D}$ in a case where the solution is known.

3.2. Examples illustrating success and failure of the AFM method. In this section we present examples which show that even for band-limited wave profiles η of arbitrarily small amplitude, the system (2.5-2.6), with the sums interpreted as infinite series, may fail to produce the Neumann data for Dirichlet data of a certain class, which includes real analytic functions, regardless of how many digits of arithmetic are used in the computation.

Both AFM and AFM* methods involve linear operators acting on the unknown Neumann data. In this section, we study the singular value decomposition (SVD) of truncations of these linear operators. By over-sampling the columns, we obtain better approximations of the SVD of the underlying quasi-matrices A and B in (2.7), whose columns are continuous functions. This leads to regularized versions of the AFM/AFM* methods, the simplest version (with no pseudo-inverse cutoff) being equivalent to performing a Gram-Schmidt orthogonalization of the basis functions $e^{ikx}\cosh(k(\eta(x) + h))$ before attempting to represent $\mathscr{D}(x)$ or $\mathscr{N}(x)$ using these functions. The regularized approach enables these methods to be used even if no series of the form (2.5) exists for $\mathscr{D}(x)$.

We now construct a function $\mathscr{D}(x)$ for which no such series exists. Consider the function

$$\phi(x,y) = \frac{1}{2}\mathrm{Im}\left\{\cot\left(\frac{x+iy}{2}\right) - \cot\left(\frac{x+iy+2ih}{2}\right)\right\}, \tag{3.23}$$

which is 2π-periodic in x and harmonic for all x, y outside of the set $(2\pi\mathbb{Z}) \times \{0, -2h\}$. Further, $\phi(x, -h-y) = \phi(x, -h+y)$ and hence $\phi_y(x, -h) = 0$ for all x.

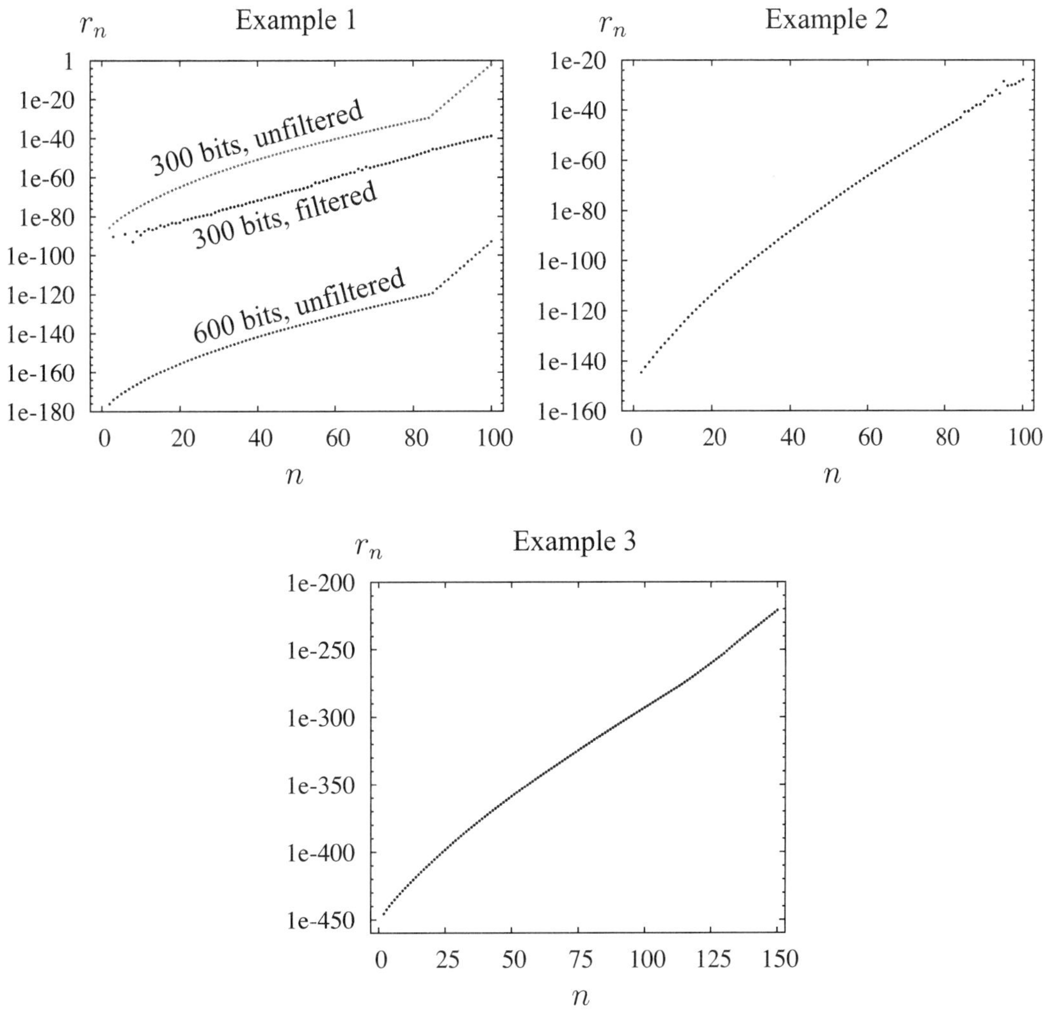

FIGURE 6. Plot of symmetry errors in the matrix $G_n(f)^\wedge$ versus n. While filtering the data improves the band-limited case significantly, the more important factor is the precision of the underlying floating point arithmetic.

Evaluating at $y = -h$, we obtain

$$\phi(x, -h) = \frac{\sinh(h)}{\cosh(h) - \cos(x)} = 1 + 2\sum_{k=1}^{\infty} e^{-kh}\cos kx = \sum_{k=-\infty}^{\infty} e^{-|k|h} e^{ikx}.$$

For values of y in the range $-2h < y < 0$, we may also write

$$\begin{aligned}\phi(x,y) &= 1 + 2\sum_{k=1}^{\infty} e^{-kh}\cosh(ky+kh)\cos kx \\ &= 1 + \sum_{k=1}^{\infty}(e^{ky} + e^{-k(y+2h)})\cos kx, \qquad (-2h < y < 0).\end{aligned} \tag{3.24}$$

In the form (3.23), ϕ is well-behaved except at the poles. However, the AFM formulation is based on representing $\mathscr{D}$ via the series (3.24), which is divergent for $y \geq 0$. Thus, we expect trouble for wave profiles $\eta(x)$ that extend above $y = 0$.

The simplest example illustrating these difficulties is the infinite depth case with

$$\eta(x) = -\varepsilon\cos(x). \tag{3.25}$$

Introducing the weights $w_k = \cosh(k(\eta_{\max} + h))$ in (2.7) and taking the limit as $h \to \infty$, the system (2.4) becomes

$$\int_0^L e^{|k|(\eta-\eta_{\max})}e^{-ikx}\mathscr{N}(x)\,dx = \int_0^L -i\,\mathrm{sgn}(k)e^{|k|(\eta-\eta_{\max})}e^{-ikx}\partial_x\mathscr{D}(x)\,dx, \tag{3.26}$$

where $\eta_{\max} = \max_{0\leq x\leq 2\pi}\eta(x) = \varepsilon$. Similarly, (2.5), (2.6) become

$$\sum_k c_k e^{|k|(\eta-\eta_{\max})}e^{ikx} = \mathscr{D}, \tag{3.27}$$

$$\mathscr{N} = (-i\partial_x)\sum_k c_k\,\mathrm{sgn}(k)e^{|k|(\eta-\eta_{\max})}e^{ikx}. \tag{3.28}$$

In our case, using $\cot\frac{x+iy}{2} = \frac{\sin x - i\sinh y}{\cosh y - \cos x}$ in (3.23), we have

$$\begin{gathered}\phi(x,y) = \frac{1}{2}\left(\frac{-\sinh y}{\cosh y - \cos x} + 1\right), \quad \mathscr{D}(x) = \frac{1}{2}\left(\frac{\sinh(\varepsilon\cos x)}{\cosh(\varepsilon\cos x) - \cos x} + 1\right), \\ \mathscr{N}(x) = \phi_y - \eta_x\phi_x = \frac{\cosh(\varepsilon\cos x)\cos x - 1 + \varepsilon(\sin^2 x)\sinh(\varepsilon\cos x)}{2(\cosh(\varepsilon\cos x) - \cos x)^2}.\end{gathered} \tag{3.29}$$

Since η dips below the poles at $x \in 2\pi\mathbb{Z}$, ϕ is harmonic on $-\infty < y < \eta(x)$. Moreover, $\mathscr{D}(x) = \phi(x,\eta(x))$ is real analytic and 2π-periodic. Nevertheless, there is no solution of (3.27) valid over the whole interval $0 \leq x \leq 2\pi$. From (3.24), we see that the coefficients

$$c_k = \begin{cases} 1 & k = 0 \\ (1/2)e^{|k|\eta_{\max}} & k \neq 0 \end{cases} \tag{3.30}$$

will work over $\{x : \eta(x) < 0\}$, but not elsewhere; see Figure 7.

There remains the possibility that for any prescribed tolerance, a choice of the c_k can be made such that (3.27) is satisfied approximately, to the specified tolerance. In other words, the left-hand side is not treated as a series, but instead as a finite linear combination of basis functions that can approximate $\mathscr{D}$ to arbitrary accuracy. Rather than just add more terms to improve accuracy, it may be necessary to change all the coefficients c_k. To explore this possibility, we construct the M by $K-1$ matrices A and B with entries

$$A_{jk} = \frac{1}{M}\exp\Big(|k|\big[\eta(x_j) - \eta_{\max}\big]\Big)e^{ikx_j}, \qquad B_{jk} = i\,\mathrm{sgn}(k)A_{jk}, \tag{3.31}$$

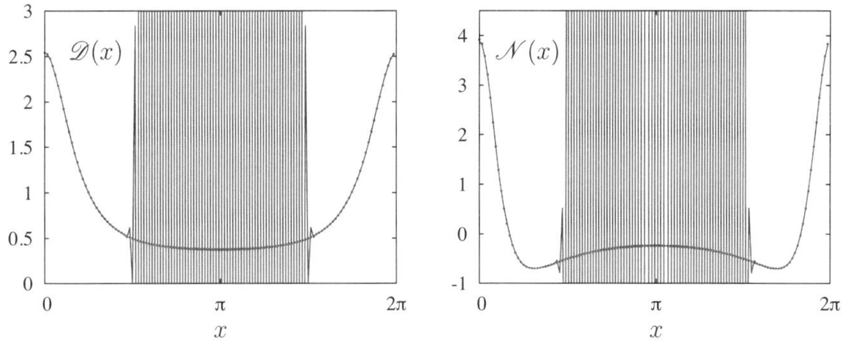

FIGURE 7. Comparison of the series representation (3.27), (3.28), (3.30) of $\mathscr{D}$ and $\mathscr{N}$, (blue curves), to the exact formulas (3.29), (red markers), for $\eta(x) = -\varepsilon \cos x$, $\varepsilon = 0.5$. The series was truncated at $|k| = 64$, and the results were plotted at 128 grid points. A centered finite difference was used to compute the derivative in (3.28). As expected, the series diverges for $\frac{\pi}{2} \le x \le \frac{3\pi}{2}$, where $\eta(x) \ge 0$.

where $-K/2 < k < K/2$, $x_j = 2\pi j/M$, and $M \ge K$. We then compute the singular value decomposition $A = USV^*$, with S and V square and U of dimension $M \times (K-1)$, and evaluate

$$\mathscr{N} = U \operatorname{pinv}(S) V^* B^* \partial_x \mathscr{D}, \qquad \text{(AFM)}, \tag{3.32}$$

$$\mathscr{N} = -\partial_x B V \operatorname{pinv}(S) U^* \mathscr{D}, \qquad \text{(AFM}^*\text{)}. \tag{3.33}$$

Here ∂_x is computed spectrally via the FFT (with no filter). The idea here is to sample each column of A and B in (2.7) with enough points that computing the SVD of the resulting matrix A in (3.31) is equivalent (up to scaling by $\sqrt{M}$ in various places) to computing the SVD of the quasi-matrix A in (2.7), whose columns are continuous, L^2 functions:

$$A = USV^*, \qquad U : \mathbb{C}^{K-1} \to L^2, \qquad S : \mathbb{C}^n \to \mathbb{C}^n, \qquad V : \mathbb{C}^n \to \mathbb{C}^n. \tag{3.34}$$

Here A has been truncated to have $K-1$ columns, U and V are unitary, S is diagonal with positive decreasing entries, and L^2 is equipped with the inner product $\langle f, g \rangle = \frac{1}{2\pi} \int_0^{2\pi} f\bar{g}\, dx$ to avoid factors of $\sqrt{2\pi}$ elsewhere. The columns of U form an orthonormal basis for the column-span of A. Once sufficient grid resolution is reached, approximating $\mathscr{D}$ and $\mathscr{N}$ at the M collocation points leads to accurate approximation throughout $(0, 2\pi)$, using trigonometric polynomials to interpolate between grid points. The columns of the matrix version of U can be thought of as sampled versions of the columns of the quasi-matrix U from (3.34), up to a factor of $\sqrt{M}$. For smaller values of M, trigonometric interpolation becomes less accurate, and the errors can be amplified significantly on division by small singular values. We remark that if the pseudo-inverse in (3.32) and (3.33) is replaced by an inverse, then one can use a QR factorization instead of the SVD to obtain an orthonormal basis U for the column span of A. This gives up some flexibility in regularizing the AFM/AFM* methods, but is cheaper and has the advantage that the leading basis functions do not change if K is increased. This QR approach is equivalent to Gram-Schmidt orthogonalization when the columns are sampled sufficiently.

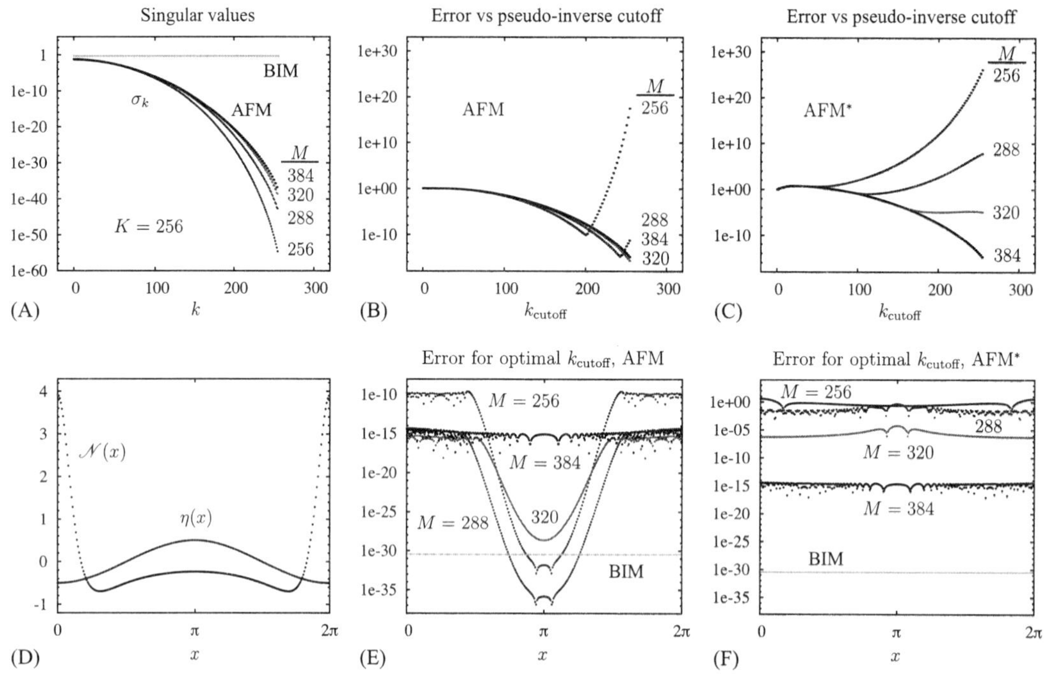

FIGURE 8. Singular values (A), exact solution (D), and effect of varying M and the pseudo-inverse cutoff, k_{cutoff}, in the AFM and AFM* methods (B,C,E,F). Errors were computed relative to the exact solution $\mathscr{N}$ given in (3.29). 360 bits of precision were used in the AFM and AFM* calculations so that aliasing and truncation errors dominate roundoff errors.

The results of computing $\mathscr{N}$ from $\mathscr{D}$ in (3.29) are shown in Figure 8. The first panel shows the singular values of A with $K = 256$ and $M \in \{256, 320, 288, 384\}$. In all four cases, the singular values decay very rapidly, with slightly slower decay when M is larger. By contrast, the singular values of A in the boundary integral approach (BIM) remain nearly constant. Panel B shows the error in $\mathscr{N}$ from (3.32),

$$\text{(3.35)} \qquad \text{error} = \sqrt{\tfrac{1}{M}\textstyle\sum_{j=0}^{M-1} |E_j|^2}, \qquad E_j = \mathscr{N}(x_j) - [U\,\text{pinv}(S)V^* B^* \partial_x \mathscr{D}]_j,$$

where $\mathscr{N}$ refers to the exact solution (3.29), plotted in Panel D. The error depends on the pseudo-inverse cutoff index, k_{cutoff}, defined by

$$\text{pinv}(S)_{ij} = \begin{cases} S_{ii}^{-1} & i = j \le k_{\text{cutoff}}, \\ 0 & \text{otherwise.} \end{cases}$$

It consists of two parts, one due to how well $\mathscr{N}$ is approximated by the leading k_{cutoff} columns of U, and one by how well the coefficients $c = UU^*\mathscr{N}$ are approximated by

$$c \approx \text{pinv}(S)V^* B^* \partial_x \mathscr{D}.$$

For all values of $M \geq K$, the error in panel B decreases initially as k_{cutoff} increases. This suggests that the leading entries of $\text{pinv}(S)V^*B^*\partial_x\mathscr{D}$ are a good approximation of those of c. When M is close to K ($M = 256$ or 288 in the figure), the error reaches a minimum at an optimal k_{cutoff}, and then increases rapidly. This occurs because the singular values σ_k become so small that the high-index entries of $\text{pinv}(S)V^*B^*\partial_x\mathscr{D}$ become large and no longer approximate the corresponding entries of c. For larger M, the singular values decay more slowly and the error curve decreases monotonically all the way to $k_{\text{cutoff}} = K - 1$.

Panel E shows the vector version of the error, namely E_j in (3.35), corresponding to the minima of the error curves in Panel B. For smaller values of M, the error E_j is largest where $\eta(x)$ is smallest. This is not surprising since the corresponding rows of A are smaller in this region due to the exponential growth of the basis functions $e^{ky}e^{ikx}$ in the y-direction. What is surprising is the extreme accuracy that is achieved by the AFM method in the region where $\eta(x) > 0$ for smaller values of M. We do not know why this occurs. This additional accuracy disappears as the errors are reduced in the region where $\eta(x) < 0$ by increasing M. Once M reaches 384, the columns of A are well-resolved in $L^2(0, 2\pi)$ as discussed above, and the error E_j is roughly uniform throughout the domain.

Panels C and F show the same results for the AFM* method. The results are very poor when $M = K = 256$, presumably a consequence of aliasing errors in sampling the columns of A being amplified on division by small singular values. As M increases, the aliasing errors become smaller and the singular values become larger. By the time M reaches 384, the error of the AFM* method is similar to that of the AFM method, of order 10^{-15}. By contrast, the boundary integral method has errors of order 10^{-30} with $K = M = 256$. There is little amplification of roundoff error in the BIM approach since A is so well-conditioned.

Figure 9 shows a similar computation to the above, but with $\eta(x)$ offset vertically so that the series (3.24) converges at all points on the curve. Specifically, we set

$$\eta(x) = -1.0 - \varepsilon\cos(x), \qquad \varepsilon = 0.5.$$

The formulas (3.29) remain nearly the same, with $\varepsilon\cos x$ replaced by $1 + \varepsilon\cos x$. The singular values in Panel A are the same as those in Figure 8 since $\eta_{\max}$ is also shifted downward by 1 in (3.31). The AFM method behaves similarly to before, achieving exceptional accuracy in the region where η is largest when M is close to K, and achieving nearly uniform accuracy once M is large enough to fully resolve the columns of the continuous version of A. Because $\mathscr{D}$ and $\mathscr{N}$ are smoother (with faster decay of Fourier modes), all three methods (AFM, AFM* and BIM) yield smaller errors in Figure 9 than in Figure 8. The AFM* method turns out to be superior to the AFM method on this example for all four choices of M. By contrast, in Figure 8, the AFM method was better for $M = 256, 288$ and 320, and the two methods were equal when $M = 384$. A partial explanation is that the left-hand side of (3.27) is a genuinely convergent series in this second example, as opposed to a finite linear combination of vectors from a dense set. Thus, the AFM* method has an easier time selecting the coefficients c_k that best represent $\mathscr{D}$. By contrast, for the AFM method, there is little difference between the two examples. In either case, the exact solutions $\mathscr{D}$ and $\mathscr{N}$ satisfy the AFM global relation, and the question of convergence comes down to how well the trapezoidal rule approximates the integrals, and how much the errors are amplified by the

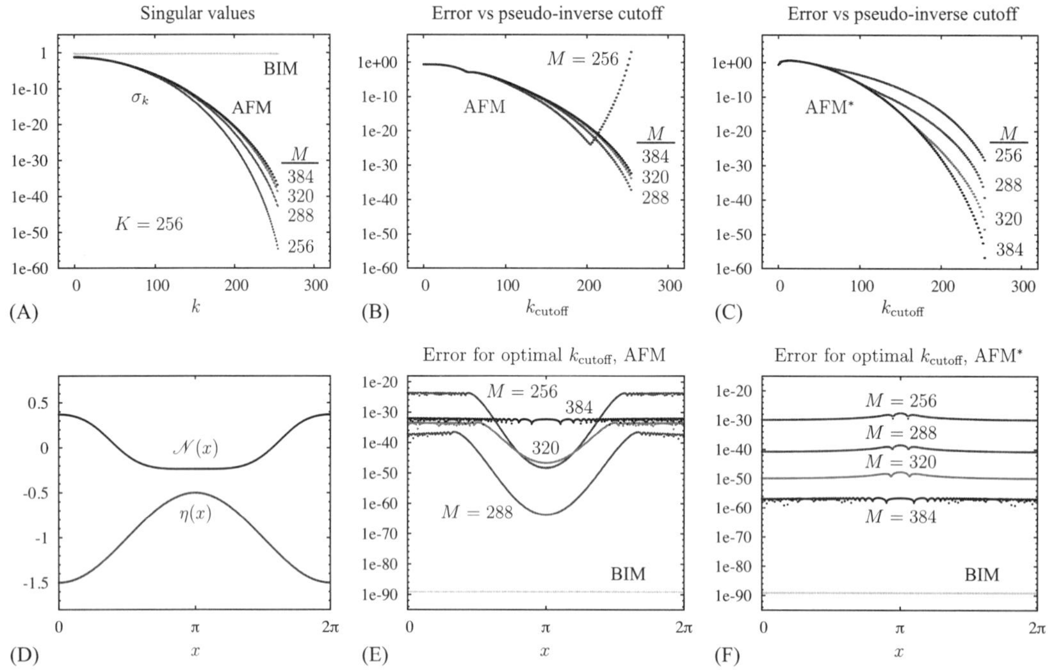

FIGURE 9. Same as Figure 8, but with a modified wave profile $\eta(x)$.

poorly-conditioned A matrix. We do not know why the AFM* method turns out to be 25 orders of magnitude more accurate once M reaches 384. As before, the BIM method is superior to both AFM methods.

In Figure 10, we check convergence of the Craig-Sulem expansion for this example. As in Figure 8, we consider $\eta(x) = -\varepsilon \cos x$, $\varepsilon = 0.5$, with Dirichlet data as in (3.29). The computations were done on a 256-point grid with 360 bits of precision. Panel A shows the errors in the partial sums, defined as

$$(3.36) \quad \|E^{(n)}\| = \sqrt{\tfrac{1}{M}\textstyle\sum_{j=0}^{M-1} \left|E^{(n)}(x_j)\right|^2}, \qquad E^{(n)}(x) = \mathscr{N}(x) - \sum_{j=0}^{n}[G_j(\eta)\mathscr{D}](x).$$

The errors decrease steadily until $n = 95$, where $\|E^{(n)}\|$ reaches 10^{-32}, the level of aliasing errors in the Fourier representation of $\mathscr{D}$ and $\mathscr{N}$ on a 256-point grid. Recall that the error in the boundary integral method with 256 points was also around 10^{-32}. Panel B shows the Fourier spectrum of $\mathscr{N}(x)$ (from the exact solution) and $E^{(0)}(x)$. The rapid decay of the Fourier modes of $E^{(0)}(x)$ show that the zeroth order approximation does an excellent job of predicting the high-frequency components of $\mathscr{N}(x)$, but not the low-frequency ones. All further corrections will only be made to the first 50 Fourier modes. In other words, we actually use

$$(3.37) \qquad \hat{E}_k^{(n)} = \begin{cases} \hat{\mathscr{N}_k} - \sum_{j=0}^{n}[G_j(\eta)\mathscr{D}]_k^\wedge, & |k| < 50, \\ \hat{\mathscr{N}_k} - [G_0(\eta)\mathscr{D}]_k^\wedge, & |k| \geq 50 \end{cases}$$

when reconstructing $E(x)$ in Panels E–H. As a result, the modes marked "unrecoverable error" in Panel B are frozen, and will not be altered by successive corrections.

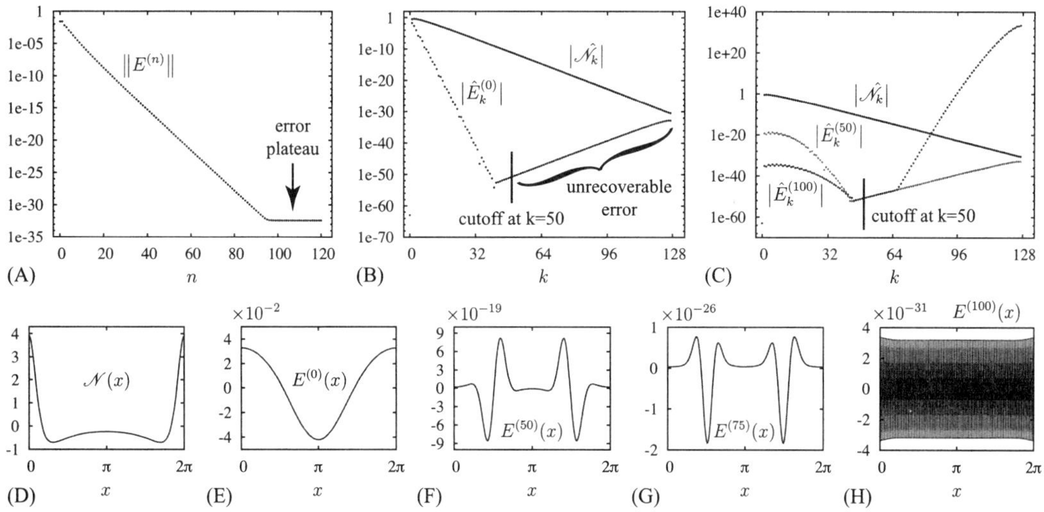

FIGURE 10. Error in the Craig-Sulem expansion at various orders. We omit the negative index Fourier modes since $\hat{\mathcal{N}}_{-k} = \overline{\hat{\mathcal{N}}_k}$ and $\hat{E}^{(n)}_{-k} = \overline{\hat{E}^{(n)}_k}$.

This is reasonable since these modes are dominated by aliasing errors in sampling $\mathcal{N}(x)$ on the 256-point grid. On a larger grid, if more accuracy were desired, the cutoff would need to be increased. The error plateau in Panel A is due to these aliasing errors.

Panel C shows the magnitude of $\hat{E}^{(n)}_k$ versus k for $n = 50$ and $n = 100$. Each successive correction reduces the error in the leading Fourier modes of the Neumann data. Note that the bulge near $k = 0$ in $\hat{E}^{(100)}_k$ is below 10^{-32}, and therefore additional corrections do not improve the global error, which is dominated by high-frequency modes at that point. The rapid growth of $\hat{E}^{(100)}_k$ above $k = 65$ is due to truncating the Fourier series to $|k| \le 128$ when computing the Craig-Sulem expansion. Analogous behavior was seen in Figures 3–6, where we observed that errors in the columns of $G_n(f)^\wedge$ propagate inward from high to low wave numbers. Due to the cutoff, these large errors do not affect the reconstruction of $\mathcal{N}(x)$. Panel D shows the exact solution $\mathcal{N}(x)$ while panels E–H show the errors in various partial sum reconstructions. Because the low-order modes are the least accurate (as seen in Panel C), the corrections are remarkably smooth, non-oscillatory curves. The highly oscillatory error in $E^{(100)}$ is due to the "unrecoverable error" in Panel B; if M were increased, $E^{(100)}$ would also be very smooth. In summary, the Craig-Sulem expansion performs well on this example, converging to the exact solution, up to aliasing errors of order 10^{-32}, in 95 iterations.

3.3. Water wave examples. We conclude with two examples in which the wave profile and velocity potential come from solutions of the water wave problem. The first is a large-amplitude standing water wave shortly before reaching maximum height. The second consists of two Stokes waves of different amplitudes

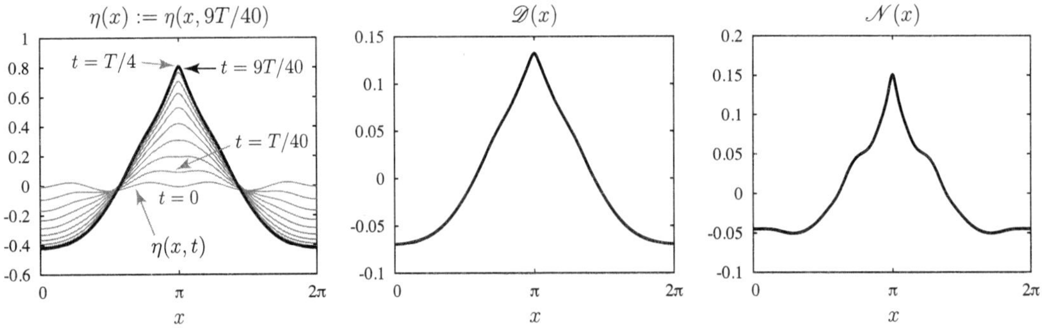

FIGURE 11. Evolution of $\eta(x,t)$ over a quarter period, $T/4$, in increments of $T/40$, along with the Dirichlet and Neumann data corresponding to $t = 9T/40$.

traveling to the right. Both examples can be evolved efficiently using the boundary integral method [**29**, **32**]. Our interest here is whether the DNO expansion method and the two AFM methods can take data (η and $\mathscr{D}$) that are only known to double- or quadruple-precision accuracy and return Neumann data ($\mathscr{N}$) with the same accuracy. We allow ourselves to use additional precision for intermediate calculations.

Figures 11–13 show the calculation of the DNO operator for a large-amplitude standing water wave. We selected the infinite depth wave corresponding to the first local maximum of wave height (half the vertical crest-to-trough distance) for this example. When wavelength is set to $L = 2\pi$ and the acceleration of gravity is $g = 1$, this wave has period $T = 6.53996$, wave height $h = 0.620173$, crest acceleration $A_c = 0.926312$, and fifth Fourier mode of φ at $t = 0$ (a good bifurcation parameter in this regime) of $\hat{\varphi}_5(0) = 0.00245499$. See [**31**, **32**] for details on how the wave was computed. Since the wave comes to rest at $t = T/4$, $\mathscr{D} \equiv 0$ at that time. Thus, to avoid a trivial DNO calculation while keeping η close to its maximum-amplitude state, we selected the wave profile at $t = 9T/40$ for this example. Figure 11 shows snapshots of $\eta(x,t)$ at equal intervals of size $\Delta t = T/40$ over a quarter-period, along with the Dirichlet and Neumann data corresponding to $t = 9T/40$.

Figure 12 shows the results of the DNO calculation using the Craig-Sulem method. The first panel shows the error as a function of order for $0 \le n \le 150$. The second panel shows the Fourier decomposition of the error at orders $n = 20$, 120 and 150, together with the Fourier modes of the "exact" solution. Unlike the results of Figure 10, the zeroth order approximation does not lead to a large improvement in the high-frequency modes (seen as a faster decay rate in $|\hat{E}_k^{(0)}|$ than $|\hat{\mathscr{N}}_k|$ in Figure 10). In fact, $|\hat{E}_k^{(0)}|$ is difficult to distinguish from $|\hat{\mathscr{N}}_k|$, so we plotted $|\hat{E}_k^{(20)}|$ instead, which has a similar decay rate but is shifted down slightly. The data for this problem, $\eta(x)$ and $\mathscr{D}(x)$, are specified via their leading 750 Fourier modes (recorded in quadruple-precision, i.e. 32 digits). This gives an approximation of the standing wave to 25 digits of accuracy, but is regarded here as specifying the DNO problem with infinite precision. The "exact" solution was computed using the boundary integral method with 2304 collocation points and 212 bits of precision, leading to approximately 40 digits of accuracy. The resulting 1152 Fourier modes of this

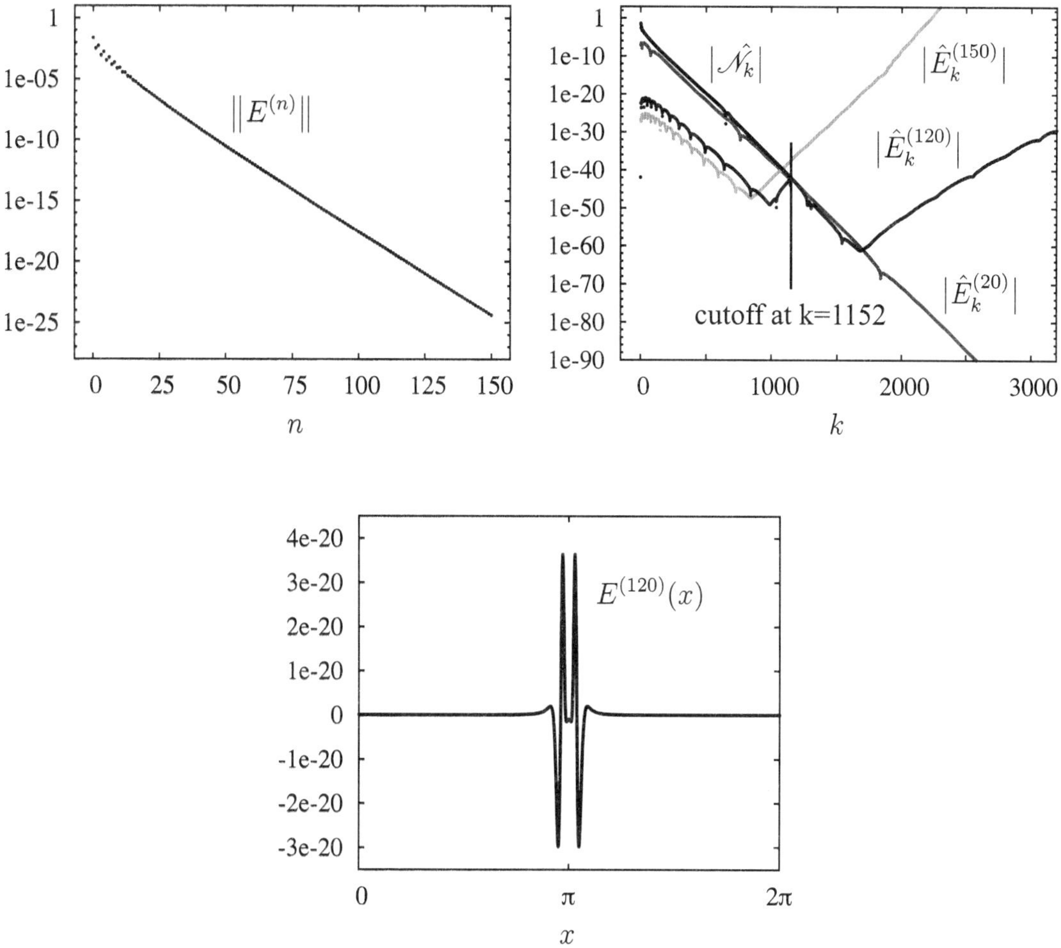

FIGURE 12. Errors in the Craig-Sulem calculation of the Neumann data for the standing wave at $t = 9T/40$. Fourier modes above $k = 1152$ are set to zero in the reconstruction of $\mathcal{N}(x)$ and the computation of $E^{(n)}(x)$.

"exact" solution are labeled $\hat{\mathcal{N}_k}$ in the second panel. The Craig-Sulem expansion was performed using $M = 16384$ grid points for the FFT and 900 bits of precision in intermediate calculations in order to achieve accurate results up to order $n = 150$ for modes $|k| < K/2$, $K = 2304$. This cutoff $|k| < 1152$ was chosen to agree with the last computed mode of the "exact" solution. The third panel shows the error in the 120th order approximation as a function of x. We see that the error is a smooth function of order 10^{-20} that is largest near the crest tip at $x = \pi$. In summary, using 900 bits (271 digits) and $M = 16384$ grid points, the 150th order Craig-Sulem method is able to achieve errors around 10^{-25}, which is comparable to the original standing wave calculation, which was done in quadruple-precision (32 digits) with 2048 grid points using the boundary integral method. Though the Craig-Sulem

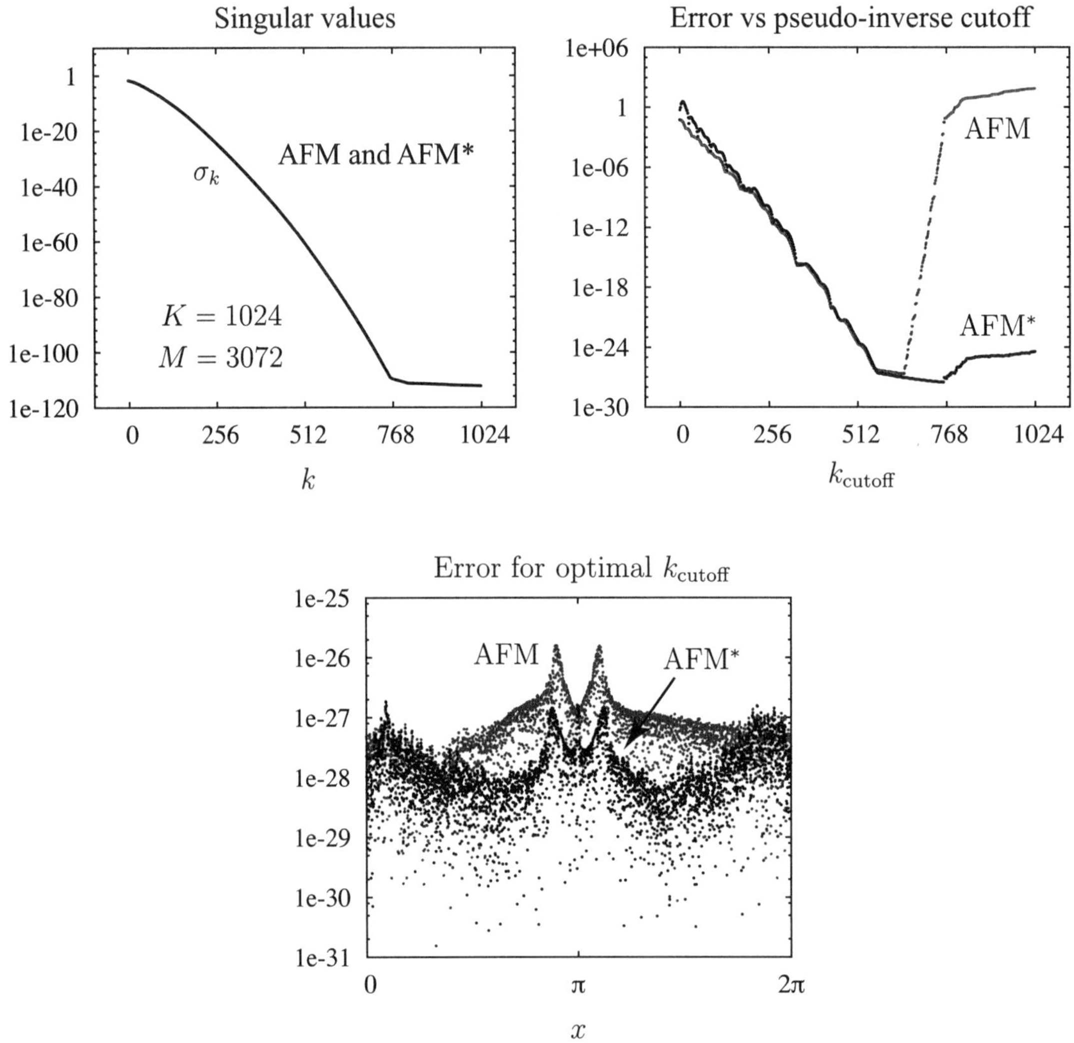

FIGURE 13. AFM and AFM* calculation of the Neumann data for the standing wave at $t = 9T/40$. The optimal cutoffs in the third panel correspond to the error minima in the second panel: $k_{\text{cutoff}} = 643$ (AFM), $k_{\text{cutoff}} = 759$ (AFM*).

method is not competitive, it is still remarkable that such a large-amplitude wave would be inside the radius of convergence of the DNO expansion.

Figure 13 shows the same calculation using the AFM and AFM* methods. The same "exact" solution as in Figure 12 was used to measure errors. The large (crest-to-trough) wave height leads to very rapidly-decaying singular values. Nevertheless, using $K = 1024$ modes and $M = 3072$ collocation points with 360 bits (108 digits) of precision, we are able to achieve 25 digits of accuracy in the solution. The AFM and AFM* methods are comparable for cutoffs up to about $k_{\text{cutoff}} = 640$. After that, the error in the AFM method grows rapidly while the AFM* method flattens out. Thus, the AFM* method is somewhat more robust in this example.

The final example of this section consists of two Stokes waves traveling right on a 2π-periodic domain. Initially, one wave is centered at $x = 0$ and the other is centered at $x = \pi$. The fluid depth is $h = 0.05$ and the parameters of the waves are

(3.38)

	$\hat{\eta}_1$	$\eta_{\max} - \eta_{min}$	c
wave 1	7.4×10^{-4}	0.028919	0.27349
wave 2	-3.0×10^{-4}	0.005202	0.23290

where c is the wave speed. Plots of $\eta(x)$, $\mathscr{D}(x)$ and $\mathscr{N}(x)$ are given in Figure 14. Figure 15 shows the results of the DNO calculation using the Craig-Sulem method. The recursion (3.14) is modified as follows to account for finite depth:

$$G_0(f) = |D| \tanh(h|D|), \qquad Y_s = \left\{ \begin{array}{ll} |D|^s, & s \text{ even} \\ |D|^s \tanh(h|D|), & s \text{ odd} \end{array} \right\},$$

$$G_n(f) = A_n(f) - \sum_{s=1}^{n-1} \frac{1}{(n-s)!} Y_{n-s} f^{n-s} G_s(f), \qquad n = 1, 2, 3, \dots,$$

$$A_n(f) = \left\{ \begin{array}{ll} (n!)^{-1} |D|^{n-1} \big(\tanh(h|D|) D f^n D - |D| f^n G_0 \big), & n \text{ even} \\ (n!)^{-1} |D|^{n-1} \big(D f^n D - G_0 f^n G_0 \big), & n \text{ odd} \end{array} \right\},$$

$$A_n(f)^{\wedge}_{kj} = \frac{jk^n}{n!} a_{nkj} (f^n)^{\wedge}_{k-j}, \qquad a_{nkj} = \left\{ \begin{array}{ll} \tanh kh - \tanh jh, & n \text{ even} \\ 1 - (\tanh kh)(\tanh jh), & n \text{ odd} \end{array} \right\}.$$

We used $M = 9216$ grid points for the FFT and computed the expansion through order $n = 100$. The "exact" solution was computed via the boundary-integral method in quadruple-precision, which is correct to about 28 digits of accuracy. The Craig-Sulem expansion reaches this level of accuracy at order $n = 90$. The errors then increase (first panel) due to the error growth region crossing the cutoff mode $k = 1750$ at 92nd order (second panel). Increasing M would delay this crossing, but is unnecessary since 90th order is sufficient to reach the accuracy of the underlying "exact" solution. The bottom panel shows the error as a function of x at order $n = 50$. The error is concentrated near $x = 0$, where $\eta(x)$ is largest. In the region not shown, the error is uniformly less than 10^{-29}, even near $x = \pi$, the location of the second Stokes wave. Thus, most of the work goes into resolving the solution near the larger peak.

Since this last example has finite-depth, we can also use the variant of the transformed field expansion method described in §2.4 to compute the Neumann data from the Dirichlet data. The first panel of Figure 16 shows the error $\|E^{(n)}\|$ from (3.36) in the Neumann data as a function of the order n, as well as the difference between the nth order terms from the TFE and CS expansions,

$$\Gamma_n = \left\| G_n^{TFE}(\eta)\mathscr{D} - G_n^{CS}(\eta)\mathscr{D} \right\|. \tag{3.39}$$

The error $\|E^{(n)}\|$ reaches a plateau of 10^{-14} for $n \geq 33$ in double-precision, and 10^{-29} for $n \geq 88$ in quadruple-precision. As before, the "exact" solution was computed in quadruple-precision using the boundary integral method. The orange markers show that the terms in the TFE expansion agree with the corresponding terms in the CS expansion to roundoff error accuracy. In particular, a plot of $E^{(50)}(x)$ for the TFE method (not shown) looks identical to that of the CS expansion in the bottom panel of Figure 15.

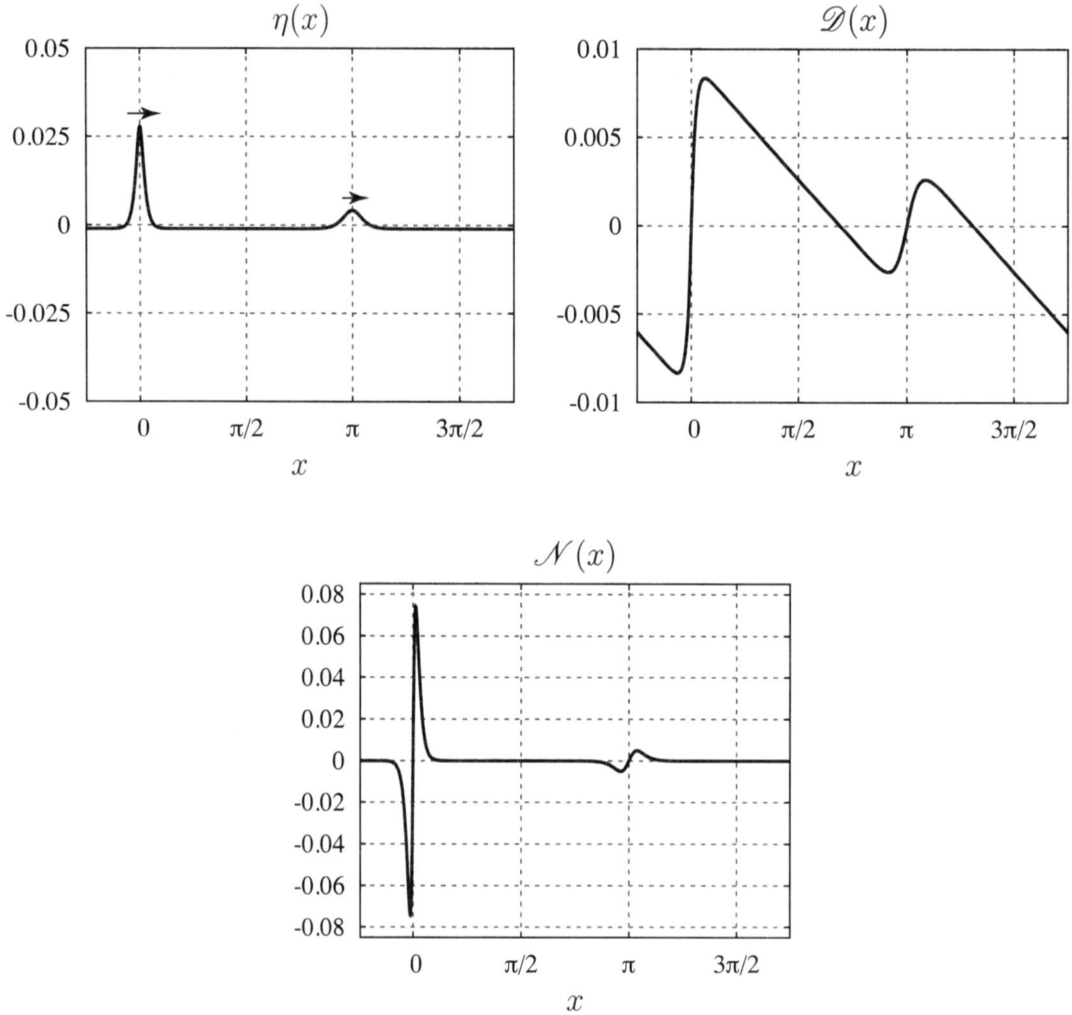

FIGURE 14. Plots of $\eta(x)$, $\mathscr{D}(x)$ and $\mathscr{N}(x)$ for the superposition of Stokes waves with parameters in (3.38) and fluid depth $h = 0.05$.

We also note in Figure 16 that Γ_n exhibits a downward trend as n increases, indicating that the terms in the TFE expansion maintain several correct digits of relative accuracy beyond the point that $\|E^{(n)}\|$ reaches the plateau region, i.e. the point where successive terms are smaller in magnitude than the absolute errors of the leading terms. In the double-precision case, most of the error in the plateau region is due to the error in the zeroth order term — Γ_0 is larger than the sum of the other Γ_n. In quadruple-precision, Γ_0 is still largest, although the Γ_n with $15 \leq n \leq 30$ are of comparable size. This may be partly due to the cutoff at $k = 1750$ used in the CS expansion in Figure 15 to eliminate high-frequency noise. Certainly, the rapid growth in Γ_n for $n \geq 80$ in Figure 16 is due to errors in the CS expansion rather than the TFE expansion. Indeed, $\|E^{(n)}\|$ in Figure 15 grows

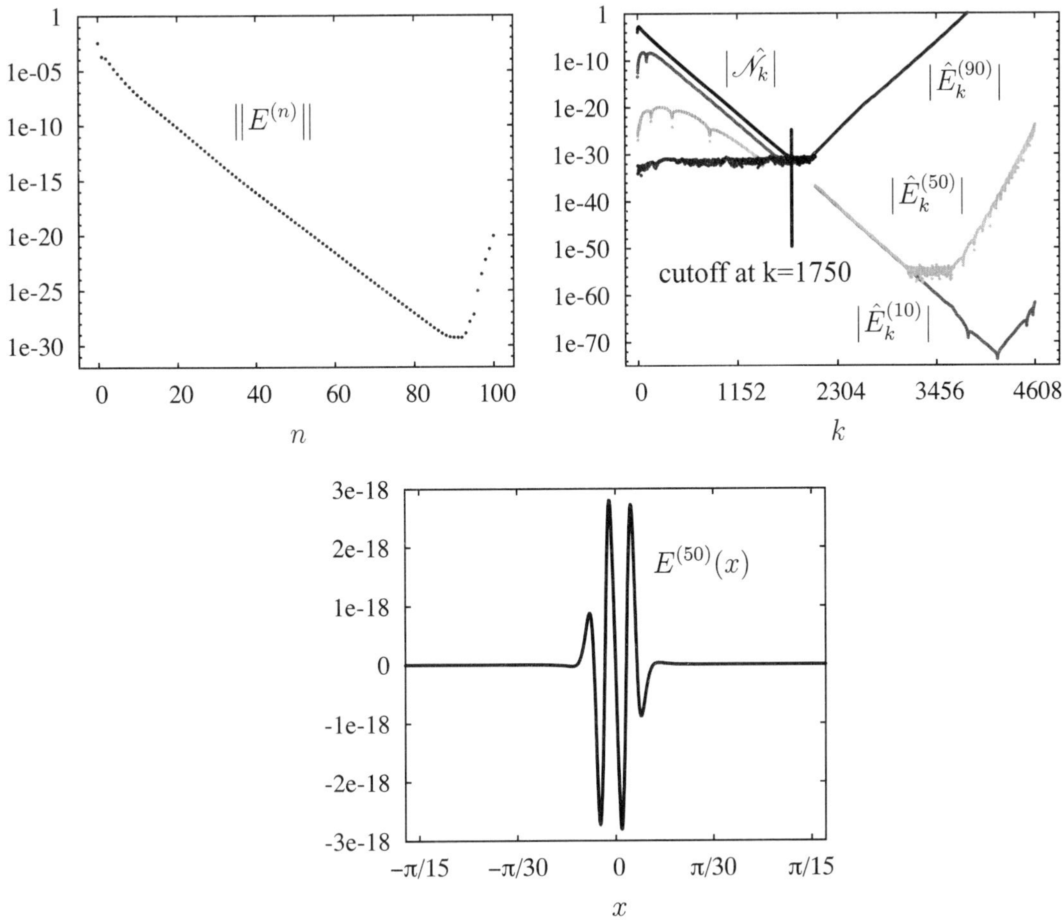

FIGURE 15. Errors in the Craig-Sulem calculation of the Neumann data for a pair of traveling Stokes waves with parameters (3.38) and fluid depth $h = 0.05$.

rapidly for $n \geq 90$ while it remains flat for $n \geq 90$ in Figure 16. Recall that a third method (the BIM method) was used for the exact solution in both plots.

The second and third panels of Figure 16 were used to decide how many grid points to use in the TFE method. We used $M = 2048$, $N = 32$ in double-precision and $M = 4096$, $N = 48$ in quadruple-precision. These plots show the partial norms

$$\kappa_{nj} = \sqrt{\textstyle\sum_k |\alpha_j^n(k)|^2}, \qquad \gamma_{nk} = \sqrt{\textstyle\sum_j |\alpha_j^n(k)|^2}, \tag{3.40}$$

where $\alpha_j^n(k)$ are the Chebyshev coefficients of the kth Fourier mode of u_n:

$$\hat{u}_n(k,y) = \sum_{j=0}^{N} \alpha_j^n(k)\, T_j(1 + 2h^{-1}y), \qquad (|k| \leq M/2,\ -h < y < 0).$$

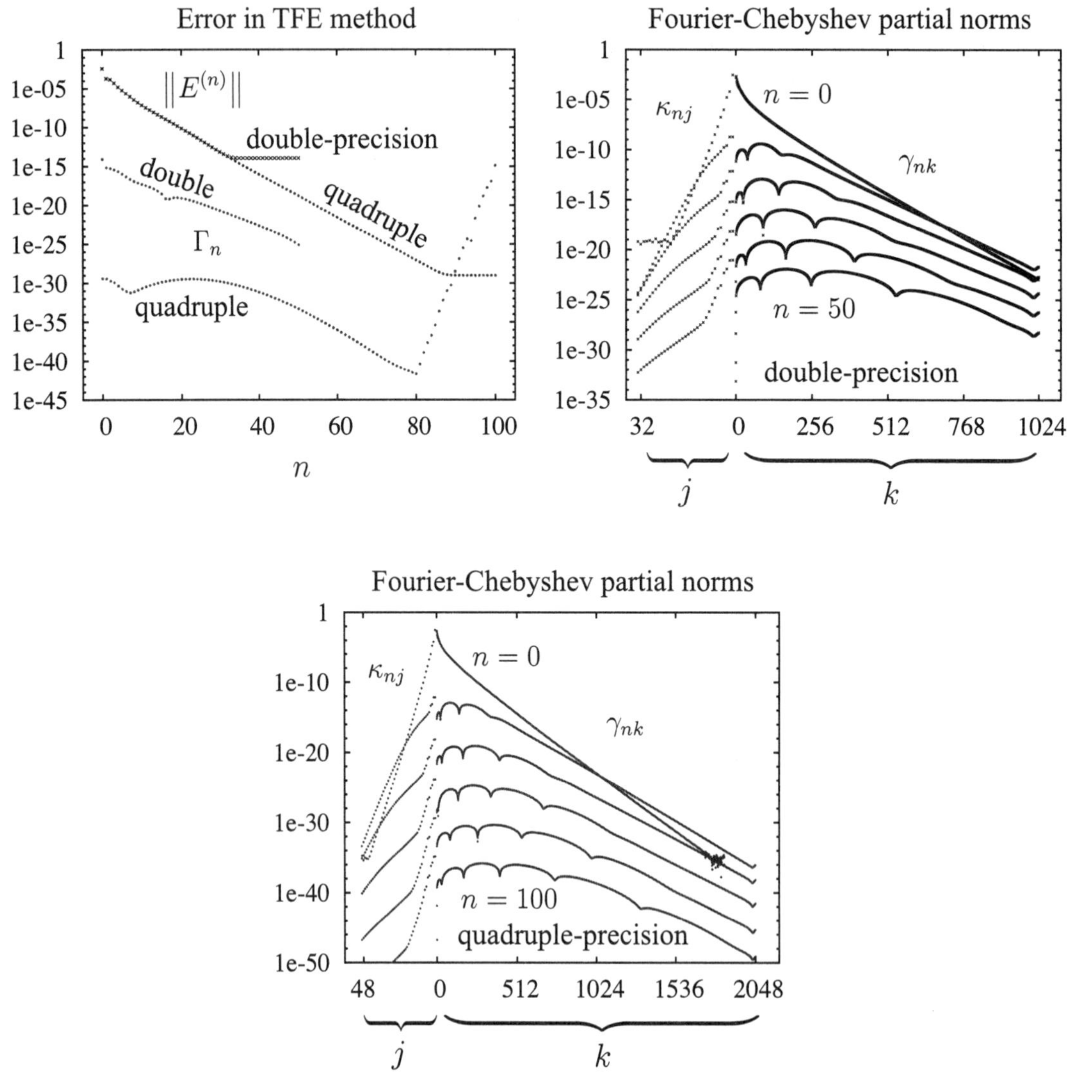

FIGURE 16. Two measures of error ($\|E^{(n)}\|$ and Γ_n) and Fourier-Chebyshev partial norms (3.40) for the TFE calculation of the Neumann data for a pair of traveling Stokes waves with parameters (3.38) and fluid depth $h = 0.05$.

Recall that $u(x, y) = \phi(x, (1 + h^{-1}\eta)y + \eta)$ is defined on a rectangle and expanded in powers of ε, $u(x, y) = \sum_n \varepsilon^n u_n(x, y)$, where $\eta = \varepsilon f$. The partial norms κ_{nj} and γ_{nk} are the norms of the rows and columns of the matrix of Fourier-Chebyshev coefficients of the function $u_n(x, y)$. To resolve the solution spectrally, the mesh needs to be large enough that the partial norms decay to the desired tolerance as $k \to M/2$ or $j \to N$. As shown in Figure 16, the meshes we selected are sufficient to reach roundoff level tolerances in these limits.

It is worth noting that many fewer grid points are needed in the y-direction than in the x-direction, so the price of discretizing the bulk fluid is not as severe as one might imagine. Moreover, as with the BIM method, intermediate calculations can be done in double or quadruple-precision arithmetic to achieve similar levels of accuracy in the solution. The overall running times (in seconds) of the various methods on this example are given in the following table:

method	BIM(d)	BIM(q)	TFE(d)	TFE(q)	CS	AFM	AFM-QR
time	0.156	3.53	2.46	24.0	2762	11222	920

The code was run on a 3.33 GHz Intel Xeon X5680 system with 12 cores. Here (d) and (q) stand for double and quadruple-precision, and the other methods were run with 360 bits (108 digits) of precision. The running time of AFM* is nearly identical to that of AFM, and the AFM-QR variant will be described below. Clearly, the BIM and TFE approaches are superior to the CS and AFM-based methods since arbitrary-precision arithmetic is not required.

We conclude with the results of the AFM and AFM* methods on this example, which we selected initially as being relevant to water waves and likely to cause difficulties for the AFM and AFM* methods. Our reasoning was that the second wave is large enough to require many Fourier modes to resolve its shape, but small enough that $\cosh(k\eta(x))$ is many orders of magnitude larger at the crest of the first wave than at the crest of the second (for large k). Thus, substantial cancellation must occur near $x = 0$ in order to resolve the behavior of ϕ near $x = \pi$. However, the singular values in the first panel of Figure 17 decay slowly in comparison to the infinite-depth standing wave case of Figure 13. This is because the waves in these two examples have similar Fourier decay rates for $\hat{\eta}_k$ and $\hat{\mathscr{D}}_k$, but the deep-water standing wave has a much larger vertical crest-to-trough height. To the extent that this is generally the case for waves in deep versus shallow water, the AFM and AFM* methods appear to be better suited for finite depth problems.

The errors in the AFM and AFM* methods are nearly identical for every choice of pseudo-inverse cutoff in the second panel of Figure 17. Thus, the methods are equally effective at computing the DNO for this example. Beyond $k_{\text{cutoff}} = 970$, the error in the second panel reaches a plateau of 3×10^{-28}. This is due to errors in the "exact" solution. Since this plateau is reached already at $k_{\text{cutoff}} = 970$, it would appear that K and possibly M can be reduced while still achieving the target accuracy of 10^{-28}. However, reducing K leads to worse results (not shown). The reason is that the high-index columns of A in

$$A_{j0} = \frac{1}{M}, \qquad \begin{pmatrix} A_{j,2k-1} \\ A_{j,2k} \end{pmatrix} = \frac{\sqrt{2}}{M} \frac{\cosh[k\eta(x_j) + h]}{\cosh[k\eta_{\max} + h]} \begin{pmatrix} \cos kx \\ \sin kx \end{pmatrix}, \qquad (1 \leq k < K/2)$$

comprise a small but important part of the low-index singular vectors U of the singular value decomposition, $A = USV^T$. A better way to understand how large K needs to be in order to achieve a prescribed accuracy is to perform a QR factorization, $A = QR$, which is a numerically robust way to perform a Gram-Schmidt orthogonalization of the columns of A. We then define the "discrete AFM transform"

$$\tilde{\mathscr{N}_0} = c_0, \qquad \tilde{\mathscr{N}_k} = c_{2k-1} + ic_{2k}, \qquad \{c\}_{k=0}^{K-2} = \frac{1}{\sqrt{M}} Q^T \{\mathscr{N}(x_j)\}_{j=0}^{M-1}$$

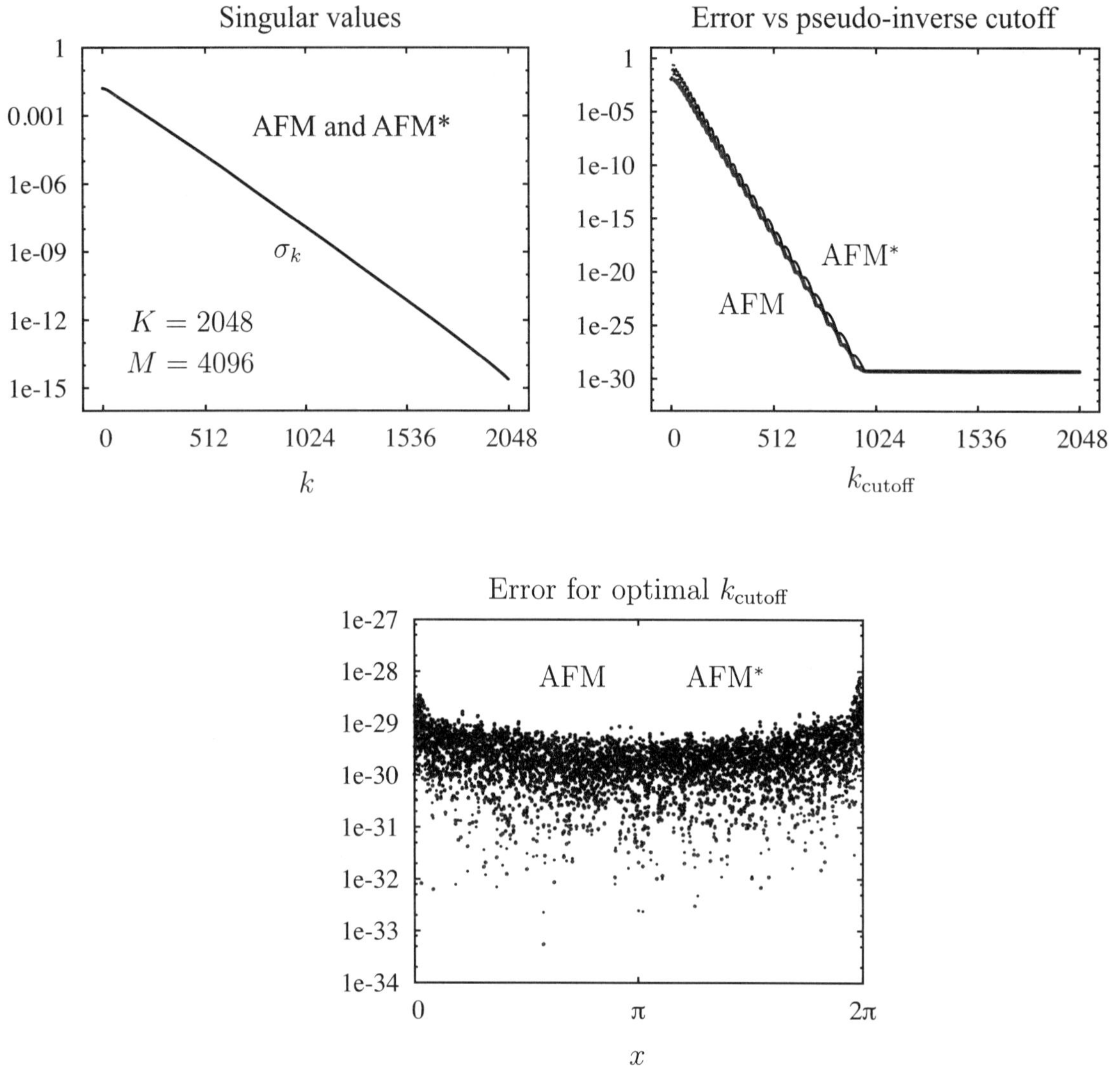

FIGURE 17. AFM and AFM* calculation of the Neumann data for the superposition of Stokes waves in Figure 14. These AFM methods are much better conditioned in this shallow water regime than the deep water cases considered above. The optimal cutoff modes at right were $k_{\text{cutoff}} = 2046$ (AFM) and 2045 (AFM*).

and, in Figure 18, compare it to the discrete Fourier transform

$$\hat{\mathscr{N}}_k = \frac{1}{M} \sum_{j=0}^{M-1} \mathscr{N}(x_j) e^{-2\pi i j k}, \qquad 0 \le k \le M/2.$$

The red curves give the Fourier modes of the input Dirichlet data, which are taken to specify the problem with infinite precision even though they only describe the

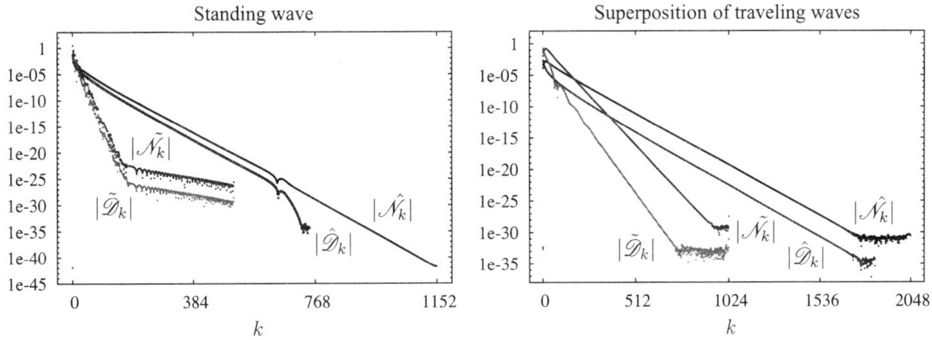

FIGURE 18. Comparison of the coefficients in an expansion of the Dirichlet and Neumann data in an orthogonalized AFM basis versus a Fourier basis.

standing wave and traveling waves to around 25 and 30 digits, respectively. The black curves give the Fourier modes of the boundary integral solutions, which were computed with 212 and 106 bits of precision, respectively. The orange and blue curves give the "AFM transform" of the Dirichlet and Neumann data. The observation that K cannot be reduced significantly below 2048 for the traveling wave problem is seen clearly in the right panel, which shows that $\tilde{\mathscr{N}}_k$ does not reach roundoff error until just below $k = 1024$.

The fact that the AFM coefficients $\tilde{\mathscr{D}}_k$ and $\tilde{\mathscr{N}}_k$ decay faster than the Fourier coefficients $\hat{\mathscr{D}}_k$ and $\hat{\mathscr{N}}_k$ suggests that the Gram-Schmidt version of the AFM basis is more efficient at representing the Dirichlet and Neumann data of many problems of physical interest (such as standing and traveling waves) than the Fourier basis. The abrupt change in slope of the orange and blue curves in the left panel occurs when the AFM method begins to resolve errors in the given Dirichlet data, which is taken to be exact even though it only agrees with physical standing waves to 25 digits. While it is interesting that the Gram-Schmidt version of the AFM basis is more efficient than a Fourier basis for these problems, computing the orthogonal AFM basis is expensive since it requires a great deal of additional precision to avoid losing all significant digits during the QR factorization process.

Conclusions

We have shown that the Craig-Sulem expansion of the DNO operator, the implicit formulation due to Ablowitz, Fokas and Musslimani, and its dual, due to Ablowitz and Haut, can all be used to compute spectrally accurate solutions of the Dirichlet-Neumann problem. All three methods involve ill-conditioned intermediate calculations when the vertical crest-to-trough distance becomes large relative to the inverse of the highest-frequency wave numbers involved in a Fourier description of the wave profile $\eta(x)$ and the Dirichlet data $\mathscr{D}(x)$. However, this ill-conditioning can be tamed using extended precision arithmetic. Most importantly, $\eta(x)$ and $\mathscr{D}(x)$ need only be specified in double-precision to obtain double-precision results for $\mathscr{N}(x)$; extended precision is only required in intermediate calculations. While it is undesirable to work in extended precision arithmetic, a spectrally accurate method requiring 64 digits of accuracy in intermediate calculations may require less work than a more traditional 16-digit calculation using a 2nd or 4th order method if high accuracy is desired.

Nevertheless, our goal in writing the paper was not to advocate the use of these methods, but to explore their limits of applicability. We originally thought that the AFM and AFM* would break down if the potential ϕ cannot be extended analytically to a strip containing the peak of the wave profile. This was the motivation for studying a potential ϕ with poles on the real axis and a wave profile that dips below the poles but also extends above them. It is surprising that the AFM basis can still be used to approximate $\mathscr{D}$ in this case to arbitrary accuracy, and that the term-by-term Neumann data turns out to be a good approximation of the correct solution. We also did not expect the CS expansion to converge for realistic large-amplitude standing waves, but it was able to achieve 25 digits of accuracy at 150th order. For the same problem, the AFM and AFM* methods achieved 25 digits of accuracy in spite of singular values dropping below 10^{-100}. For waves in shallow water, the condition numbers are much more reasonable, making the AFM methods more appealing.

As noted in [**23**], the CS expansion suffers from significant cancellations requirements. Our numerical investigations confirm this behavior. We note that a suitable rearrangement of the series permits one to account for some of these cancellations, though not all. One of the goals of the present work was to understand whether this behavior is present in the AFM formulation and to what extent as the AFM method may be interpreted as a certain summation of the CS expansion [**2**]. Our investigations reveal the AFM method is similarly ill-conditioned. Indeed the rapid decay of singular values in the AFM (and AFM*) method mirrors the cancellation properties observed in the CS expansion. Given the delicate nature of CS, AFM and AFM* we recommend the BIM, especially for two-dimensional flows. The TFE version of the CS expansion also performs remarkably well without the need for extended precision arithmetic in intermediate calculations.

Regarding extension to three dimensions, the CS and TFE expansions work with little change. Since the latter can be run effectively in double-precision, it is a viable method. Although the bulk fluid must be discretized, the system of equations that must be solved decouples into many one-dimensional boundary value problems rather than a large system of equations such as arise in finite element methods. The AFM and AFM* methods will lead to large, highly ill-conditioned linear systems that are easy to set up but nearly impossible to solve. Performing the SVD or QR factorization of such a large matrix in extended precision arithmetic would be extremely costly, and iterative methods such as GMRES will not converge when the condition number grows to 10^{15} or higher. By contrast, the boundary integral method yields condition numbers close to 1; hence, GMRES converges in just a few iterations. Although it is difficult to implement boundary integral methods in three dimensions due to the slowly decaying lattice sums involved, techniques such as Ewald summation [**3, 13, 18**] are available to accelerate convergence.

Acknowledgments

The author was supported in part by the Director, Office of Science, Computational and Technology Research, U.S. Department of Energy under Contract No. DE-AC02-05CH11231, and by the National Science Foundation through grant

DMS-0955078. Any opinions, findings, and conclusions or recommendations expressed in this material are those of the authors and do not necessarily reflect the views of the funding sources.

References

[1] M. J. Ablowitz, A. S. Fokas, and Z. H. Musslimani, *On a new non-local formulation of water waves*, J. Fluid Mech. **562** (2006), 313–343, DOI 10.1017/S0022112006001091. MR2263547 (2007k:76013)

[2] M. J. Ablowitz and T. S. Haut, *Spectral formulation of the two fluid Euler equations with a free interface and long wave reductions*, Anal. Appl. (Singap.) **6** (2008), no. 4, 323–348, DOI 10.1142/S0219530508001213. MR2459114 (2010a:76016)

[3] David M. Ambrose, Michael Siegel, and Svetlana Tlupova, *A small-scale decomposition for 3D boundary integral computations with surface tension*, J. Comput. Phys. **247** (2013), 168–191, DOI 10.1016/j.jcp.2013.03.045. MR3066169

[4] Gregory Baker and André Nachbin, *Stable methods for vortex sheet motion in the presence of surface tension*, SIAM J. Sci. Comput. **19** (1998), no. 5, 1737–1766 (electronic), DOI 10.1137/S1064827595296562. MR1617821 (99c:76073)

[5] Gregory R. Baker, Daniel I. Meiron, and Steven A. Orszag, *Generalized vortex methods for free-surface flow problems*, J. Fluid Mech. **123** (1982), 477–501, DOI 10.1017/S0022112082003164. MR687014 (84a:76002)

[6] Gregory R. Baker and Chao Xie, *Singularities in the complex physical plane for deep water waves*, J. Fluid Mech. **685** (2011), 83–116, DOI 10.1017/jfm.2011.283. MR2844303

[7] Oscar P. Bruno and Fernando Reitich, *Solution of a boundary value problem for the Helmholtz equation via variation of the boundary into the complex domain*, Proc. Roy. Soc. Edinburgh Sect. A **122** (1992), no. 3-4, 317–340, DOI 10.1017/S0308210500021132. MR1200203 (94e:35131)

[8] Claudio Canuto, M. Yousuff Hussaini, Alfio Quarteroni, and Thomas A. Zang, *Spectral methods in fluid dynamics*, Springer Series in Computational Physics, Springer-Verlag, New York, 1988. MR917480 (89m:76004)

[9] I. M. Cohen and P. K. Kundu, *Fluid Mechanics*, Academic Press, 2004.

[10] Walter Craig, Philippe Guyenne, David P. Nicholls, and Catherine Sulem, *Hamiltonian long-wave expansions for water waves over a rough bottom*, Proc. R. Soc. Lond. Ser. A Math. Phys. Eng. Sci. **461** (2005), no. 2055, 839–873, DOI 10.1098/rspa.2004.1367. MR2121939 (2006e:76019)

[11] W. Craig and C. Sulem, *Numerical simulation of gravity waves*, J. Comput. Phys. **108** (1993), no. 1, 73–83, DOI 10.1006/jcph.1993.1164. MR1239970 (94h:76064)

[12] Bernard Deconinck and Katie Oliveras, *The instability of periodic surface gravity waves*, J. Fluid Mech. **675** (2011), 141–167, DOI 10.1017/S0022112011000073. MR2801039 (2012j:76057)

[13] Z.-H. Duan and R. Krasny, *An Ewald summation based multipole method*, J. Chem. Phys. **113** (2000), no. 9, 3492–3495.

[14] A.I. Dyachenko, E.A. Kuznetsov, M.D. Spector, and V.E. Zakharov, *Analytical description of the free surface dynamics of an ideal fluid (canonical formalism and conformal mapping)*, Phys. Letters A **221** (1996), 73 – 79.

[15] Laurent Fousse, Guillaume Hanrot, Vincent Lefèvre, Patrick Pélissier, and Paul Zimmermann, *MPFR: a multiple-precision binary floating-point library with correct rounding*, ACM Trans. Math. Software **33** (2007), no. 2, Art. 13, 15, DOI 10.1145/1236463.1236468. MR2326955 (2008e:65157)

[16] Robert Krasny, *A study of singularity formation in a vortex sheet by the point-vortex approximation*, J. Fluid Mech. **167** (1986), 65–93, DOI 10.1017/S0022112086002732. MR851670 (87g:76028)

[17] David Lannes, *Well-posedness of the water-waves equations*, J. Amer. Math. Soc. **18** (2005), no. 3, 605–654 (electronic), DOI 10.1090/S0894-0347-05-00484-4. MR2138139 (2005m:76021)

[18] D. Lindbo and A.-K. Tornberg, *Fast and spectrally accurate Ewald summation for 2-periodic electrostatic systems*, J. Chem. Phys. **136** (2012), 164111.

[19] M. S. Longuet-Higgins and E. D. Cokelet, *The deformation of steep surface waves on water. I. A numerical method of computation*, Proc. Roy. Soc. London Ser. A **350** (1976), no. 1660, 1–26. MR0411355 (53 #15091)

[20] G. N. Mercer and A. J. Roberts, *Standing waves in deep water: their stability and extreme form*, Phys. Fluids A **4** (1992), no. 2, 259–269, DOI 10.1063/1.858354. MR1146134 (92k:76014)

[21] G. N. Mercer and A. J. Roberts, *The form of standing waves on finite depth water*, Wave Motion **19** (1994), no. 3, 233–244, DOI 10.1016/0165-2125(94)90056-6. MR1276940

[22] N. I. Muskhelishvili, *Singular integral equations*, Dover Publications, Inc., New York, 1992. Boundary problems of function theory and their application to mathematical physics; Translated from the second (1946) Russian edition and with a preface by J. R. M. Radok; Corrected reprint of the 1953 English translation. MR1215485 (94a:45001)

[23] David P. Nicholls and Fernando Reitich, *A new approach to analyticity of Dirichlet-Neumann operators*, Proc. Roy. Soc. Edinburgh Sect. A **131** (2001), no. 6, 1411–1433, DOI 10.1017/S0308210500001463. MR1869643 (2003b:35216)

[24] David P. Nicholls and Fernando Reitich, *Stable, high-order computation of traveling water waves in three dimensions*, Eur. J. Mech. B Fluids **25** (2006), no. 4, 406–424, DOI 10.1016/j.euromechflu.2005.11.003. MR2241379 (2007d:76026)

[25] K. L. Oliveras, V. Vasan, B. Deconinck, and D. Henderson, *Recovering the water-wave profile from pressure measurements*, SIAM J. Appl. Math. **72** (2012), no. 3, 897–918, DOI 10.1137/110853285. MR2968755

[26] Chris H. Rycroft and Jon Wilkening, *Computation of three-dimensional standing water waves*, J. Comput. Phys. **255** (2013), 612–638, DOI 10.1016/j.jcp.2013.08.026. MR3109807

[27] D. H. Smith and A. J. Roberts, *Branching behavior of standing waves—the signatures of resonance*, Phys. Fluids **11** (1999), no. 5, 1051–1064, DOI 10.1063/1.869976. MR1683714 (99m:76024)

[28] Vishal Vasan and Bernard Deconinck, *The inverse water wave problem of bathymetry detection*, J. Fluid Mech. **714** (2013), 562–590, DOI 10.1017/jfm.2012.497. MR3008073

[29] J. Wilkening, *Traveling-standing water waves*, (2014), (submitted).

[30] J. Wilkening, A. Cerfon, and M. Landreman, *Projected dynamics of kinetic equations with energy diffusion in spaces of orthogonal polynomials*, (2014), (submitted).

[31] Jon Wilkening, *Breakdown of self-similarity at the crests of large-amplitude standing water waves*, Phys. Rev. Lett. **107** (2011), 184501.

[32] Jon Wilkening and Jia Yu, *Overdetermined shooting methods for computing standing water waves with spectral accuracy*, Comput. Sci. Disc. **5** (2012), no. 1, 014017.

[33] V. E. Zakharov, *Stability of periodic waves of finite amplitude on the surface of a deep fluid*, Zhurnal Prikladnoi Mekhaniki i Tekhnicheskoi Fiziki **8** (1968), 86–94.

Department of Mathematics, University of California, Berkeley, California 94720-3840

Department of Mathematics, Pennsylvania State University, University Park, Pennsylvania 16802

Selected Published Titles in This Series

635 **Christopher W. Curtis, Anton Dzhamay, Willy A. Hereman, and Barbara Prinari, Editors,** Nonlinear Wave Equations, 2015

632 **Gohar Kyureghyan, Gary L. Mullen, and Alexander Pott, Editors,** Topics in Finite Fields, 2015

631 **Siddhartha Bhattacharya, Tarun Das, Anish Ghosh, and Riddhi Shah, Editors,** Recent Trends in Ergodic Theory and Dynamical Systems, 2015

630 **Pierre Albin, Dmitry Jakobson, and Frédéric Rochon, Editors,** Geometric and Spectral Analysis, 2014

629 **Milagros Izquierdo, S. Allen Broughton, Antonio F. Costa, and Rubí E. Rodríguez, Editors,** Riemann and Klein Surfaces, Automorphisms, Symmetries and Moduli Spaces, 2014

628 **Anita T. Layton and Sarah D. Olson, Editors,** Biological Fluid Dynamics: Modeling, Computations, and Applications, 2014

627 **Krishnaswami Alladi, Frank Garvan, and Ae Ja Yee, Editors,** Ramanujan 125, 2014

626 **Veronika Furst, Keri A. Kornelson, and Eric S. Weber, Editors,** Operator Methods in Wavelets, Tilings, and Frames, 2014

625 **Alexander Barg and Oleg R. Musin, Editors,** Discrete Geometry and Algebraic Combinatorics, 2014

624 **Karl-Dieter Crisman and Michael A. Jones, Editors,** The Mathematics of Decisions, Elections, and Games, 2014

623 **Pramod N. Achar, Dijana Jakelić, Kailash C. Misra, and Milen Yakimov, Editors,** Recent Advances in Representation Theory, Quantum Groups, Algebraic Geometry, and Related Topics, 2014

622 **S. Ejaz Ahmed, Editor,** Perspectives on Big Data Analysis, 2014

621 **Ludmil Katzarkov, Ernesto Lupercio, and Francisco J. Turrubiates, Editors,** The Influence of Solomon Lefschetz in Geometry and Topology, 2014

620 **Ulrike Tillmann, Søren Galatius, and Dev Sinha, Editors,** Algebraic Topology: Applications and New Directions, 2014

619 **Gershon Wolansky and Alexander J. Zaslavski, Editors,** Variational and Optimal Control Problems on Unbounded Domains, 2014

618 **Abba B. Gumel, Editor,** Mathematics of Continuous and Discrete Dynamical Systems, 2014

617 **Christian Ausoni, Kathryn Hess, Brenda Johnson, Wolfgang Lück, and Jérôme Scherer, Editors,** An Alpine Expedition through Algebraic Topology, 2014

616 **G. L. Litvinov and S. N. Sergeev, Editors,** Tropical and Idempotent Mathematics and Applications, 2014

615 **Plamen Stefanov, András Vasy, and Maciej Zworski, Editors,** Inverse Problems and Applications, 2014

614 **James W. Cogdell, Freydoon Shahidi, and David Soudry, Editors,** Automorphic Forms and Related Geometry, 2014

613 **Stephan Stolz, Editor,** Topology and Field Theories, 2014

612 **Patricio Cifuentes, José García-Cuerva, Gustavo Garrigós, Eugenio Hernández, José María Martell, Javier Parcet, Keith M. Rogers, Alberto Ruiz, Fernando Soria, and Ana Vargas, Editors,** Harmonic Analysis and Partial Differential Equations, 2014

611 **Robert Fitzgerald Morse, Daniela Nikolova-Popova, and Sarah Witherspoon, Editors,** Group Theory, Combinatorics, and Computing, 2014